VOLUME ONE HUNDRED AND FIVETEEN

ADVANCES IN BOTANICAL RESEARCH

African plant-based products as a source of potent drugs to overcome cancers and their chemoresistance

Part 3 - Potential pharmaceuticals to overcome cancers and their chemoresistance

ADVANCES IN BOTANICAL RESEARCH

Series Editor

Jean-Pierre Jacquot
Membre de l'Institut Universitaire de France, Unité Mixte de Recherche INRA, UHP 1136 "Interaction Arbres Microorganismes", Université de Lorraine, Faculté des Sciences, Vandoeuvre, France

VOLUME ONE HUNDRED AND FIFTEEN

ADVANCES IN BOTANICAL RESEARCH

African plant-based products as a source of potent drugs to overcome cancers and their chemoresistance

Part 3 - Potential pharmaceuticals to overcome cancers and their chemoresistance

Edited by

VICTOR KUETE

Professor, Faculty of Science, Department of Biochemistry,
University of Dschang, Dscahng, Cameroon

Academic Press is an imprint of Elsevier
125 London Wall, London, EC2Y 5AS, United Kingdom
50 Hampshire Street, 5th Floor, Cambridge, MA 02139, United States
525 B Street, Suite 1650, San Diego, CA 92101, United States

First edition 2025

ISBN: 978-0-443-29334-4
ISSN: 0065-2296

Publisher: Zoe Kruze
Acquisitions Editor: Mariana L. Kuhl
Editorial Project Manager: Sneha Apar
Production Project Manager: James Selvam
Cover Designer: Greg Harris

Typeset by MPS Limited, India

Contents

Contributors

Daniel Buyinza
Department of Chemistry, Kabale University, Kabale, Uganda

Appolinaire Kene Dongmo
Department of Chemistry, Faculty of Science, University of Dschang, Dschang, Cameroon

Hugues Fouotsa
Department of Process Engineering, National Higher Polytechnic School of Douala, University of Douala, Douala, Cameroon

Ibrahim Hashim
Department of Soil Science, College of Agriculture and Life Sciences, University of Wisconsin-Madison, Madison, United States

Robert V.T. Kepdieu
Department of Chemistry, Faculty of Science, University of Dschang, Dschang, Cameroon

Jenifer R.N. Kuete
Department of Chemistry, Faculty of Science, University of Dschang, Dschang, Cameroon

Victor Kuete
Department of Biochemistry, Faculty of Science, University of Dschang, Dschang, Cameroon

Armelle T. Mbaveng
Department of Biochemistry, Faculty of Science, University of Dschang, Dschang, Cameroon

Justus Mukavi
Institute of Pharmaceutical Biology and Phytochemistry, University of Münster, Münster, Germany

Vaderament-A. Nchiozem-Ngnitedem
Institute of Chemistry, University of Potsdam, Potsdam-Golm, Germany

Leonidah K. Omosa
Department of Chemistry, Faculty of Science and Technology, University of Nairobi, Nairobi, Kenya

Julio Issah Mawouma Pagna
Department of Organic Chemistry, Faculty of Science, University of Yaoundé I, Yaoundé, Cameroon

Rémy B. Teponno
Department of Chemistry, Faculty of Science, University of Dschang, Dschang, Cameroon

Leonel Donald Feugap Tsamo
Department of Chemistry, Faculty of Science, University of Dschang, Dschang, Cameroon

Preface

Globally, the number of cancer deaths increases from 7.1 million in 2002, 7.62 million deaths in 2007, 9.56 million in 2017 to 10 million cancer deaths in 2020 and is projected to reach 11.5 million in 2030. Despite the progress achieved in managing malignant diseases, deficiencies and room for improvement remain. Chemotherapy or palliative chemotherapy is applied in all types of cancers. Nonetheless, side effects in chemotherapy exist for most drugs. Plant-based therapy may reduce adverse side effects. The usefulness of medicinal plants as a source of cytotoxic agents is well known, though a few plant products are used in cancer chemotherapy. Some established anticancer drugs from plant source include vinblastine, vincristine, combretastatins, paclitaxel, camptothecin, and homoharringtonine. The library of herbal cytotoxic agents has increased enormously in recent years, giving hope that the discovery of new chemotherapeutic agents will flourish in the future. This is the case with African medicinal plants that have shown exceptional cytotoxic potential in vitro and in vivo on a large panel of human cancer cell lines, including resistant phenotypes. Nonetheless, the main shortcoming of studies on medicinal plants with anticancer potential remains the scarcity of clinical studies. This book is a unique tool to analyze the most prominent data on the cytotoxic potential of the botanicals and phytochemicals from the flora of Africa and highlight the potential phytomedicine to undergo clinical studies to overcome cancer and cancer drug resistance. This book covers several aspects of pharmacology, cell biology, molecular biology, phytochemistry, medicinal chemistry, alternative medicine, and drug discovery with relevant to academic use, scientific research, and possible pharmaceutical application. The complete book is made of three volumes (volumes 113-115) published in *Advances in Botanical Research* as follows:

- Volume 113. African plant-based products as a source of potent drugs to overcome cancers and their chemoresistance. Part 1. Cancer chemoresistance, screening methods, and the updated cutoff points for the classification of natural cytotoxic products.
- Volume 114. African plant-based products as a source of potent drugs to overcome cancers and their chemoresistance. Part 2. Potent botanicals to overcome cancers and their chemoresistance.
- Volume 115. African plant-based products as a source of potent drugs to overcome cancers and their chemoresistance. Part 3. Potential pharmaceuticals to overcome cancers and their chemoresistance.

In the present volume 115, more than 296 plants and 606 plant constituents or other molecules with relevance to cancer chemotherapy are mentioned, to highlight the impressive potential of the plant kingdom in general, and the flora of Africa in particular, as a source of cytotoxic agents to tackle both drug-sensitive and refractory cancers. These compounds are pooled according to the main groups or classes of cytotoxic secondary metabolites, including terpenoids (Chapter 1), Coumarins (Chapter 2), Quinones (Chapter 3), Benzophenones (Chapter 4), Flavonoids and Isoflavonoids (Chapters 5 and 6), lignans, neolignans, and stilbenes (Chapter 7), xanthones (Chapter 8), and Alkaloids (Chapter 9). In each Chapter, the synopsis of the biosynthetic pathway of the class of cytotoxic secondary metabolites is also discussed, to provide a necessary background to appreciate the structure-activity relationship of naturally occurring agents. This book is preceded by volume 113 as indicated above, which provides the necessary background to appreciate the significance of the antiproliferative potential of the documented phytochemicals, as well as volume 114 which also highlights the best cytotoxic African medicinal plants. The book also opens the door for future volumes, which will update the various aspects according to the evolution of research at the continental level in the future, with a special emphasis on clinical studies.

I am very grateful to Mariana Kühl Leme, the Acquisitions Editor, Consolação, São Paulo-SP, Brazil, Jhon Michael Peñano, the Developmental Editor at Elsevier, Quezon City, Philippines, Sneha Apar, the Editorial Project Manager, and Jean-Pierre Jacquot, the Serial Editor, for their support and fruitful collaboration.

Victor Kuete
University of Dschang, Cameroon

CHAPTER ONE

Terpenoids, steroids, and saponins from African medicinal plants as potential pharmaceuticals to fight cancers, and their refractory phenotypes

Jenifer R.N. Kuete[a], Robert V.T. Kepdieu[a], Rémy B. Teponno[a], and Victor Kuete[b,*]

[a]Department of Chemistry, Faculty of Science, University of Dschang, Dschang, Cameroon
[b]Department of Biochemistry, Faculty of Science, University of Dschang, Dschang, Cameroon
*Corresponding author. e-mail address: kuetevictor@yahoo.fr

Contents

Advances in Botanical Research, Volume 115
ISSN 0065-2296, https://doi.org/10.1016/bs.abr.2024.02.011

Abstract

Terpenoids are the largest and most diverse group of normally occurring exacerbates generally found in plants with a variety of pharmacological activities. In this chapter, we have identified 111 terpenoids isolated from African medicinal plants with cytotoxic effects on various human cancer cell lines. They include five sesquiterpenoids, 25 diterpenoids, 29 steroids and steroidal saponins, and 52 triterpenoids and triterpene saponins. The most potent terpenoids include sesquiterpenoids 2*α*-hydroxyalantolactone (**1**), damsin (**2**), neoambrosin (**3**), vernomelitensin (**4**), and vernopicrin (**5**), diterpenes *ent*-trachyloban-3*β*-ol (**16**), *ent*-trachyloban-3-one (**17**), and salvimulticanol (**28**), steroids and steroidal saponins were 16*β*-formyloxymelianthugenin (**31**), 16*β*-hydroxybersaldegenin 1,3,5-orthoacetate (**32**), 2*β*-acetoxy-3,5-di-*O*-acetylhellebrigenin (**33**), 2*β*-acetoxy-5*β*-*O*-acetylhellebrigenin (**34**), 2*β*-acetoxymelianthusigenin (**35**), 2*β*-hydroxy-3*β*,5*β*-di-*O*-acetylhellebrigenin (**36**), stigmasterol (**39**), (25*R*)-17α-hydroxyspirost-5-en-3*β*-yl *O*-α-$_L$-rhamnopyranosyl-(1-2)-*O*-[*O*-*α*-$_L$-rhamnopyranosyl-(1-4)-*α*-$_L$-rhamnopyranosyl-(1-4)]-*β*-$_D$-glucopyranoside (**42**), balanitin 4 (**46**), balanitin 6 (**47**), balanitin 7 (**48**), and progenin III (**55**), triterpenoids and triterpene saponins include 2*α*, 3*α*, 19α,20*β*, 23-pentahydroxyurs-12-en-28-oic acid (**65**), 3*β*-hydroxy-urs-11-en13(28)-olide (**67**), betulin (**70**), dichapetalin X (**73**), erythrodiol (**75**), ardisiacrispin B (**99**), and lebbeckosides A (**4**) and B (**103**). Those having the ability to combat cancer drug resistance were identified as sesquiterpenoids **1**, **2**, and **3**, diterpenoid **28**, sterols **31**, **33**, **34**, **35**, and **36**, and steroidal saponin **55**, triterpenoids **70** and **71**, and triterpene saponins α-hederin (**98**) and olean-12-en-3-*ß*-*O*-$_D$-glucopyranoside (**105**). The most active terpenoids should be further explored to develop novel cytotoxic products to fight cancer.

1. Introduction

Terpenoids also called isoprenoids are the largest and most diverse group of normally occurring exacerbates generally found in plants; however, larger classes of terpenes, such as sterols and squalene, can be found in animals, microbes, and marine organisms. They are responsible for the smell, taste, and color of plants. Terpenes are characterized according to the association and number of isoprene units they contain (Yang et al., 2012). They occur in various chemical structures in a usual assortment of linear hydrocarbons or chiral carbocyclic backbones with different chemical modifications such as hydroxyl groups, ketones, aldehydes, and peroxides. The main classes of terpenoids include hemiterpenoids (C_5H_{10}), monoterpenoids ($C_{10}H_{16}$), sesquiterpenoids ($C_{15}H_{24}$), diterpenoids ($C_{20}H_{32}$), sesterterpenoids ($C_{25}H_{40}$), triterpenoids ($C_{30}H_{48}$), tetraterpenoids or caratenoids ($C_{40}H_{64}$), and polyisoprenoids ((C_5H_8)n) (Sell, 2007). Different terpene molecules are reported to have antimicrobial, antifungal, antiviral, antiparasitic, antihyperglycemic, antiallergic, anti-inflammatory, antispasmodic, immunomodulatory, and chemotherapeutic properties

(Cowan, 1999; Kuete, 2010; Kuete, 2013; Kuete, 2023a, 2023b; Kuete, Tangmouo, Penlap Beng, Ngounou, & Lontsi, 2006; Omosa et al., 2016). They can also be used as natural insecticides and protective substances in the storage of agricultural products. This diversity of structures and functions of terpenoids has led to great interest in their medicinal use and commercial applications such as flavors, fragrances, and spices. In addition, terpenoids have recently become important players in the biofuel market. Terpenoids with established medical applications include the antimalarial drug artemisinin and the anticancer drug taxol (Ajikumar et al., 2008; Guan et al., 2015; Thoppil & Bishayee, 2011; Wang, Tang, Bidigare, Zhang, & Demain, 2005). Some well-known cytotoxic terpenoids include sesquiterpenoids artemisinin and parthenolide, diterpenoids oridonin and triptolide, and triterpenoids alisol, betulinic acid, oleanolic acid, platycodin D, and ursolic acid (El-Baba et al., 2021). In the present Chapter, the synopsis of the cytotoxic potential of terpenoids isolated from African medicinal plants against human cancer cell lines including the multidrug-resistant (MDR) phenotypes will be provided. These terpenoids will include sesquiterpenoids, diterpenoids, sterol, and their saponins, and triterpenes and their saponins. The biogenesis of naturally occurring terpenoids as well as the terpene saponins will also be summarized.

2. Biosynthesis of terpenoids

Terpenoids are the largest, most structurally varied class of natural products with a spectacular abundance (~30,000 known compounds) and an extremely rich and long biochemical history. They are the oldest known natural products, as they are found in fossils and sediments of various ages (Talapatra & Talapatra, 2015). All terpenoids are oligomeric/polymeric products of isopentenyl pyrophosphate (IPP) and dimethylallyl pyrophosphaste (DMAPP) in which DMAPP serves as the primer. The origin of the isoprene units, IPP and DMAPP can be mevalonoid or non-mevalonoid pathways.

2.1 Biosynthesis of terpenoids

Terpenes are a group of compounds widely distributed in the plant kingdom. Their most important structural feature is the presence in their skeleton of isoprenic units with 5 carbon atoms (C_5H_8). This isoprene is the basis of the concept of the "isoprenic rule" or Ruzika's rule. This rule considers the isopentenyl diphosphate, designated as active isoprene, as the true precursor of the terpene molecule; hence the name isoprenoids by which they are also designated (Lamarti, Badoc, Deffieux, & Carde, 1994). IPP can be formed in two biological ways: the mevalonic acid pathway and the non-mevalonic or methylerythritol phosphate pathway, the most current is the mevalonic acid pathway (Scheme 1).

The condensation of IPP and DMAPP through the enzymatic activity of geranyl diphosphate (GPP) synthase (GPS) or farnesyl diphosphate synthase (FPS) provides the prenyl diphosphate substrates such as C10 (GPP) or C15 (farnesyl diphosphate (FPP)). GPP and FPP are the universal precursors of monoterpenoids (C10) and sesquiterpenoids (C15), respectively (Ludwiczuk, Skalicka-Woźniak, Georgiev, Badal, & Delgoda, 2017). Geranylgeranyl diphosphate (GGPP) synthase catalyzes the condensation of FPP with IPP, which results in the formation of the C20 precursor of the diterpenes, GGPP; or tail-to-tail coupling while dimerization of two FPP molecules and removal of the diphosphate groups through the activity of squalene synthase which results in the biosynthesis of squalene (C30) (Scheme 2). Squalene monooxygenase or epoxidase adds an oxygen group to the squalene, resulting in the production of 2,3-oxidosqualene, the precursor of triterpenoids (C30) as well as steroids in plants. Dimerization of two GGPP molecules and elimination of the two diphosphate groups by phytoene synthase results in the formation of a C40 compound: phytoene, the precursor of the tetraterpenoids or carotenoids (Table 1) (Ludwiczuk et al., 2017; Sell, 2003).

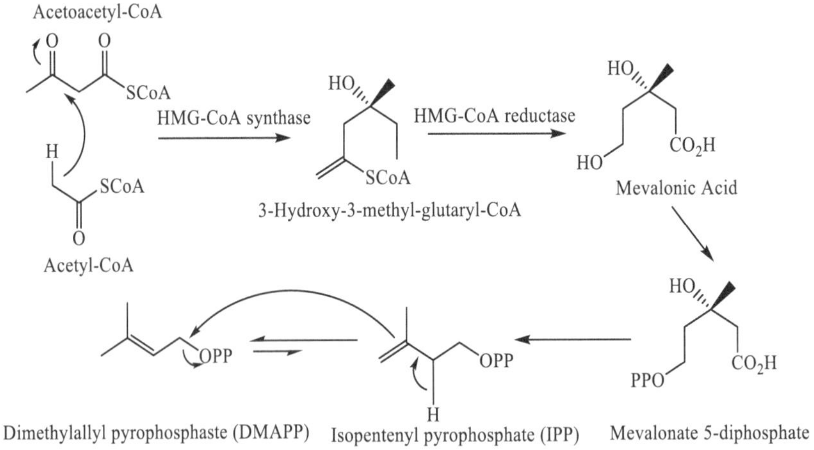

Scheme 1 Formation of IPP and DMAPP from the mevalonic acid pathway (Lamarti et al., 1994) HMG-CoA: β-Hydroxy β-methylglutaryl- coenzyme A.

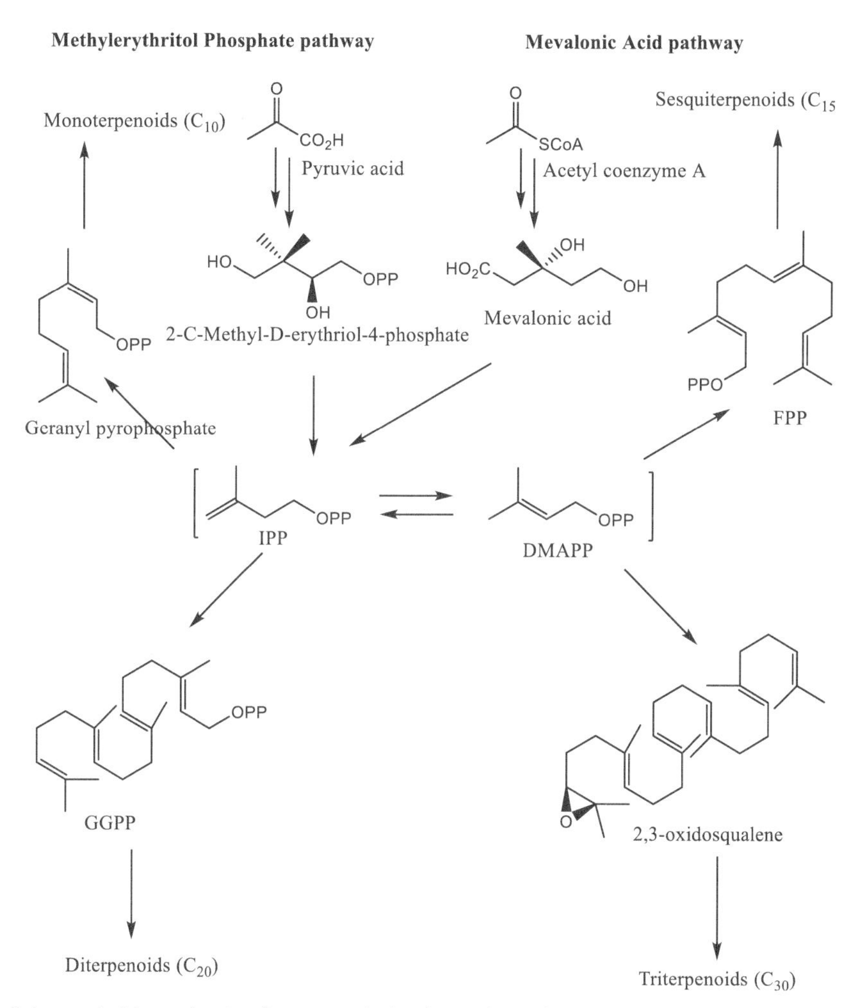

Scheme 2 Biosynthesis of terpenoids (Ludwiczuk et al., 2017).

2.2 Biogenesis of monoterpenes

The precursor molecule for the biosynthetic formation of all acyclic and cyclic monoterpenes is geranyl pyrophosphate formed by the head-to-tail coupling of IPP and DMAPP catalyzed by GPS (Talapatra & Talapatra, 2015). Monoterpenes are volatile, can be carried away by water vapor, often have a pleasant odor, and represent most of the constituents of essential oils (Lamarti et al., 1994). They are predominantly products of the secondary metabolism of plants, although specialized classes occur in some animals and microorganisms, and are usually isolated from the oils obtained

Table 1 Classification of terpenoids (Sell, 2003).

Name	Number of isoprene units	Number of carbon atoms	General formula
Hemiterpenoids	1	5	C_5H_{10}
Monoterpenoids	2	10	$C_{10}H_{16}$
Sesquiterpenoids	3	15	$C_{15}H_{24}$
Diterpenoids	4	20	$C_{20}H_{32}$
Sesterterpenoids	5	25	$C_{25}H_{40}$
Triterpenoids	6	30	$C_{30}H_{48}$
Tetraterpenoids/ caratenoids	8	40	$C_{40}H_{64}$
Polyisoprenoids	>8	>40	$(C_5H_8)n$

by steam distillation or solvent extraction of leaves, fruits, some heartwoods, and rarely roots and bark (Tchimene, Okunji, Iwu, Kuete, & Kuete, 2013). Monoterpenoids consist of a 10-carbon backbone (2 isoprene units) structure and can be divided into three subgroups: acyclic, monocyclic, and bicyclic (Scheme 3).

2.3 Biogenesis of sesquiterpenes

Sesquiterpenoids are produced from FPP through the catalytic activity of sesquiterpene synthases (Schemes 4–6). Sesquiterpenoids are derived from three isoprene units and exist in a wide variety of forms, including linear, monocyclic, bicyclic, and tricyclic frameworks. They are the most diverse group of terpenoids (Kashkooli, Krol, & Bouwmeester, 2018). They are also constituents of essential oils found particularly in higher plants and in other many living systems such as marine organisms and fungi. They occur in nature as hydrocarbons or in oxygenated forms including lactones, alcohols, acids, aldehydes, and ketones. Sesquiterpenes also include essential oils as well as aromatic components from plants (Awouafack, Tane, Kuete, & Eloff, 2013). Sesquiterpene lactones are a sub-class with over 4000 different known structures, chemically distinct from other sesquiterpenoids by the presence of a γ-lactone system, and can be distinguished in three major types: germacranolides, eudesmanolides, and guaianolides. Sesquiterpene lactones are mainly colorless and bitter compounds found mainly in plant species in the Asteraceae family (Kashkooli et al., 2018; Ludwiczuk et al., 2017).

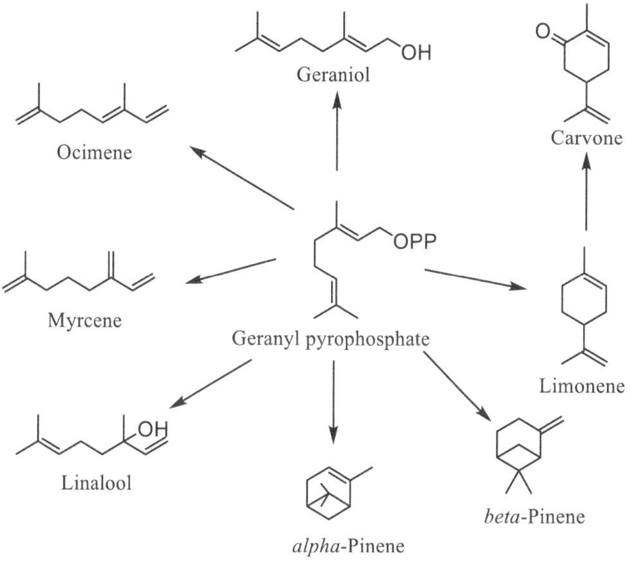

Scheme 3 Some monoterpenes formed from GPP in plants (Kang & Lee, 2016).

2.4 Biogenesis of diterpenes

Diterpenes contain 20 atoms of carbon (C_{20}), based of four isoprene units resulting from the condensation of IPP and DMAPP. However, the formation of diterpenoids starts with the condensation of DMAPP and IPP in a head-to-tail process by prenyltransferases to form GPP. Furthermore, GPP condenses with two more IPP isomerases, yielding through a farnesyl diphosphate intermediate, GGPP, which is the common precursor of diterpenes (Sandjo & Kuete, 2013a). They can be classified as linear (phytanes), bicyclic (labdanes, halimanes, clerodanes), tricyclic (pimaranes, cassanes, rosanes), tetracyclic (kauranes, gibberellanes), or macrocyclic diterpenes (taxanes, daphnanes) depending on their skeletal cores and are commonly found in a polyoxygenated form with keto and hydroxyl groups, often esterified by small-sized aliphatic or aromatic acids (Scheme 7) (Ludwiczuk et al., 2017).

2.5 Biogenesis of sesterterpenes

Sesterterpenoids are a relatively small group of terpenoids isolated from fungi, bacteria, lichens, higher plants, insects, and various marine invertebrate organisms, especially sponges. They are pentaprenyl terpenoids whose often complex polycyclic structures are derived from the linear precursor

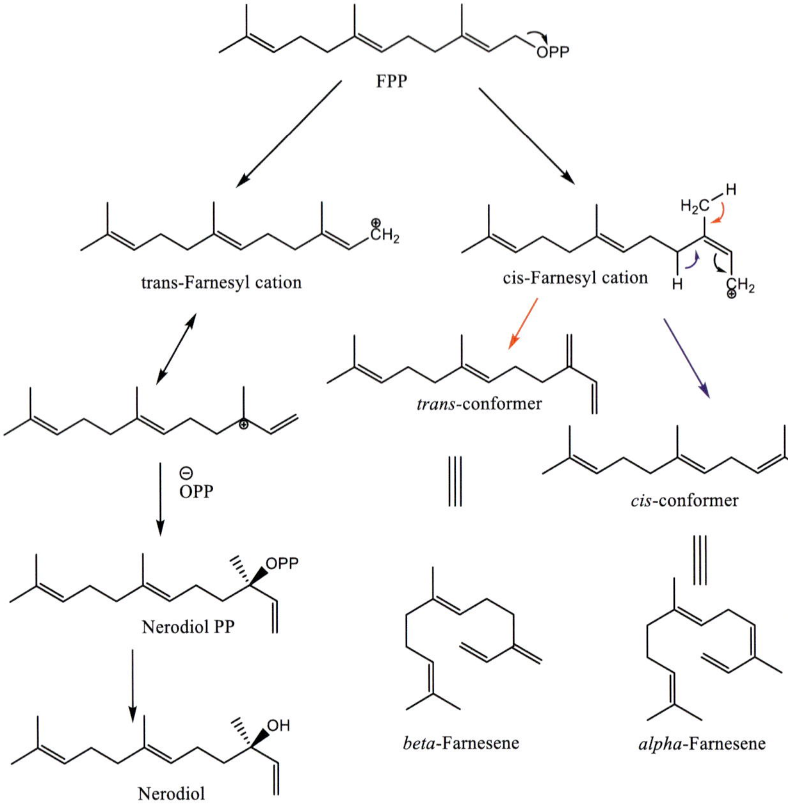

Scheme 4 Formation of acyclic sesquiterpenoids (Talapatra & Talapatra, 2015).

geranylfarnesyl diphosphate (GFPP) (Li & Gustafson, 2021). Plants have physically colocalized prenyl transferase (PT) and terpene synthase (TPS) gene pairs, which facilitates genome mining and subsequent analysis. Initial biosynthetic studies of sesterterpenes in plants focused on the PT enzymes responsible for the production of the C25 prenyl diphosphate precursor GFPP (Li & Gustafson, 2021). LCMS/MS studies revealed that isoprenyl diphosphate synthases from *Arabidopsis thaliana* preferentially catalyze the formation of GFPP. It is interesting to note that the production of GFPP was only detected in the roots, not in other plant parts. In addition, it was shown that the isopentenyl diphosphate synthases, which condense the basic C5 isoprene building blocks into the larger linear substrates for TPS, localized to distinct subcellular compartments in *A. thaliana* including plastids and mitochondria (Nagel et al., 2015).

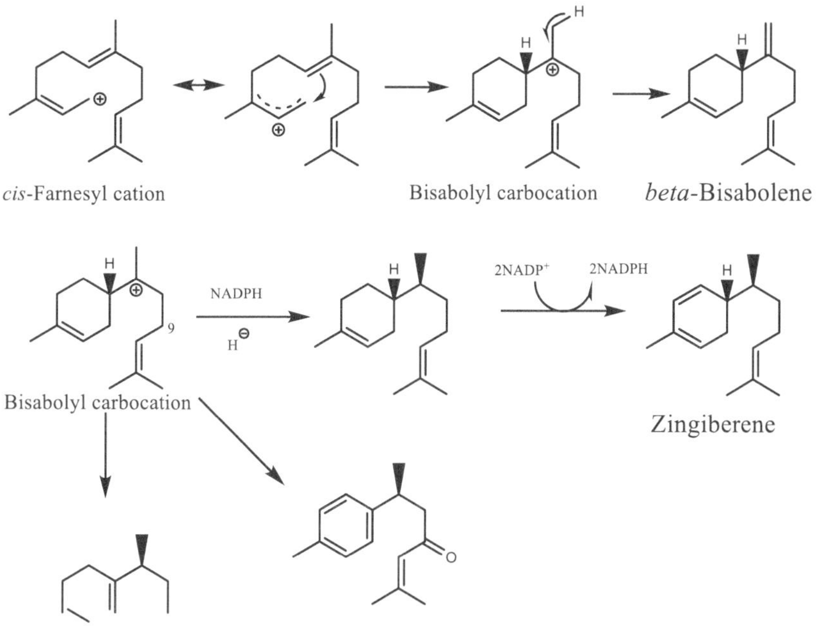

cis-Farnesyl cation

Bisabolyl carbocation

beta-Bisabolene

Bisabolyl carbocation

NADPH

2NADP$^+$ 2NADPH

Zingiberene

Scheme 5 Monocyclic sesquiterpenoids.

2.6 Biogenesis of triterpenes and steroids

With more than 20.000 identified compounds, triterpenoids are a large and diversified class of compounds including pentacyclic triterpenes, tetracyclic triterpenes, steroids, and saponins (glycosylated triterpenes or steroids) (Yan, Xia, Qiu, Li Sheng, & Xu, 2013). They derive from the tail-to-tail coupling of two units of FPP to form squalene (C_{30}), the precursor of triterpenoids. Squalene is then converted into (3*S*)-2,3-oxidosqualene by squalene epoxidase and some triterpene synthases convert 2,3-oxidosqualene by oxidosqualene cyclases, in a chair-boat-chair conformation or the chair-chair-chair conformation to give different triterpene skeletons (Scheme 8) (Thimmappa, Geisler, Louveau, O'Maille, & Osbourn, 2014). If epoxysqualene adopts a chair-chair-chair conformation, cyclization leads to a protostanyl cation that is a precursor to cycloartanes, lanostanes, and cucurbitanes that will lead to steroids such as ergosterol in fungi, cholesterol in animals, and β-sitosterol, stigmasterol or brassinosteroids in plants. If on the other hand, epoxysqualene adopts the chair-chair-chair conformation, cyclization gives rise to a dammaranyl cation. This will lead to tetracyclic triterpenes with a dammarane, euphane, and tirucallane skeletons when the cyclization is incomplete and most often it leads to pentacyclic triterpenes

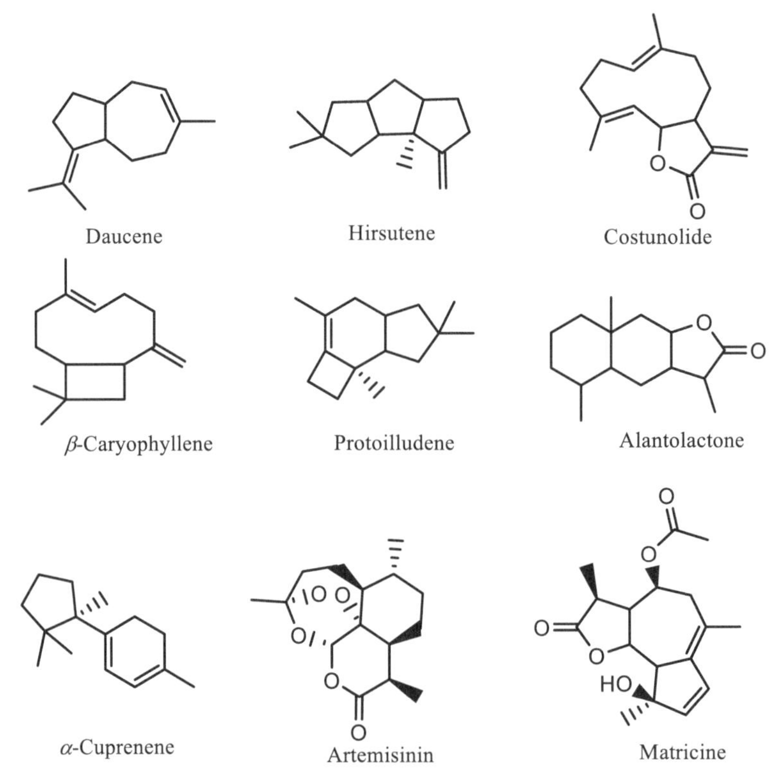

Scheme 6 Some bicyclic and tricyclic sesquiterpenes from farnesyl cation (Ludwiczuk et al., 2017; Quin, Flynn, & Dannert, 2014).

such as oleanane, ursane, hopane, lupane, taraxerane *etc.* when the cyclization is complete (Thimmappa et al., 2014). The most common form of triterpenoids is tetracyclic and pentacyclic. Steroids are considered like tetracyclic triterpenes which have lost three methyl groups, generally at C-4 and C-14. The transformation of a tetracyclic triterpene into a steroid occurs in several steps (Mercer, 1984).

- A methylation at C-24 thanks to the cofactor *S*-adenosylmethionine which constitutes a methyl donor in several biochemical reactions. This reaction leads to the formation of an exo-methylene if it is incomplete and to the formation of a double bond at C-22 if it is complete.
- A demethylation in C-14 and C-4: it results in a series of oxidations and decarboxylations, leading to the loss of these different methyl groups.
- And finally, there is the formation or rearrangement of double bonds in the molecule depending on the subclass of the steroid (Scheme 9).

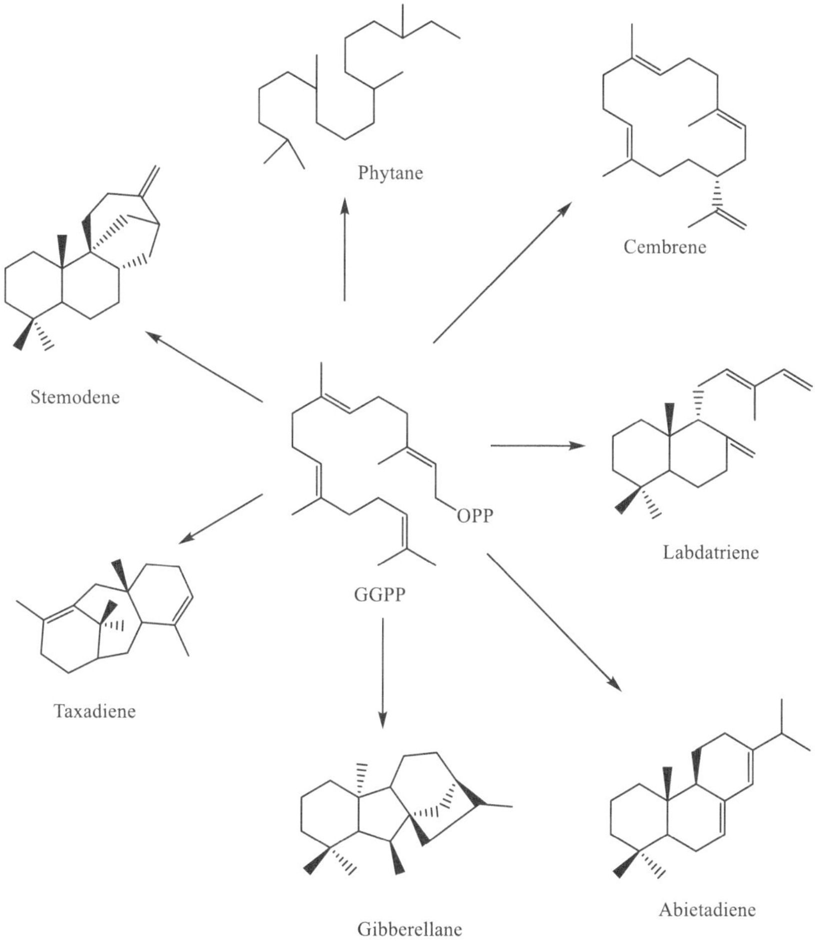

Phytane
Cembrene
Stemodene
OPP
GGPP
Labdatriene
Taxadiene
Gibberellane
Abietadiene

Scheme 7 Some diterpenoids skeletons from geranylgeranyl pyrophosphate (Sandjo & Kuete, 2013a).

3. Cell lines used to screen the cytotoxicity of terpenoids isolated from African medicinal plants

Several human and animal cancer and normal cell lines were used to assess the cytotoxicity of terpenoids isolated from African medicinal plants. They include cell lines from breast cancer (MCF-7, MDA-MB-231-*pcDNA*, and MDA-MB-231-*BCRP*), cervical cancer (HeLa), colon cancer (Caco-2, DLD-1, HCT116 $p53^{+/+}$, HCT116 $p53^{-/-}$, and HT-29), epidermoid carcinoma (A431), fibroblasts (BJ), glioblastoma (U87MG and

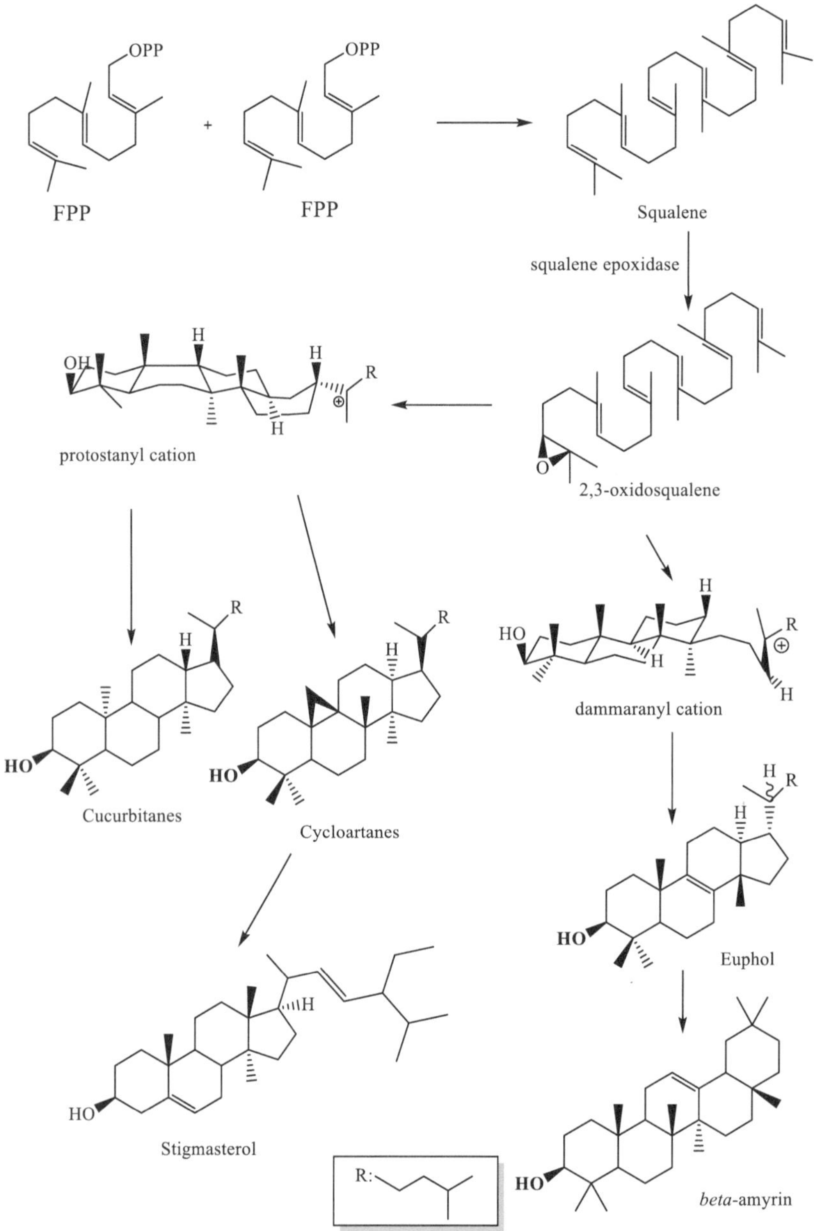

Scheme 8 Biosynthesis of tetracyclic triterpenes, pentacyclic triterpenes, and steroids.

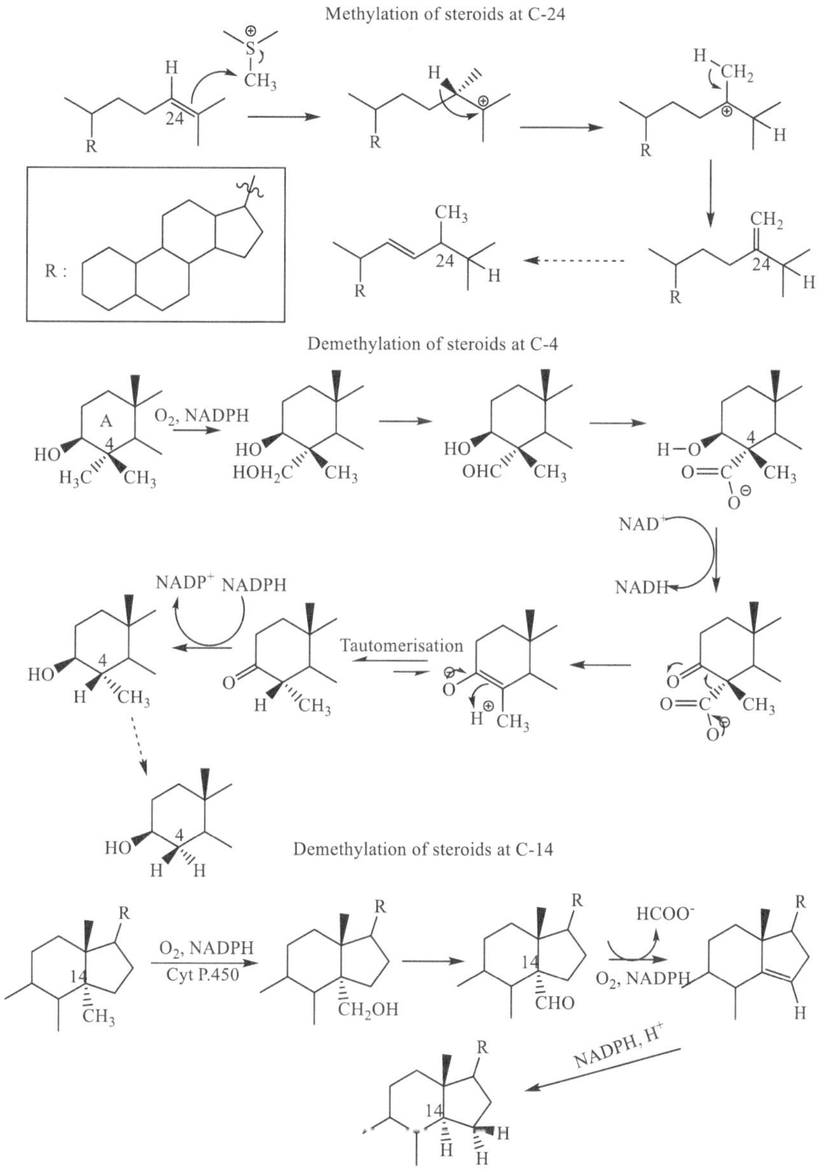

Scheme 9 Conversion of tetracyclic triterpenes to steroids.

U87MG.Δ*EGFR*, TG1 GSC, and U373), hepatocarcinoma (HepG2), human umbilical vein endothelial cell line (ECV-304), leukemia (CCRF-CEM, CEM/ADR5000, HL60, Jurkat, THP-1, and P388 (from mouse)), and lung cancer (A549 and SPC212), melanoma (A375 and B16-F1 (mouse)), pancreatic cancer (AsPC-1, Panc-1, and MiaPaCa-2), and prostate cancer (DU-45 and PC-3) cell lines (Table 2).

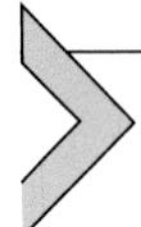

4. Cytotoxic terpenoids from African medicinal plants towards drug sensitive and multidrug-resistant cancer cells

A total of 111 cytotoxic terpenoids belonging to sesquiterpenoids, diterpenoids, steroids and steroidal saponins, and triterpenoids and triterpene saponins have been isolated from African medicinal plants. Their effects on various human cancer cell lines are summarized in Table 2. The degree of resistance (D.R.) is determined as the ratio of the IC_{50} value in resistant cells *versus* the IC_{50} values in the corresponding sensitive cell line: CEM/ADR5000 cells *vs* CCRF-CEM cells, MDA-MB-231-*BCRP* *vs* MDA-MB-231-*pcDNA3*, HCT116 $p53^{-/-}$ cells *vs* HCT116 $p53^{+/+}$ cells, and U87MG.Δ*EGFR* *vs* U87MG, is also depicted in Table 2 (Mbaveng, Noulala, et al., 2020). In the laboratory scale, collateral sensitivity is achieved when the D.R. is defined as the ratio of the IC_{50} values of the cytotoxic agent towards the resistant cell line *versus* that in its sensitive counterpart is below 1 whilst the D.R. above 1 defines the normal sensitivity. It has been established that the D.R. < 0.9 define the collateral sensitivity, whilst D.R. between 0.9 and 1.2 define the normal sensitivity. The cross-resistance is noted if the cytotoxic agent is more active in the sensitive cell line than its resistant subline, with D.R above 1.2 (Efferth et al., 2020; Efferth et al., 2021; Mbaveng, Kuete, & Efferth, 2017). Collateral or normal sensitivities should be achieved for samples with the ability to combat drug resistant cancer cells (Efferth et al., 2020; Kuete & Efferth, 2015). In this section, the activity of these flavonoids will be discussed according to the cut-off points of cytotoxic phytochemicals defined as follows: outstanding activity ($IC_{50} \leq 0.5\ \mu M$), excellent activity ($0.5 < IC_{50} \leq 2\ \mu M$), very good activity ($2 < IC_{50} \leq 5\ \mu M$), good activity ($5 < IC_{50} \leq 10\ \mu M$), average activity ($10 < IC_{50} \leq 20\ \mu M$), weak activity ($20 < IC_{50} \leq 60\ \mu M$), very weak activity ($60 < IC_{50} \leq 150\ \mu M$), and not active ($IC_{50} > 150\ \mu M$) (Kuete, 2025). In this section, emphasis will be focused on the promising compounds with cytotoxic activity ranging from outstanding to good and those that can combat cancer drug resistance.

Table 2 Cytotoxic terpenoids from African medicinal plants and their effects on sensitive and drug-resistant cancer cell lines.

Compounds names	Source	Country	Cancer cell lines and IC_{50} values (μM) and degree of resistance in bracket	References
Sesquiterpenoids				
2α-Hydroxyalantolactone (**1**)	*Pulicaria undulata*	Egypt	0.36 (CCRF-CEM) *vs* 0.27 (CEM/ADR5000) [0.75]; 12.16 (MDA-MB-231-*pcDNA*) *vs* 40.48 (MDA-MB-231-*BCRP*) [3.32]; 5.91 (HCT116 $p53^{+/+}$) *vs* 6.15 (HCT116 $p53^{-/-}$) [1.04]; 15.15 (U87MG) *vs* 49.22 (U87MG.Δ*EGFR*) [3.24]	Hegazy et al. (2021)
Damsin (**2**)	*Ambrosia maritima*	Sudan	4.8 (CCRF-CEM) *vs* 4.7 (CEM/ADR5000) [1]; 33.8 (MDA-MB-231-*pcDNA*) *vs* 46.4 (MDA-MB-231-*BCRP*) [1.4]; 32.5 (HCT116 $p53^{+/+}$) *vs* 133.6 (HCT116 $p53^{-/-}$) [4.1]; 154.4 (U87MG) *vs* 24.1 (U87MG.Δ*EGFR*) [0.15]	Saeed et al. (2015)
Neoambrosin (**3**)	*Ambrosia maritima*	Sudan	4.5 (CCRF-CEM) *vs* 4.8 (CEM/ADR5000) [1.07]; 29.9 (MDA-MB-231-*pcDNA*) *vs* 43.3 (MDA-MB-231-*BCRP*) [1.4]; 26.5 (HCT116 $p53^{+/+}$) *vs* 39.4 (HCT116 $p53^{-/-}$) [1.5]; 132.2 (U87MG) *vs* 24.1 (U87MG.Δ*EGFR*) [0.18]	Saeed et al. (2015)
Vernomelitensin (**4**)	*Vernonia guineensis*	Cameroon	1.12 (HL60); 0.367 (MDA-MB-231-*BCRP*); 1.56 (MCF-7); 0.14 (HCT116); 0.64 (PC-3); 0.41 (DU-45); 0.25 (MiaPaCa-2); 1.13 (A549)	Toyang et al. (2013)

(*continued*)

Table 2 Cytotoxic terpenoids from African medicinal plants and their effects on sensitive and drug-resistant cancer cell lines. (*cont'd*)

Compounds names	**Source**	**Country**	**Cancer cell lines and IC_{50} values (μM) and degree of resistance in bracket**	**References**
Vernopicrin (**5**)	*Vernonia guineensis*	Cameroon	1.55 (HL60); 1.01 (MDA-MB-231-*BCRP*); 0.67 (MCF-7); 0.45 (HCT116); 1.00 (PC-3); 0.49 (DU-45); 0.42 (MiaPaCa-2); 2.04 (A549)	Toyang et al. (2013)
Diterpenoids				
2,18,19-Trachylobanetriol (**6**)	*Psiadia punctulata*	Cameroon	99.74 (A549); 36.96 (SPC212)	Kuete, Omosa, Midiwo, Karaosmanoğlu, and Sivas (2019)
2,6,19-Trachylobanetriol (ent-2α,6α)-form (**7**)	*Psiadia punctulata*	Kenya	108.73 (MCF-7); 94.84 (Caco-2); 104.08 (DLD-1); 104.08 (HepG2); 86.19 (A549); 18.91 (SPC212)	Kuete et al. (2019)
2-Oxocandesalvone (**8**)	*Salvia multicaulis*	Egypt	11.58 (CCRF-CEM) *vs* 11.37 (CEM/ADR5000) [0.98]; 20.55 (MDA-MB-231-*pcDNA*) *vs* 14.53 (MDA-MB-231-*BCRP*) [0.70]; 19.82 (HCT116 $p53^{+/+}$) *vs* 19.94 (HCT116 $p53^{-/-}$) [1.00]; 19.95 (U87MG) *vs* 23.84 (U87MG.Δ*EGFR*) [1.19]	Hegazy, Hamed, El-Halawany, et al. (2018)
4,12,20-Trideoxyphorbol-13-(2,3-dimethyl) butyrate (**9**)	*Euphorbia sanctae-catharinae*	Egypt	29.4 (Caco-2)	Hegazy, Hamed, Ibrahim, et al. (2018)

4,20-Dideoxy (4α) phorbol-12-benzoate-13-isobutyrate (**10**)	*Euphorbia Sanctae-Catharinae*	Egypt	31.3 (A549); 26.1 (Caco-2)	Hegazy, Hamed, Ibrahim, et al. (2018)
6-β-Hydroxycandesalvone B (**11**)	*Salvia multicaulis*	Egypt	21.54 (CCRF-CEM) *vs* 10.77 (CEM/ADR5000) [0.5]; 46.46 (HCT116 $p53^{+/+}$) *vs* 77.15 (HCT116 $p53^{-/-}$) [1.66]	Hegazy, Hamed, El-Halawany, et al. (2018)
7β,13β,17-*O*-Triacetyl-5α-*O*-(2-methylbutyryl)-3β-*O*-propanoyl14-oxopremyrsinol (**12**)	*Euphorbia sanctae-catharinae*	Egypt	43.5 (Caco-2); 50.1 (A549)	Hegazy, Hamed, Ibrahim, et al. (2018)
Candesalvone B (**13**)	*Salvia multicaulis*	Egypt	31.52 (CCRF-CEM) *vs* 19.61 (CEM/ADR5000) [0.62]	Hegazy, Hamed, El-Halawany, et al. (2018)
Candesalvone B methyl ester (**14**)	*Salvia multicaulis*	Egypt	20.95 (CCRF-CEM) *vs* 4.13 (CEM/ADR5000) [0.19]; 89.15 (MDA-MB-231-*pcDNA*) *vs* 60.44 (MDA-MB-231-*BCRP*) [0.67]; 41.02 (HCT116 $p53^{+/+}$) *vs* 70.43 (HCT116 $p53^{-/-}$) [1.71]	Hegazy, Hamed, El-Halawany, et al. (2018)
Ent-18-hydroxy-trachyloban-3-one (**15**)	*Croton zambesicus*	Benin	12.2 (HeLa)	Block et al. (2004)

(continued)

Table 2 Cytotoxic terpenoids from African medicinal plants and their effects on sensitive and drug-resistant cancer cell lines. (*cont'd*)

Compounds names	Source	Country	Cancer cell lines and IC_{50} values (μM) and degree of resistance in bracket	References
Ent-Trachyloban-3β-ol (**16**)	*Croton zambesicus*	Benin	7.3 (HeLa)	Block et al. (2002)
Ent-Trachyloban-3-one (**17**)	*Croton zambesicus*	Benin	9.6 (HeLa)	Block et al. (2004)
Esopimara-7,15-dien-3β-ol (**18**)	*Croton zambesicus*	Benin	25.3 (HeLa)	Block et al. (2004)
Euphosantianane A (**19**)	*Euphorbia sanctae-catharinae*	Egypt	75.8 (Caco-2)	Hegazy, Hamed, Ibrahim, et al. (2018)
Euphosantianane B (**29**)	*Euphorbia sanctae-catharinae*	Egypt	40.5 (Caco-2); 48.5 (A549)	Hegazy, Hamed, Ibrahim, et al. (2018)
Euphosantianane C (**21**)	*Euphorbia sanctae-catharinae*	Egypt	31.0 (Caco-2); 21.5 (A549)	Hegazy, Hamed, Ibrahim, et al. (2018)
Euphosantianane D (**22**)	*Euphorbia sanctae-catharinae*	Egypt	33.2 (Caco-2); 32.8 (A549)	Hegazy, Hamed, Ibrahim, et al. (2018)

Galanal A (**23**)	*Aframomum arundinaceum*	Cameroon	17.32 (CCRF-CEM); 18 (Jurkat); 27.99 (MDA-MB-231-*BCRP*) [<0.70]	Kuete, Ango, et al. (2014); Miyoshi et al. (2003)
Galanal B (**24**)	*Aframomum arundina-ceum*	Cameroon	19.81 (CCRF-CEM); 32 (Jurkat); 83.69 (MDA-MB-231-*pcDNA*) *vs* 74.66 (MDA-MB-231-*BCRP*) [0.89]	Kuete, Ango, et al., (2014); Miyoshi et al. (2003)
Hautriwaic acid lactone (**25**)	*Dodonaea angustifolia*	Kenya	78.91 (MCF-7); 64.38 (Caco-2); 85.41 (DLD-1); 85.41 (HepG2)	Kuete et al. (2019)
Premyrsinol-3-propanoate-5-isobutyrate-7,13,17-triacetate (**26**)	*Euphorbia sanctae-catharinae*	Egypt	33.3 (Caco-2); 33.1 (A549)	Hegazy, Hamed, Ibrahim, et al. (2018)
Premyrsinol-3-propanoate-5-isobutyrate-7,13-triacetate-17-nicotinate (**27**)	*Euphorbia Sanctae-Catharinae*	Egypt	60.3 (A549); 40.3 (Caco-2)	Hegazy, Hamed, Ibrahim, et al. (2018)
Salvimulticanol (**28**)	*Salvia multicaulis*	Egypt	15.32 (CCRF-CEM) *vs* 8.36 (CEM/ADR5000) [0.54]; 32.01 (MDA-MB-231-*pcDNA*) *vs* 26.40 (MDA-MB-231-*BCRP*) [0.82]; 30.91 (HCT116 *p53*$^{+/+}$) *vs* 38.95 (HCT116 *p53*$^{-/-}$) [1.26]; 30.33 (U87MG) *vs* 35.31 (U87MG.Δ*EGFR*) [1.16]	Hegazy, Hamed, El-Halawany, et al. (2018)

(continued)

Table 2 Cytotoxic terpenoids from African medicinal plants and their effects on sensitive and drug-resistant cancer cell lines. (*cont'd*)

Compounds names	Source	Country	Cancer cell lines and IC_{50} values (μM) and degree of resistance in bracket	References
Salvimulticaoic acid (**29**)	*Salvia multicaulis*	Egypt	28.28 (CCRF-CEM) *vs* 21.6 (CEM/ADR5000) [0.76]	Hegazy, Hamed, El-Halawany, et al. (2018)
Trans-phytol (**30**)	*Croton zambesicus*	Benin	13.8 (HeLa)	Block et al., (2004)
16α-Hydroxy-*ent*-kauran-19-oic acid (**61**)	*Xylopia aethiopica*	Cameroon	72.28 (CCRF-CEM) *vs* 101.59 (CEM/ADR5000) [1.41]; 85.86 (HepG2)	Kuete, Sandjo, Mbaveng, Zeino, and Efferth, (2015)
Steroids and steroidal saponins				
16β-Formyloxymelianthugenin (**31**)	*Melianthus comosus*	South Africa	0.07 (CCRF-CEM) *vs* 0.06 (CEM/ADR5000) [0.86]; 0.36 (MCF-7)	Bedane et al. (2020a)
16β-Hydroxybersaldegenin 1,3,5-orthoacetate (**32**)	*Melianthus comosus*	South Africa	0.63 (CCRF-CEM) *vs* 0.80 (CEM/ADR5000) [1.27]; 3.09 (MCF-7)	Bedane et al. (2020a)
2β-Acetoxy-3,5-di-*O*-acetylhellebrigenin (**33**)	*Melianthus major*	South Africa	0.1 (CCRF-CEM) *vs* 0.1 (CEM/ADR5000) [1]; 0.3 (MCF-7)	Bedane et al., (2020b)
2β-Acetoxy-5β-*O*-acetylhellebrigenin (**34**)	*Melianthus comosus*	South Africa	0.13 (CCRF-CEM) *vs* 0.08 (CEM/ADR5000) [0.62]; 0.53 (MCF-7)	Bedane et al. (2020a)
2β-Acetoxymelianthusigenin (**35**)	*Melianthus comosus*	South Africa	1.44 (CCRF-CEM) *vs* 1.08 (CEM/ADR5000) [0.75]; 8.54 (MCF-7)	Bedane et al. (2020a)

2β-Hydroxy-3β,5β-di-O-acetylhellebrigenin (**36**)	*Melianthus comosus*	South Africa	0.19 (CCRF-CEM) *vs* 0.17 (CEM/ADR5000) [0.89]; 0.65 (MCF-7)	Bedane et al. (2020a)
Daucosterol (**37**)	*Crateva adansonii*	Cameroon	30.25 (PC-3); 50.71 (DU-45)	Zingue et al. (2020)
Sitosterol (**38**)	*Citrus reticulata; Raphia vinifera*	Cameroon	38.03 (CCRF-CEM) *vs* 76.40 (CEM/ADR5000) [2.00]; 232.61 (A549)	Chi et al. (2020); Tahsin et al. (2017)
Stigmasterol (**39**)	*Citrus reticulata; Drimia maritima*	Cameroon; Egypt	2.0 (PC-3); 51.21 (A549)	Mohamed, Ibrahim, Shaala, Alshali, & Youssef (2014); Tahsin et al. (2017)
β-Spinasterol (**40**)	*Garcinia epunctata*	Cameroon	69.81 (Caco-2)	Kuete, Fokou, Karaosmanoğlu, Beng, and Sivas (2017)
(20*R*)-*O*-(3)-β-D-glucopyranosyl(1-2)-α-L-arabinopyranosyl-pregn-5-en-3β,20-diol (**41**)	*Brucea antidysenterica*	Cameroon	31.99 (CCRF-CEM); 32.16 (HCT116 $p53^{-/-}$); 31.17 (HepG2)	Youmbi et al. (2023)

(continued)

Table 2 Cytotoxic terpenoids from African medicinal plants and their effects on sensitive and drug-resistant cancer cell lines. (*cont'd*)

Compounds names	Source	Country	Cancer cell lines and IC_{50} values (µM) and degree of resistance in bracket	References
(25*R*)-17α-Hydroxyspirost-5-en-3β-yl *O*-α-L-rhamnopyranosyl-(1-2)-*O*-[*O*-α-L-rhamnopyranosyl-(1-4)-α-L-rhamnopyranosyl-(1-4)]-β-D-glucopyranoside **(42)**	*Dioscorea preussii*	Cameroon	2.17 (HCT116); 1.64 (HT-29)	Tabopda et al. (2014)
(25*R*)-17α-Hydroxyspirost-5-en-3β-yl *O*-α-L-rhamnopyranosyl-(1-4)-*O*-α-L-rhamnopyranosyl-(1-4)-β-D-glucopyranoside **(43)**	*Dioscorea preussii*	Cameroon	37.41 (HCT116); 42.43 (HT-29)	Tabopda et al. (2014)
(25*R*)-spirost-5-ene-3β, 22β-3-*O*-β-D-glucopyranosyl(1-2)-*O*-α-L-rhamnopyranoside (**44**)	*Raphia vinifera*	Cameroon	3.55 (CCRF-CEM) *vs* 9.19 (CEM/ADR5000) [7.30]; 13.42 (MDA-MB-231-*pcDNA*) *vs* 17.56 (MDA-MB-231-*BCRP*) [1.31]; 12.64 (HCT116 $p53^{+/+}$) *vs* 22.52 (HCT116 $p53^{-/-}$) [1.78]; 18.08 (U87MG) *vs* 17.48 (U87MG.Δ*EGFR*) [0.97]; 21.10 (HepG2)	Chi et al. (2020)
26-*O*-β-D-Glucopyranosyl-(25*R*)-5-en-furost-3β,17α,22α,26-tetraol-3-*O*-α-L-rhamnopyranosyl-(1-4)-α-L-rhamnopyranosyl-(1-4)-[α-L-rhamnopyranosyl-(1-2)]-β-D-glucopyranoside **(45)**	*Dioscorea bulbifera*	Cameroon	14.3 (ECV-304)	Tapondjou, Jenett-siems, Böttger, and Melzig (2013)

Balanitin 4 **(46)**	*Balanites aegyptica*	Burkina Faso	0.41 (P388)	Pettit, Doubek, and Herald, (1991)
Balanitin 6 **(47)**	*Balanites aegyptica*	Burkina Faso	0.3 (A549)	Gnoula et al. (2008)
Balanitin 7 **(48)**	*Balanites aegyptica*	Burkina Faso	0.5 (U373)	Gnoula et al. (2008)
Diosgenin (**49**)	*Raphia vinifera*	Cameroon	51.90 (CCRF-CEM) *vs* >100 (CEM/ADR5000) [>1.93]; 88.47 (HepG2)	Chi et al. (2020)
Diospreussinosides E **(50)**	*Dioscorea preussii*	Cameroon	48.70 (HCT116); 31.0 (HT-29)	Tabopda et al. (2014)
Fruticoside H **(51)**	*Cordyline fruticosa*	Cameroon	69.63 (MDA-MB 231); 37.83 (A375); 39.80 (HCT116)	Fouedjou et al. (2014)
Fruticoside I **(52)**	*Cordyline fruticosa*	Cameroon	50.45 (MDA-MB 231); 46.59 (A375); 59.97 (A375)	Fouedjou et al. (2014)
Parquispiroside **(53)**	*Cestrum parqui*	Egypt	7.7 (HeLa); 7.2 (HepG2); 14.1 (MCF-7); 3.3 (U87)	Mosad et al. (2017)

(*continued*)

Table 2 Cytotoxic terpenoids from African medicinal plants and their effects on sensitive and drug-resistant cancer cell lines. (*cont'd*)

Compounds names	Source	Country	Cancer cell lines and IC_{50} values (μM) and degree of resistance in bracket	References
Pennogenin 3-*O*-α-L-rhamnopyranosyl-(1-4)-α-L-rhamnopyranosyl-(1-4)-[α-L-rhamnopyranosyl-(1-2)]-β-D-glucopyranoside **(54)**	*Dioscorea bulbifera*	Cameroon	8.5 (ECV-304)	Tapondjou et al. (2013)
Progenin III (**55**)	*Raphia vinifera*	Cameroon	1.59 (CCRF-CEM) *vs* 1.70 (CEM/ADR5000) [1.07]; 3.17 (MDA-MB-231-*pcDNA*) *vs* 4.22 (MDA-MB-231-*BCRP*) [1.33]; 3.43 (HCT116 $p53^{+/+}$) *vs* 3.69 (HCT116 $p53^{-/-}$) [1.08]; 3.13 (U87MG) *vs* 4.77 (U87MG.Δ*EGFR*) [1.52]; 10.24 (HepG2)	Mbaveng, Chi, Nguenang, et al. (2020)
Raphvinin 1 (**56**)	*Raphia vinifera*	Cameroon	21.62 (CCRF-CEM) *vs* >100 (CEM/ADR5000) [4.63]	Chi et al. (2020)
Raphvinin 3 (**57**)	*Raphia vinifera*	Cameroon	7.14 (CCRF-CEM) *vs* 12.29 (CEM/ADR5000) [1.72]; 33.34 (MDA-MB-231-*pcDNA*) *vs* 28.12 (MDA-MB-231-*BCRP*) [0.84]; 22.78 (HCT116 $p53^{+/+}$) *vs* 20.23 (HCT116 $p53^{-/-}$) [0.89]; 33.46 (U87MG) *vs* 27.39 (U87MG.Δ*EGFR*) [0.82]; 29.41 (HepG2)	Chi et al. (2020)
Spiroconazole A **(58)**	*Dioscorea bulbifera*	Cameroon	5.8 (ECV-304)	Tapondjou et al. (2013)

Trillin (**59**)	*Raphia vinifera*	Cameroon	56.88 (CCRF-CEM) *vs* >100 (CEM/ADR5000) [1.76]	Chi et al. (2020)
Triterpenoids and triterpene saponins				
11-Oxo-α-amyryl acetate (**60**)	*Uapaca togoensis*	Cameroon	4.53 (CCRF-CEM) *vs* 78.93 (CEM/ADR5000) [17.42]; 56.89 (HepG2)	Kuete et al. (2015); Seukep, Sandjo, Ngadjui, and Kuete (2016)
16β-Hydroxylupeol (**62**)	*Garcinia epunctata*	Cameroon	9.12 (MCF-7); 20.76 (Caco-2); 15.41 (DLD-1); 35.64 (HepG2)	Kuete et al. (2018)
1β, 2α, 3α, 5, 19α, 24-Hexahydroxyurs-11(12), 20(30)-dien-28-oic acid (**63**)	*Manilkara pellegriniana*	Cameroon	59.18 (A549); 12.39 (SPC212)	Mogue et al. (2019)
23-Hydroxyursolic acid (**64**)	*Cussonia bancoensis*	Cameroon	30.04 (A375); 35.30 (MDA-MB 231); 23.84 (HCT116)	Ponou et al. (2014)
2α, 3α, 19α,20β, 23-Pentahydroxyurs-12-en-28-oic acid (**65**)	*Manilkara pellegriniana*	Cameroon	72.81 (A549); 0.52 (SPC212)	Mogue et al. (2019)
2β,3β,13α,22α-Tetrahydroxy oleanane-23,28-dioic acid (**66**)	*Gladiolus segetum*	Egypt	16.51 (HepG2); 4.87 (PC-3)	Abd El-Kader et al. (2020)

(continued)

Table 2 Cytotoxic terpenoids from African medicinal plants and their effects on sensitive and drug-resistant cancer cell lines. (*cont'd*)

Compounds names	Source	Country	Cancer cell lines and IC_{50} values (μM) and degree of resistance in bracket	References
3-*O*-β-D-Glucopyranosyl-(1-3)-β-D-glucopyranosyl-27-hydroxyolean-12-en-28-oic acid (**87**)	*Tetrapleura tetraptera*	Cameroon	20.07 (CCRF-CEM) *vs* 16.15 (CEM/ADR5000) [0.80]; 24.11 (MDA-MB-231-*pcDNA*) *vs* 28.54 (MDA-MB-231-*BCRP*) [1.18]; 23.20 (HCT116 $p53^{+/+}$) *vs* 20.22 (HCT116 $p53^{-/-}$) [0.87]; 29.23 (U87MG) *vs* 23.16 (U87MG.Δ*EGFR*) [0.79]; 17.91 (HepG2)	Mbaveng, Chi, Bonsou, et al. (2020)
3-*O*-β-D-Glucopyranosyl-(1-6)-β-D-glucopyranosylurs-12-en-28-oic acid (**88**)	*Tetrapleura tetraptera*	Cameroon	15.38 (CCRF-CEM) *vs* 37.95 (CEM/ADR5000) [2.47]; 8.19 (MDA-MB-231-*pcDNA*) *vs* 10.98 (MDA-MB-231-*BCRP*) [1.19]; 12.30 (HCT116 $p53^{+/+}$) *vs* 10.23 (HCT116 $p53^{-/-}$) [0.83]; 10.86 (U87MG) *vs* 11.12 (U87MG.Δ*EGFR*) [1.03]; 13.03 (HepG2)	Mbaveng, Chi, Bonsou, et al. (2020)
3-*O*-[β-D-Galactopyranosyl-(1-4)-β-D-galactopyranosyl]-oleanolic acid (**89**)	*Acacia polyacantha*	Cameroon	35.16 (CCRF-CEM)	Fotso et al. (2019)
3-*O*-[β-D-Glucopyranosyl-(1-2)-β-D-glucopyranosyl-(1-4)-β-D-xylopyranosyl]-2β,3β,16α-trihydroxyolean-12-en-23,28-dioic acid-28-*O*-α-L-rhamnopyranosyl-(1-4)-α-L-rhamnopyranosyl-(1-2)-β-D-glucopyranosyl-(1-2)-α-L-arabinopyranoside (**90**)	*Gladiolus segetum*	Egypt	2.67 (MCF-7); 3.78 (HepG2); 2.18 (PC-3)	Abd El-Kader et al. (2020)

3-*O*-{β-D-Glucopyranosyl-(1-2)-[α-L-arabinopyranosyl-(1-3)]-β-D-glucuronopyranosyl}-21-*O*-(2-methylbutyroyl)-22-*O*-acetyl-R1-barrigenol (**91**)	*Hydrocotyle bonariensis*	Cameroon	24.1 (HT-29); 24.0 (HCT116)	Tabopda et al. (2012)
3-*O*-{β-D-Glucopyranosyl-(1-2)-[α-L-arabinopyranosyl-(1-3)]-β-D-glucuronopyranosyl}-21-*O*-(2-methylbutyroyl)-28-*O*-acetyl-R1-barrigenol (**92**)	*Hydrocotyle bonariensis*	Cameroon	83.6 (HT-29); 83.0(HCT116)	Tabopda et al. (2012)
3-*O*-α-L-Arabinopyranosyl-23-hydroxyursolic acid (**93**)	*Cussonia bancoensis*	Cameroon	48.44(A375); 50.12 (MDA-MB 231); 40.02 (HCT116)	Ponou et al. (2014)
3-*O*-β-D-Galactopyranosyl-(1-2)-β-D-glucuronopyranosyloleanolic acid (**94**)	*Cussonia bancoensis*	Cameroon	87.92 (HCT116)	Ponou et al. (2014)
3-*O*-β-D-Glucopyranosyl-23-hydroxyursolic acid (**95**)	*Cussonia bancoensis*	Cameroon	92.16 (A375); 99.05 (MDA-MB 231); 66.03 (HCT116)	Ponou et al. (2014)
3β-Hydroxy-urs-11-en13(28)-olide (**67**)	*Callistemon citrinus*	Egypt	0.4 (Panc-1)	Tawila et al. (2020)

(continued)

Table 2 Cytotoxic terpenoids from African medicinal plants and their effects on sensitive and drug-resistant cancer cell lines. (*cont'd*)

Compounds names	Source	Country	Cancer cell lines and IC_{50} values (µM) and degree of resistance in bracket	References
7-Hydroxydichapetalin P (**68**)	*Dichapetalum pallidum*	Ghana	29.72 (HL60); 2.66 (Jurkat)	Osei-Safo et al. (2017)
Albidoside E (**96**)	*Acacia albida*	Cameroon	18.6 (HeLa); 13.3 (HL60)	Tchoukoua et al. (2017)
Albidoside F (**97**)	*Acacia albida*	Cameroon	46.7 (HeLa); 15.8 (HL60)	Tchoukoua et al. (2017)
Alpha-hederin (**98**)	*Beilschmiedia acuta*	Cameroon	6.30 (CCRF-CEM) *vs* 7.43 (CEM/ADR5000) [1.18]; 21.35 (MDA-MB-231-*pcDNA*) *vs* 19.80 (MDA-MB-231-*BCRP*) [0.93]; 14.98 (HCT116 $p53^{+/+}$) *vs* 18.92 (HCT116 $p53^{-/-}$) [1.26]; 21.45 (U87MG) *vs* 43.98 (U87MG.Δ*EGFR*) [2.05]; 23.63 (HepG2)	Kuete, Tankeo, et al. (2014)
Ardisiacrispin B (**99**)	*Ardisia kivuensis*	Cameroon	1.20 (CCRF-CEM) *vs* 4.42 (CEM/ADR5000) [3.68]; 4.34 (MDA-MB-231-*pcDNA*) *vs* 4.46 (MDA-MB-231-*BCRP*) [1.03]; 6.60 (HCT116 $p53^{+/+}$) *vs* 6.63 (HCT116 $p53^{-/-}$) [0.94]; 2.40 (U87MG) *vs* 4.12 (U87MG.Δ*EGFR*) [1.72]; 6.76 (HepG2)	Mbaveng, Ndontsa, et al. (2018)
Beta-amyrin (**69**)	*Citrus reticulata*	Cameroon	174.18 (A549)	Tahsin et al. (2017)

Betulin (**70**)	*Garcinia epunctata; Callistemon citrinus*	Cameroon; Egypt	25.08 (CCRF-CEM) *vs* 10.15 (CEM/ADR5000) [0.40]; 9.68 (MDA-MB-231-*pcDNA*) *vs* 5.12 (MDA-MB-231-*BCRP*) [0.53]; 12.33 (HCT116 $p53^{+/+}$) *vs* 10.96 (HCT116 $p53^{-/-}$) [0.89]; 8.20 (U87MG) *vs* 16.18 (U87MG.Δ*EGFR*) [1.97]; 35.10 (HepG2); 0.7 (Panc-1)	Mbaveng, Fotso, et al., (2018); Tawila et al. (2020)
Betulinic acid (**71**)	*Dichrostachys cinerea; Callistemon citrinus*	Cameroon; Egypt	8.80 (CCRF-CEM) *vs* 7.65 (CEM/ADR5000) [0.87]; 38.83 (MDA-MB-231-*pcDNA*) *vs* 24.91 (MDA-MB-231-*BCRP*) [0.64]; 24.91 (U87MG) *vs* 13.92 (U87MG.Δ*EGFR*) [0.56]; 44.17 (HepG2); 15.2 (Panc-1)	Mbaveng et al. (2019); Tawila et al. (2020)
Dichapetalin A (**72**)	*Dichapetalum pallidum*	Ghana	11.19 (HL60); 2.97 (Jurkat)	Osei-Safo et al. (2017)
Dichapetalin X (**73**)	*Dichapetalum pallidum*	Ghana	5.56 (HL60); 1.80 (Jurkat)	Osei-Safo et al. (2017)
Elatumic acid (**74**)	*Omphalocarpum elatum*	Cameroon	16.60 (CCRF-CEM) *vs* 67.91 (CEM/ADR5000) [4.09]; 46.73 (MDA-MB-231-*pcDNA*); 31.46 (HCT116 $p53^{+/+}$) *vs* 17.07 (HCT116 $p53^{-/-}$) [0.54]	Sandjo et al. (2014)
Erythrodiol (**75**)	*Callistemon citrinus*	Egypt	0.2 (Panc-1)	Tawila et al. (2020)

(continued)

Table 2 Cytotoxic terpenoids from African medicinal plants and their effects on sensitive and drug-resistant cancer cell lines. (*cont'd*)

Compounds names	Source	Country	Cancer cell lines and IC_{50} values (μM) and degree of resistance in bracket	References
Friedelanol (**76**)	*Desbordesia glaucescens*	Cameroon	35.37 (Caco-2); 22.24 (HepG2); 80.79 (A549)	Kuete, Mafodong, et al. (2017)
Friedelanone (**77**)	*Desbordesia glaucescens*	Cameroon	82.39 (Caco-2); 60.61 (MCF-7)	Kuete, Mafodong, et al. (2017)
Glaberrimoside A **(100)**	*Albizia glaberrima*	Cameroon	35 (AsPC-1)	Noté, Azouaou, et al. (2016)
Glaberrimoside B **(101)**	*Albizia glaberrima*	Cameroon	60 (THP-1)	Noté, Azouaou, et al. (2016)
Glaberrimoside C **(102)**	*Albizia glaberrima*	Cameroon	433 (BJ)	Noté, Azouaou, et al. (2016)
Lanosta-7,24-dien-3-one (**78**)	*Desbordesia glaucescens*	Cameroon	43.63 (HepG2); 82.43 (A549)	Kuete, Mafodong, et al. (2017)
Lebbeckoside B **(103)**	*Albizia lebbeck*	Cameroon	2.10 (U87MG); 2.24 (TG1 GSC)	Noté et al. (2015)
Lebbeckoside A **(104)**	*Albizia lebbeck*	Cameroon	3.46 (U87MG); 1.36 (TG1 GSC)	Noté et al. (2015)

Lupeol (**79**)	*Citrus reticulata; Zanthoxylum gilletii*	Cameroon	15.82 (CCRF-CEM) *vs* 69.39 (CEM/ADR5000) [4.38]; 27.96 (MDA-MB-231-*pcDNA*) *vs* 22.36 (MDA-MB-231-*BCRP*) [0.8]; 22.31 (HCT116 $p53^{+/+}$) *vs* 19.64 (HCT116 $p53^{-/-}$) [0.88]; 20.58 (U87MG) *vs* 13.60 (U87MG.Δ*EGFR*) [1.51]; 33.53 (HepG2); 49.30 (A549)	Nyaboke et al., (2018); Tahsin et al. (2017)
Medicagenic acid (**80**)	*Gladiolus segetum*	Egypt	28.69 (HepG2); 21.31 (PC-3)	Abd El-Kader et al. (2020)
Olean-12-en-3-*β*-*O*-D-glucopyranoside (**105**)	*Tetrapleura tetraptera*	Cameroon	4.76 (CCRF-CEM) *vs* 10.65 (CEM/ADR5000) [2.24]; 5.35 (MDA-MB-231-*pcDNA*) *vs* 12.58 (MDA-MB-231-*BCRP*) [2.35]; 11.57 (HCT116 $p53^{+/+}$) *vs* 9.50 (HCT116 $p53^{-/-}$) [0.82]; 7.99 (U87MG) *vs* 10.91 (U87MG.Δ*EGFR*) [1.37]; 12.92 (HepG2)	Mbaveng, Chi, Bonsou, et al. (2020)
Oleanolic acid (**81**)	*Callistemon citrinus*	Egypt	20.5 (Panc-1)	Tawila et al. (2020)
Parkioside A (**106**)	*Butyrospermum parkii*	Cameroon	23.68 (T98G); 35.97 (MDA-MB 231); 27.32 (A375); 55.44 (HCT116)	Tapondjou et al. (2011)
Parkioside B (**107**)	*Butyrospermum parkii*	Cameroon	2.74 (T98G); 9.62 (MDA-MB 231)	Tapondjou et al. (2011)

(continued)

Table 2 Cytotoxic terpenoids from African medicinal plants and their effects on sensitive and drug-resistant cancer cell lines. (*cont'd*)

Compounds names	Source	Country	Cancer cell lines and IC_{50} values (μM) and degree of resistance in bracket	References
Platanic acid (**82**)	*Callistemon citrinus*	Egypt	14.2 (Panc-1)	Tawila et al. (2020)
Polyacanthoside A (**108**)	*Acacia polyacantha*	Cameroon	8.90 (CCRF-CEM) *vs* 14.06 (CEM/ADR5000) [1.58]; 19.26 (MDA-MB-231-*pcDNA*) *vs* 16.95 (MDA-MB-231-*BCRP*) [0.88]; 18.65 (HCT116 $p53^{+/+}$) *vs* 17.92 (HCT116 $p53^{-/-}$) [0.96]; 18.98 (U87MG) *vs* 17.65 (U87MG.Δ*EGFR*) [0.92]; 35.21 (HepG2)	Fotso et al. (2019)
Rheediinoside B **(109)**	*Entada rheedii*	Cameroon	20.48 (A431); 34.78 (PC3); 75.74 (B16-F1); 75.96 (T98G)	Nzowa et al. (2010)
Trichadonic acid (**83**)	*Hypericum roeperianum*	Cameroon	14.44 (CCRF-CEM) *vs* 18.27 (CEM/ADR5000) [1.26]; 16.47 (MDA-MB-231-*pcDNA*) *vs* 14.95 (MDA-MB-231-*BCRP*) [0.91]; 17.36 (HCT116 $p53^{+/+}$) *vs* 44.20 (HCT116 $p53^{-/-}$) [2.55]; 16.16 (U87MG) *vs* 14.69 (U87MG.Δ*EGFR*) [0.91]; 21.68 (HepG2)	Guefack et al. (2020)
Ursolic acid (**84**)	*Callistemon citrinus*	Egypt	13.7 (Panc-1)	Tawila et al. (2020)

Uvaol (**85**)	*Callistemon citrinus*	Egypt	5.2 (Panc-1)	Tawila et al. (2020)
Zanhic acid (**86**)	*Gladiolus segetum*	Egypt	30.31 (HepG2); 20.85 (PC-3)	Abd El-Kader et al. (2020)
Zygiaoside A **(110)**	*Albizia zygia*	Cameroon	9 (A431)	Noté, Simo, et al. (2016)
Zygiaoside B **(111)**	*Albizia zygia*	Cameroon	249 (A431)	Noté, Simo, et al. (2016)

4.1 Cytotoxic sesquiterpenoids

Potent cytotoxic sesquiterpenoids identified from African medicinal plants include 2α-hydroxyalantolactone (**1**), damsin (**2**), neoambrosin (**3**), vernomelitensin (**4**), and vernopicrin (**5**). Their chemical structures are shown in Fig. 1 meanwhile the cytotoxic activities in human cancer cell lines are summarized in Table 2. Outstanding cytotoxic activity ($IC_{50} < 0.5\,\mu M$) was reported with sesquiterpene **1** against CCRF-CEM and CEM/ADR5000 cells, **4** against MDA-MB-231-*BCRP* cells, HCT116 cells, DU-45 cells, and MiaPaCa-2 cells, and **5** against HCT116 cells, DU-45 cells, and MiaPaCa-2 cells. Excellent cytotoxic activity ($0.5 < IC_{50} \leq 2\,\mu M$) sesquiterpene **4** against PC-3 cells, MCF-7 cells, and A549 cells, and **5** against HL60 cells, MDA-MB-231-*BCRP* cells, MCF-7 cells, PC-3 cells, meanwhile very good cytotoxic effect ($2 < IC_{50} \leq 5\,\mu M$) was reported with and **4** against A549 cells and **5** against A549 cells. Good antiproliferative activity ($5 < IC_{50} \leq 10\,\mu M$) was obtained with sesquiterpenes **1** against HCT116 $p53^{+/+}$ cells and HCT116 $p53^{-/-}$ cells, **2** and **3** against CCRF-CEM cells and CEM/ADR5000 cells, and **5** against A549 cells. Collateral sensitivity of CEM/ADR5000 cells *vs* CCRF-CEM cancer cells (D.R. 0.75) to compound **1** was achieved while normal sensitivity of CEM/ADR5000 cells *vs* CCRF-CEM cancer cells to **2** (D.R. 1), **3** (D.R. 1.07), and HCT116 $p53^{-/-}$ *vs* HCT116 $p53^{+/+}$ (D.R. 1.04) to **1** was also obtained.

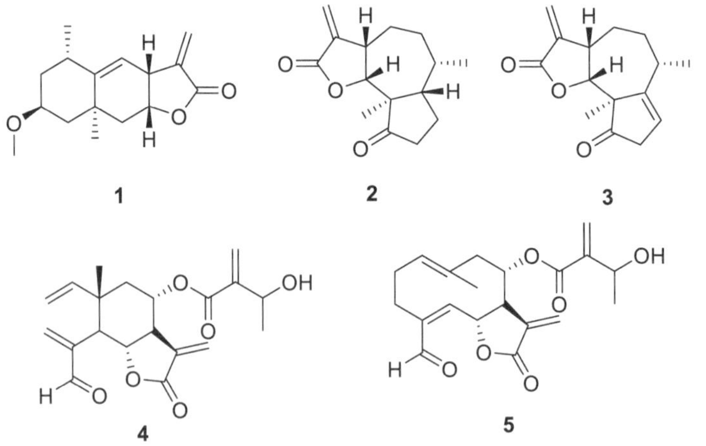

Fig. 1 Cytotoxic sesquiterpenoids isolated from African medicinal plants. **1:** 2α-hydroxyalantolactone; **2:** damsin; **3:** neoambrosin; **4:** vernomelitensin; **5:** vernopicrin.

4.2 Cytotoxic diterpenoids

Twenty-six (26) diterpenoids with antiproliferative effects in human cancer cell lines were identified in African medicinal plants. They include 2,18,19-trachylobanetriol (**6**), 2,6,19-trachylobanetriol (ent-2*α*,6*α*)-form (**7**), 2-oxo-candesalvone (**8**), 4,12,20-trideoxyphorbol-13-(2,3-dimethyl) butyrate (**9**), 4,20-dideoxy (4*α*) phorbol-12-benzoate-13-isobutyrate (**10**), 6-*β*-hydroxycandesalvone B (**11**), 7*β*,13*β*,17-*O*-triacetyl-5*α*-*O*-(2-methylbutyryl)-3*β*-*O*-propanoyl14-oxopremyrsinol (**12**), candesalvone B (**13**), candesalvone B methyl ester (**14**), *ent*-18-hydroxy-trachyloban-3-one (**15**), *ent*-trachyloban-3*β*-ol (**16**), *ent*-trachyloban-3-one (**17**), esopimara-7,15-dien-3*β*-ol (**18**), euphosantianane A (**19**), euphosantianane B (**29**), euphosantianane C (**21**), euphosantianane D (**22**), galanal A (**23**), galanal B (**24**), hautriwaic acid lactone (**25**), premyrsinol-3-propanoate-5-isobutyrate-7,13,17-triacetate (**26**), premyrsinol-3-propanoate-5-isobutyrate-7,13-triacetate-17- nicotinate (**27**), salvimulticanol (**28**), salvimulticaoic acid (**29**), *trans*-phytol (**30**), and 16*α*-hydroxy-*ent*-kauran-19-oic acid (**61**). Their chemical structures are depicted in Fig. 2 meanwhile the cytotoxic activities in human cancer cell lines are shown in Table 2. Their cytotoxic effects rather vary from average ($10 < IC_{50} \leq 20\ \mu M$) to weak ($20 < IC_{50} \leq 60\ \mu M$) (Table 2). However, very good antiproliferative activity was obtained with diterpenes **16** and **17** against HeLa cells, and **28** against CEM/ADR5000 cells. Interestingly, a collateral sensitivity of CEM/ADR5000 cells *vs* CCRF-CEM cancer cells (D.R. 0.54) to compound **28** was achieved.

4.3 Cytotoxic steroids and steroidal saponins

Twenty-nine (29) steroids and steroidal saponins with cytotoxic activities against human cancer cell lines were reported from African medicinal plants. They include 16*β*-formyloxymelianthugenin (**31**), 16*β*-hydroxybersaldegenin 1,3,5-orthoacetate (**32**), 2*β*-acetoxy-3,5-di-*O*-acetylhellebrigenin (**33**), 2*β*-acetoxy-5*β*-*O*-acetylhellebrigenin (**34**), 2*β*-acetoxymelianthusigenin (**35**), 2*β*-hydroxy-3*β*,5*β*-di-*O*-acetylhellebrigenin (**36**), daucosterol (**37**), sitosterol (**38**), stigmasterol (**39**), *β*-spinasterol (**40**), (20*R*)-*O*-(3)-*β*-D-glucopyranosyl(1-2)-*α*-L-arabinopyranosyl-pregn-5-en-3*β*,20-diol (**41**), (25*R*)-17α-hydroxyspirost-5-en-3*β*-yl *O*-α-L-rhamnopyranosyl-(1-2)-*O*-[*O*-*α*-L-rhamnopyranosyl-(1-4)-*α*-L-rhamnopyranosyl-(1-4)]-*β*-D-glucopyranoside (**42**), (25*R*)-17α-hydroxyspirost-5-en-3*β*-yl *O*-*α*-L-rhamnopyranosyl-(1-4)-*O*-*α*-L-rhamnopyranosyl-(1-4)-*β*-D-glucopyranoside (**43**), (25*R*)-spirost-5-ene-3*β*, 22*β*-3-*O*-*β*-D-glucopyranosyl(1-2)-*O*-*α*-L-rhamnopyranoside (**44**),

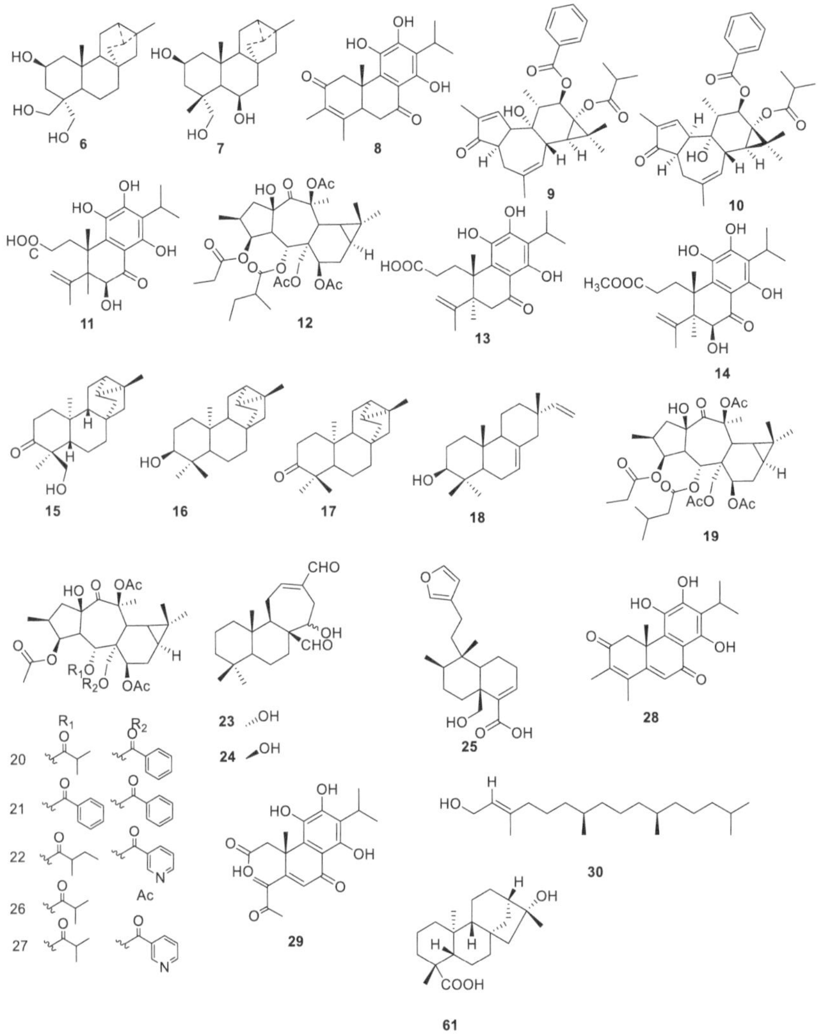

Fig. 2 Cytotoxic diterpenoids isolated from African medicinal plants. **6:** 2,18,19-trachylobanetriol; **7:** 2,6,19-trachylobanetriol (ent-2α,6α)-form; **8:** 2-oxocandesalvone; **9:** 4,12,20-trideoxyphorbol-13-(2,3-dimethyl) butyrate; **10:** 4,20-dideoxy (4α) phorbol-12-benzoate-13-isobutyrate; **11:** 6-β-hydroxycandesalvone B; **12:** 7β,13β,17-*O*-triacetyl-5α-*O*-(2-methylbutyryl)-3β-*O*-propanoyl14-oxopremyrsinol; **13:** candesalvone B; **14:** candesalvone B methyl ester; **15:** *ent*-18-hydroxy-trachyloban-3-one; **16:** *ent*-trachyloban-3β-ol; **17:** *ent*-Trachyloban-3-one; **18:** esopimara-7,15-dien-3β-ol; **19:** euphosantianane A; **20:** euphosantianane B; **21:** euphosantianane C; **22:** euphosantianane D; **23:** galanal A; **24:** galanal B; **25:** hautriwaic acid lactone; **26:** premyrsinol-3-propanoate-5-isobutyrate-7,13,17-triacetate; **27:** premyrsinol-3-propanoate-5-isobutyrate-7,13-triacetate-17- nicotinate; **28:** salvimulticanol; **29:** salvimulticaoic acid; **30:** *trans*-phytol; **61:** 16α-hydroxy-*ent*-kauran-19-oic acid.

26-*O*-*β*-D-glucopyranosyl-(25*R*)-5-en-furost-3*β*,17α,22α,26-tetraol-3-*O*-*α*-L-rhamnopyranosyl-(1-4)-*α*-L-rhamnopyranosyl-(1-4)-[α-L-rhamnopyranosyl-(1-2)]-*β*-D-glucopyranoside (**45**), balanitin 4 (**46**), balanitin 6 (**47**), balanitin 7 (**48**), diosgenin (**49**), diospreussinosides B (**50**), fruticoside H (**51**), fruticoside I (**52**), parquispiroside (**53**), pennogenin 3-*O*-*α*-L-rhamnopyranosyl-(1-4)-*α*-L-rhamnopyranosyl-(1-4)-[*α*-L-rhamnopyranosyl-(1-2)]-*β*-D-glucopyranoside (**54**), progenin III (**55**), raphvinin 1 (**56**), raphvinin 3 (**57**), spiroconazole A (**58**), and trillin (**59**). The chemical structures of the reported steroids and steroidal saponins are shown in Fig. 3. Their cytotoxic activities in human cancer cell lines are given in Table 2. Amongst them, outstanding cytotoxic activities were obtained with sterol **31, 33, 34,** and **36** against CCRF-CEM and CEM/ADR5000 cells, and **31** and **33** against MCF-7 cells, as well as steroidal saponins **46** against P388 cells, **47** against A549 cells, and **48** against U373 cells. Excellent cytotoxic activities were reported with sterol **32** and **35** against CCRF-CEM and CEM/ADR5000 cells, **34** and **36** against MCF-7 cells, and **39** against PC-3 cells, as well as steroidal saponins **42** against HT-29 cells and **55** against CCRF-CEM and CEM/ADR5000 cells. Very good antiproliferative activities were obtained with sterol **32** against MCF-7 cells, steroidal saponins **42** against HCT116 cells, **44** against CCRF-CEM cells, **53** against U87 cells, and **55** against MDA-MB-231 cells, MDA-MB-231-*BCRP* cells, HCT116 $p53^{+/+}$ cells, HCT116 $p53^{-/-}$ cells, U87MG, and U87MG.Δ*EGFR*. Good antiproliferative activities were obtained with sterol **35** against MCF-7 cells, steroidal saponins **44** and **57** against CCRF-CEM cells, **53** against HeLa cells and HepG2 cells, **54** and **58** against ECV-304 cells, and **57** against CEM/ADR5000 cells. Importantly, collateral sensitivities of CEM/ADR5000 cells *vs* CCRF-CEM cancer cells to compounds **31** (D.R. 0.86), **34** (D.R. 0.62), **35** (D.R. 0.75), and **36** (D.R. 0.89) were achieved. Also, a normal sensitivity of CEM/ADR5000 cells *vs* CCRF-CEM cells to **33** (D.R. 1) and steroidal saponins **55** (D.R. 1.07), as well as HCT116 $p53^{-/-}$ *vs* HCT116 $p53^{+/+}$ to **55** (D.R. 1.08) were obtained.

4.4 Cytotoxic triterpenoids and triterpene saponins

Fifty-one (51) triterpenoids and triterpene saponins with cytotoxic activities against human cancer cell lines were reported from African medicinal plants. They are 11-oxo-α-amyryl acetate (**60**), 16*β*-hydroxylupeol (**62**), 1*β*, 2*α*, 3*α*, 5, 19*α*, 24-hexahydroxyurs-11(12), 20(30)-dien-28-oic acid (**63**), 23-hydroxyursolic acid (**64**), 2*α*, 3*α*, 19α,20*β*, 23-pentahydroxyurs-12-en-28-oic acid (**65**), 2*β*,3*β*,13*α*,22*α*-tetrahydroxy oleanane-23,28-dioic

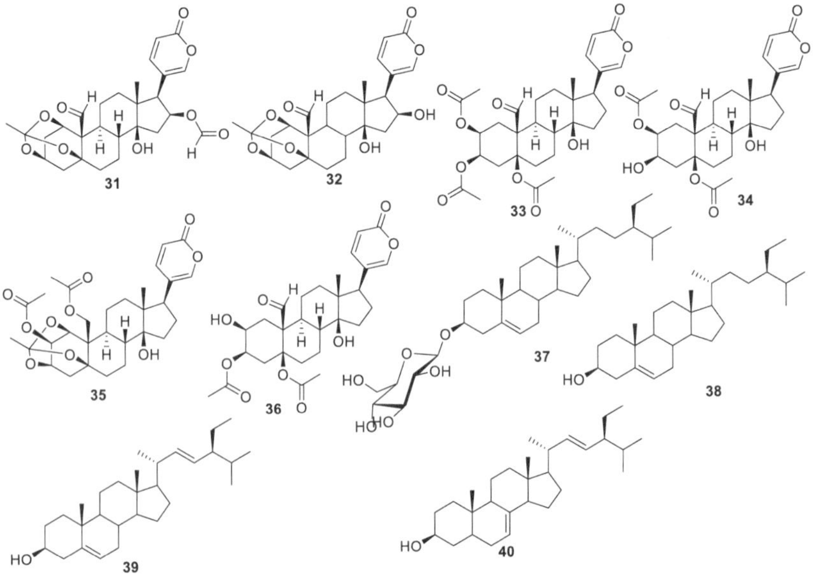

Fig. 3 Cytotoxic steroids and steroidal saponins isolated from African medicinal plants. **31:** 16*β*-formyloxymelianthugenin; **32:** 16*β*-hydroxybersaldegenin 1,3,5-orthoacetate; **33:** 2*β*-acetoxy-3,5-di-*O*-acetylhellebrigenin; **34:** 2*β*-acetoxy-5*β*-*O*-acetylhellebrigenin; **35:** 2*β*-acetoxymelianthusigenin; **36:** 2*β*-hydroxy-3*β*,5*β*-di-*O*-acetylhellebrigenin; **37:** daucosterol; **38:** sitosterol; **39:** stigmasterol; **40:** *β*-spinasterol; **41:** (20*R*)-*O*-(3)-*β*-D-glucopyranosyl (1-2)-*α*-L-arabinopyranosyl-pregn-5-en-3*β*,20-diol; **42:** (25*R*)-17α-hydroxyspirost-5-en-3*β*-yl *O*-α-L-rhamnopyranosyl-(1-2)-*O*-[*O*-*α*-L-rhamnopyranosyl-(1-4)-*α*-L-rhamnopyranosyl-(1-4)]-*β*-D-glucopyranoside; **43:** (25*R*)-17α-hydroxyspirost-5-en-3*β*-yl *O*-*α*-L-rhamnopyranosyl-(1-4)-*O*-*α*-L-rhamnopyranosyl-(1-4)-*β*-D-glucopyranoside; **44:** (25*R*)-spirost-5-ene-3*ß*, 22*ß*-3-*O*-*ß*-D-glucopyranosyl(1-2)-*O*-*α*-L-rhamnopyranoside; **45:** 26-*O*-*ß*-D-glucopyranosyl-(25*R*)-5-en-furost-3*ß*,17α,22α,26-tetraol-3-*O*-*α*-L-rhamnopyranosyl-(1-4)-*α*-L-rhamnopyranosyl-(1-4)-[α-L-rhamnopyranosyl-(1-2)]-*β*-D-glucopyranoside; **46:** balanitin 4; **50:** diospreussinoside B; **54:** pennogenin 3-*O*-*α*-L-rhamnopyranosyl-(1-4)-*α*-L-rhamnopyranosyl-(1-4)-[*α*-L-rhamnopyranosyl-(1-2)]-*β*-D-glucopyranoside; **47:** balanitin 6; **48:** balanitin 7; **49:** diosgenin; **51:** fruticoside H; **52:** fruticoside I; **53:** parquispiroside; **55:** progenin III; **56:** raphvinin 1; **57:** raphvinin 3; **58:** spiroconazole A; **59:** trillin.

acid (**66**), 3*β*-hydroxy-urs-11-en13(28)-olide (**67**), 7-hydroxydichapetalin P (**68**), beta-amyrin (**69**), betulin (**70**), betulinic acid (**71**), dichapetalin A (**72**), dichapetalin X (**73**), elatumic acid (**74**), erythrodiol (**75**), friedelanol (**76**), friedelanone (**77**), lanosta-7,24-dien-3-one (**78**), lupeol (**79**), medicagenic acid (**80**), oleanolic acid (**81**), platanic acid (**82**), trichadonic acid (**83**), ursolic acid (**84**), uvaol (**85**), zanhic acid (**86**), 3-*O*-*β*-D-glucopyranosyl-(1-3)-*β*-D-glucopyranosyl-27-hydroxyolean-12-en-28-oic acid (**87**),

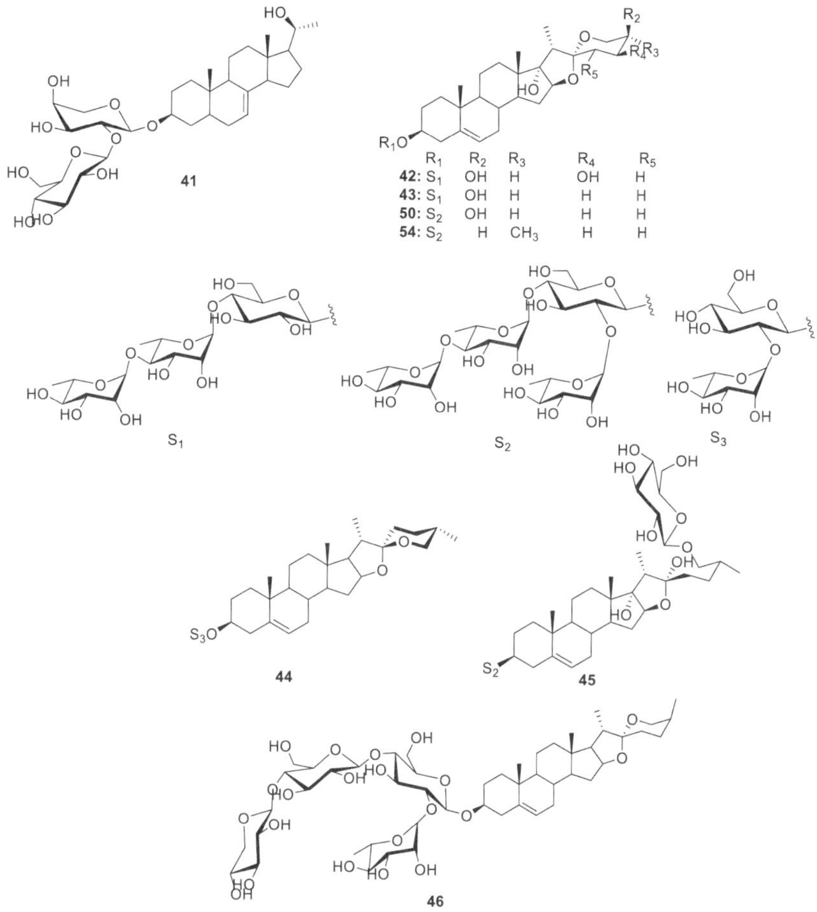

	R1	R2	R3	R4	R5
42:	S1	OH	H	OH	H
43:	S1	OH	H	H	H
50:	S2	OH	H	H	H
54:	S2	H	CH3	H	H

Fig. 3 (*Continued*)

3-*O*-β-D-glucopyranosyl-(1-6)-β-D-glucopyranosylurs-12-en-28-oic acid (**88**), 3-*O*-[β-D-galactopyranosyl-(1-4)-β-D-galactopyranosyl]-oleanolic acid (**89**), 3-*O*-[β-D-glucopyranosyl-(1-2)-β-D-glucopyranosyl-(1-4)-β-D-xylopyranosyl]-2β, 3β,16α-trihydroxyolean-12-en-23,28-dioic acid-28-*O*-α-L-rhamnopyranosyl-(1-4)-α-L-rhamnopyranosyl-(1-2)-β-D-glucopyranosyl-(1-2)-α-L-arabinopyranoside (**90**), 3-*O*-{β-D-glucopyranosyl-(1-2)-[α-L-arabinopyranosyl-(1-3)]-β-D-glucuronopyranosyl}-21-*O*-(2-methylbutyroyl)-22-*O*-acetyl-R1-barrigenol (**91**), 3-*O*-{β-D-glucopyranosyl-(1-2)-[α-L-arabinopyranosyl-(1-3)]-β-D-glucuronopyranosyl}-21-*O*-(2-methylbutyroyl)-28-*O*-acetyl-R1-barrigenol (**92**), 3-*O*-α-L-arabinopyranosyl-23-hydroxyursolic acid (**93**), 3-*O*-β-D-galactopyranosyl-(1-2)-β-D-glucuronopyranosyloleanolic acid (**94**),

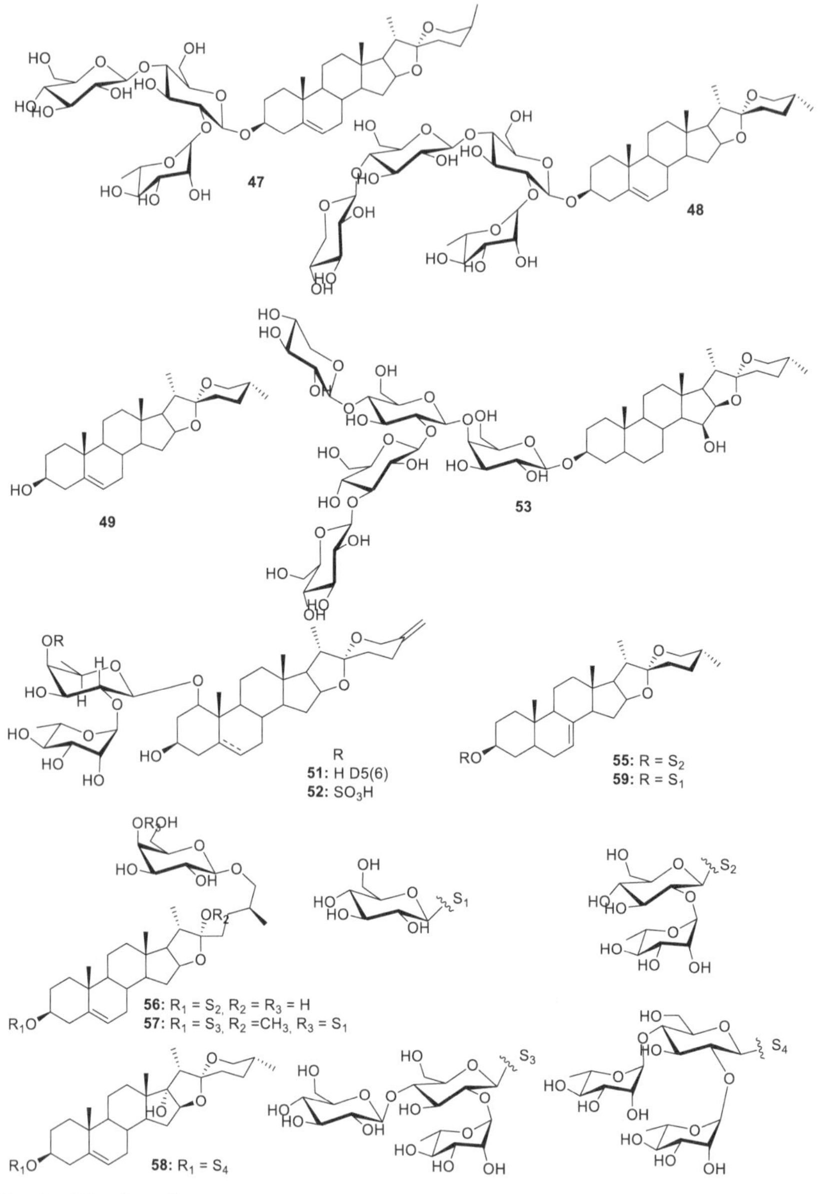

Fig. 3 (*Continued*)

3-*O*-*β*-D-glucopyranosyl-23-hydroxyursolic acid (**95**), albidoside E (**96**), albidoside F (**97**), alpha-hederin (**98**), ardisiacrispin B (**99**), glaberrimoside A (**100**), glaberrimoside B (**101**), glaberrimoside C (**102**), lebbeckoside B (**103**), lebbeckoside A (**104**), olean-12-en-3-*β*-*O*-D-glucopyranoside (**105**), parkioside A

(**106**), parkioside B (**107**), polyacanthoside A (**108**), rheediinoside B (**109**), zygiaoside A (**110**), and Zygiaoside B **(111)**. The chemical structures of the reported steroids and sterol saponins are shown in Fig. 4. Their cytotoxic activities in human cancer cell lines are given in Table 2. Amongst them, outstanding cytotoxic activity was obtained with triterpenoids **67** and **75** against Panc-1 cells while excellent cytotoxic activity was obtained with triterpenoids **65** against SPC212 cells, **70** against Panc-1 cells, and **73** against Jurkat cells, and triterpene saponins **99** against CCRF-CEM cells, **99** and **103** against U87MG cells, and **103** and **104** against TG1 GSC cells. Very good antiproliferative activity was obtained with triterpenoids **60** against CCRF-CEM cells, **62** against MCF-7 cells, and **66** against PC-3 cells, and triterpene saponins **90** against MCF-7 cells, HepG2 cells, and PC-3 cells, **99** against MDA-MB-231-*pcDNA* cells, MDA-MB-231 cells, HCT116 *p53*$^{+/+}$ cells, and HepG2 cells, **104** and **105** against U87MG cells, **99** and **105** against MDA-MB-231 cells, **99** and **105** HCT116 *p53*$^{-/-}$ cells, **105** against CCRF-CEM cells, and **107** against T98G cells. Good antiproliferative activity was obtained with triterpenoids **70** and **71** against CCRF-CEM cells, **70** against CEM/ADR5000 cells, MDA-MB-231-*pcDNA* cells, MDA-MB-231-*BCRP* cells, and U87MG cells, **73** against HL60 cells, and triterpene saponins **98, 99** and **108** against CCRF-CEM cells, **98** against CEM/ADR5000 cells, **107** against MDA-MB-231 cells, and **110** against A431 cells. Interestingly, collateral sensitivities of CEM/ADR5000 cells *vs* CCRF-CEM cancer cells to compounds **70** (D.R. 0.40) and **71** (D.R. 0.87), and MDA-MB-231-*BCRP* cells *vs* MDA-MB-231 cells (D.R. 0.53) and HCT116 *p53*$^{-/-}$ *vs* HCT116 *p53*$^{+/+}$ to **105** (D.R. 0.79) were achieved while normal sensitivity of CEM/ADR5000 cells *vs* CCRF-CEM cells to **98** (D.R. 1.18), was also obtained.

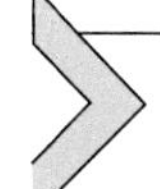

5. Modes of action of the cytotoxic terpenoids from the African medicinal plants

5.1 Terpenoids inducing apoptosis

Several bioactive terpenoids including monoterpenes, sesquiterpenes, diterpenes, triterpenes, and steroids have been isolated from African medicinal plants (Awouafack et al., 2013; Kuete, 2013; Sandjo & Kuete, 2013a, 2013b; Tchimene et al., 2013). Several cytotoxic terpenoids and saponins isolated from African medicinal plants induced apoptosis in cancer cell lines.

Triterpenes. Betulinic acid (**71**) displayed antiproliferative activity against a panel of nine cancer cell lines (CCRF-CEM, CEM/ADR5000, HCT116

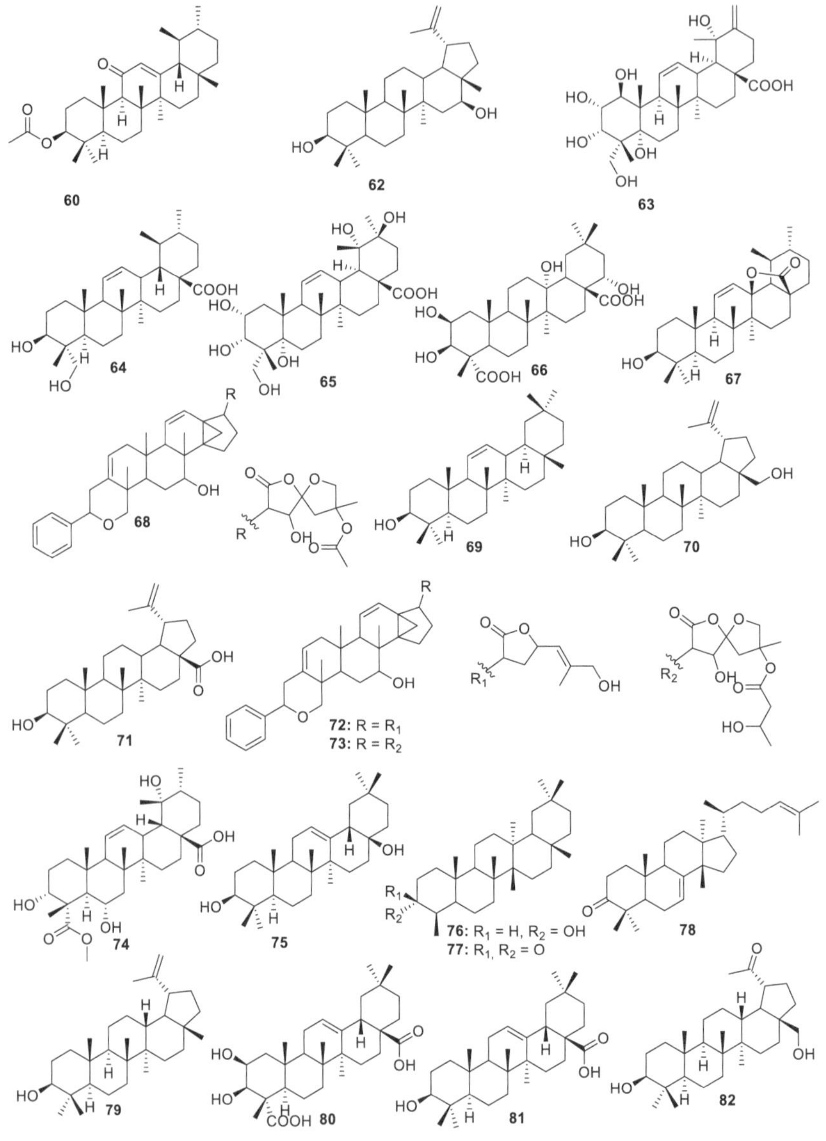

Fig. 4 Cytotoxic triterpenoids and triterpene saponins isolated from African medicinal plants. **60:** 11-oxo-α-amyryl acetate; **62:** 16*β*-hydroxylupeol; **63:** 1*β*, 2*α*, 3*α*, 5, 19*α*, 24-hexahydroxyurs-11(12), 20(30)-dien-28-oic acid; **64:** 23-hydroxyursolic acid; **65:** 2*α*, 3*α*, 19α,20*β*, 23-pentahydroxyurs-12-en-28-oic acid; **66:** 2*β*,3*β*,13*α*,22*α*-tetrahydroxy oleanane-23,28-dioic acid; **67:** 3*β*-hydroxy-urs-11-en13(28)-olide; **68:** 7-hydroxydichapetalin P; **69:** beta-amyrin; **70:** betulin; **60:** 11-oxo-α-amyryl acetate; **61:** 16*α*-hydroxy-*ent*-kauran-19-oic acid; **62:** 16*β*-hydroxylupeol; **63:** 1*β*, 2*α*, 3*α*, 5, 19*α*, 24-hexahydroxyurs-11(12), 20(30)-dien-28-oic acid; **64:** 23-hydroxyursolic acid; **65:** 2*α*, 3*α*, 19α,20*β*, 23-pentahydroxyurs-12-en-28-oic acid; **66:** 2*β*,3*β*,13*α*,22*α*-tetrahydroxy

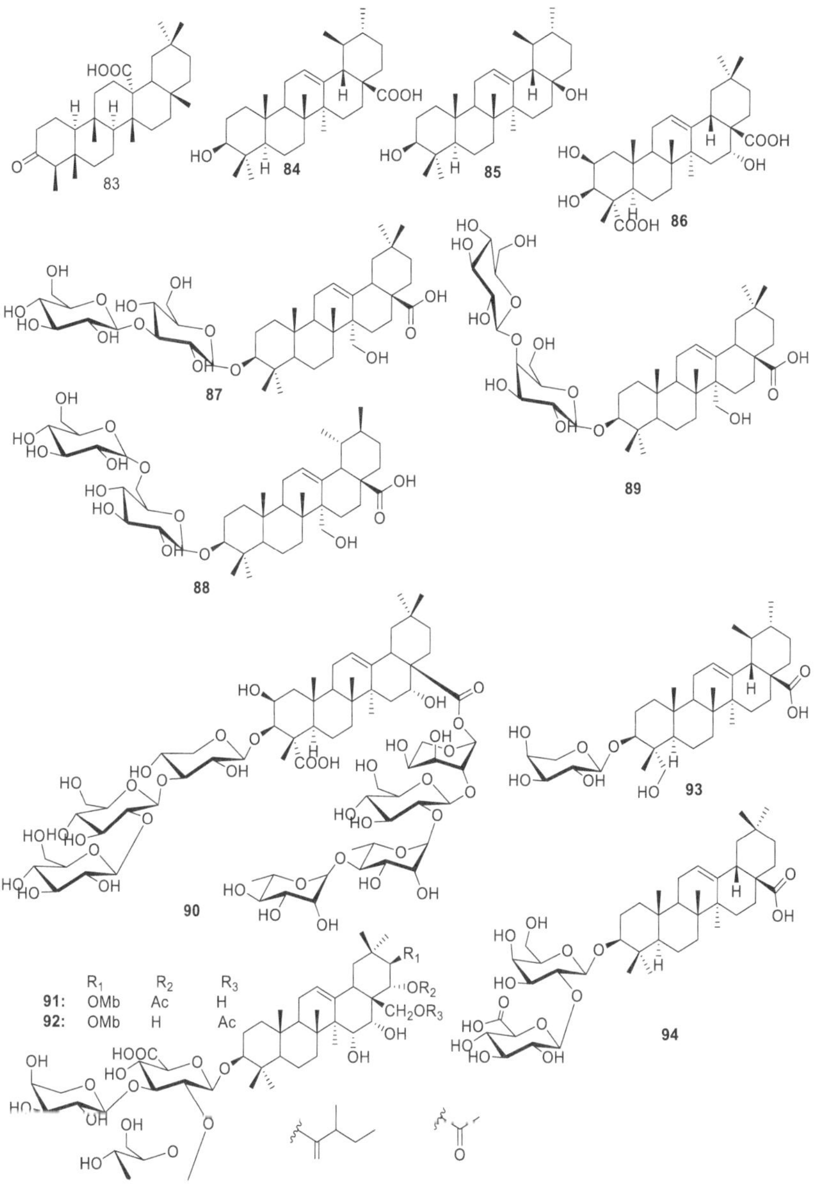

Fig. 4 (*Continued*)

oleanane-23,28-dioic acid; **67:** 3*β*-hydroxy-urs-11-en13(28)-olide; **68:** 7-hydroxydichapetalin P; **69:** beta-amyrin; **70:** betulin; **71:** betulinic acid; **72:** dichapetalin A; **73:** dichapetalin X; **74:** elatumic acid; **75:** erythrodiol; **76:** friedelanol; **77:** friedelanone; **78:** lanosta-7,24-dien-3-one; **79:** lupeol; **80:** medicagenic acid; **81:** oleanolic acid; **82:**

(Continued)

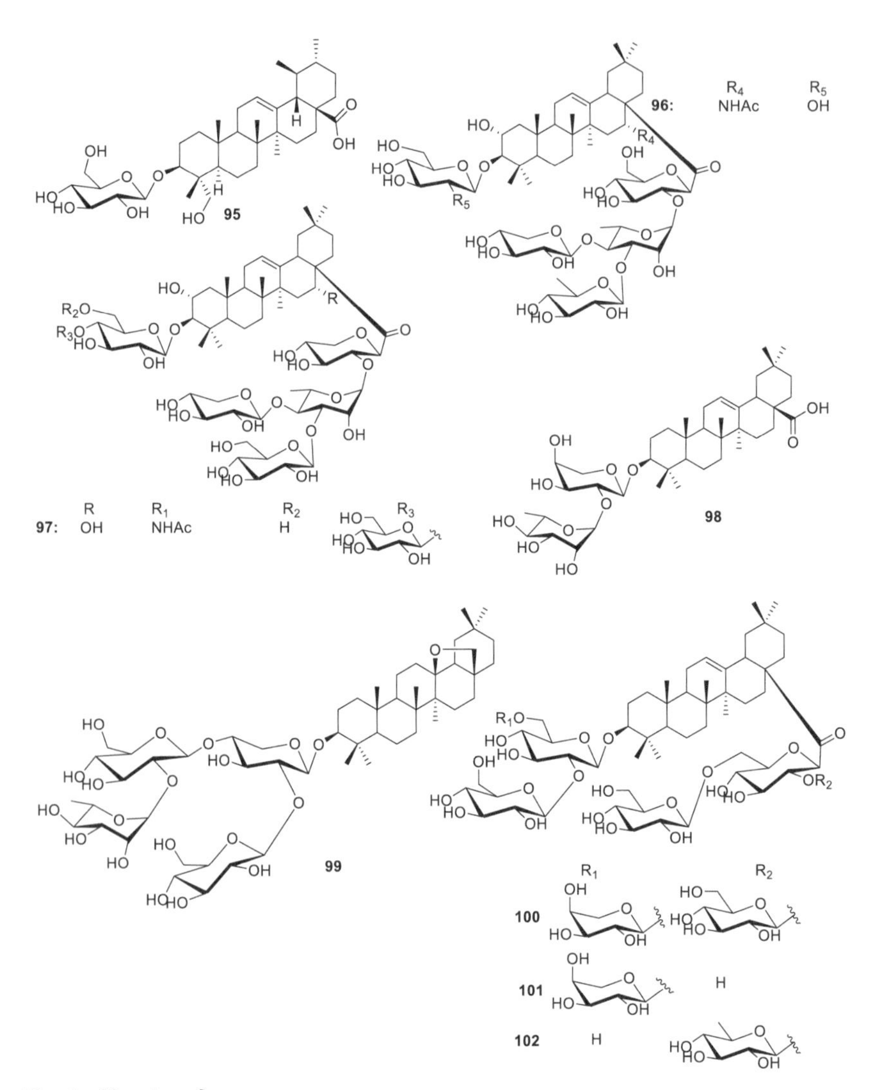

Fig. 4 (*Continued*)

Fig. 4—Cont'd platanic acid; **83:** trichadonic acid; **84:** ursolic acid; **85:** uvaol; **86:** zanhic acid; **87:** 3-*O*- ß-D-glucopyranosyl-(1-3)-ß-D-glucopyranosyl-27-hydroxyolean-12-en-28-oic acid; **88:** 3-*O*-ß-D-glucopyranosyl-(1-6)-ß-D-glucopyranosylurs-12-en-28-oic acid; **89:** 3-*O*-[*β*-D-galactopyranosyl-(1-4)-*β*-D-galactopyranosyl]-oleanolic acid; **90:** 3-*O*-[*β*-D-glucopyranosyl-(1-2)-*β*-D-glucopyranosyl-(1-4)-*β*-D-xylopyranosyl]-2*β*,3*β*,16*α*-trihydroxyolean-12-en-23, 28-dioic acid-28-*O*-*α*-L-rhamnopyranosyl-(1-4)-*α*-L-rhamnopyranosyl-(1-2)-*β*-D-glucopyranosyl-(1-2)-*α*-L-arabinopyranoside; **91:** 3-*O*-{*β*-D-glucopyranosyl-(1-2)-[*α*-L-arabinopyranosyl-(1-3)]-*β*-D-glucuronopyranosyl}-21-*O*-(2-methylbutyroyl)-22-*O*-acetyl-R1-barrigenol; **92:** 3-*O*-{*β*-D-glucopyranosyl-(1-2)-[α-L-arabinopyranosyl-(1-3)]-*β*-D-glucuronopyranosyl}-

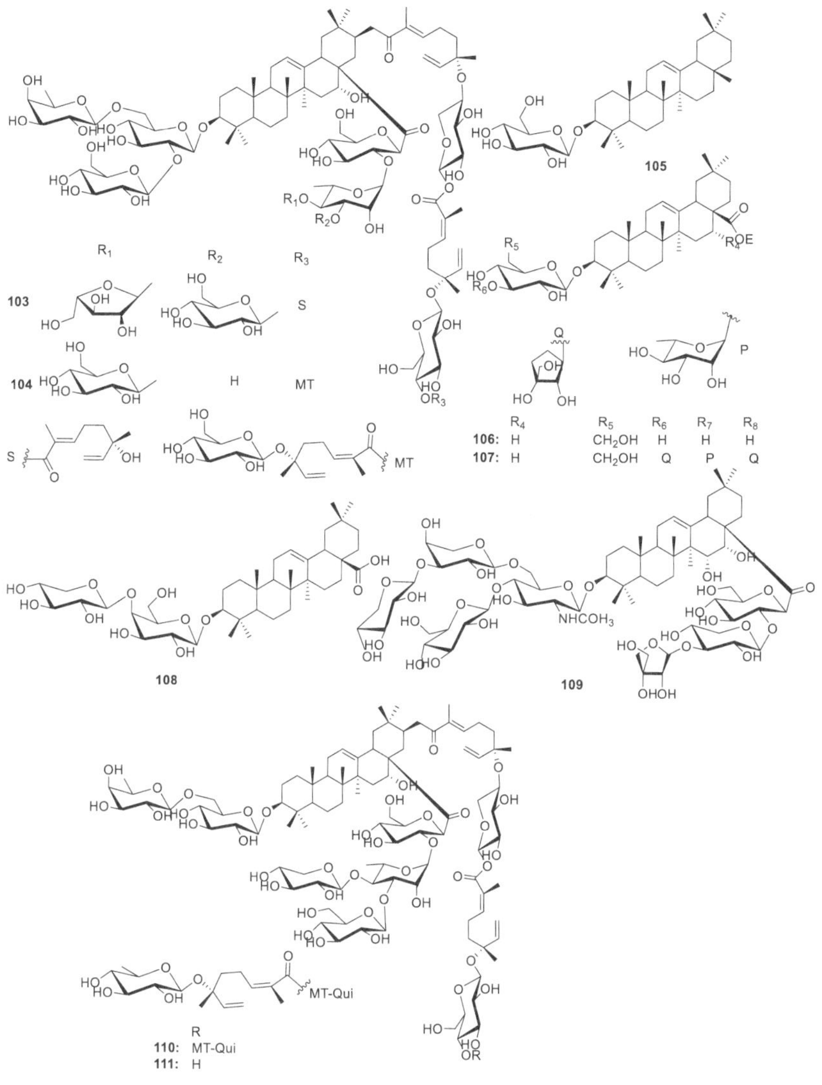

Fig. 4 (*Continued*)

21-*O*-(2-methylbutyroyl)-28-*O*-acetyl-R1-barrigenol; **93:** 3-*O*-α-L-arabinopyranosyl-23 hydroxyursolic acid. **94:** 3-*O*-β-D-galactopyranosyl-(1-2)-β-D-glucuronopyranosyloleanolic acid; **95:** 3-*O*-β-D-glucopyranosyl-23-hydroxyursolic acid; **96:** albidoside E; **97:** albidoside F; **98:** alpha-hederin; **99:** ardisiacrispin B; **100:** glaberrimoside A; **101:** glaberrimoside B; **102:** glaberrimoside C; **103:** lebbeckoside B; **104:** lebbeckoside A; **105:** olean-12-en-3-β-*O*-D-glucopyranoside; **106:** parkioside A; **107:** parkioside B; **108:** polyacanthoside A; **109:** rheediinoside B; **110:** zygiaoside A; **111:** zygiaoside B.

p53$^{+/+}$, HCT116 *p53*$^{-/-}$, U87. MG, U87. MGΔ*EGFR*, MDA-MB-231-pcDNA3, MDA-MB-231-*BCRP*, and HepG2) by the resazurin reduction assay (RRA) (Mbaveng et al., 2019). The obtained IC_{50} values varied from 7.65 μM (towards MDR CEM-ADR5000 leukemia cells) to 44.17 μM (against HepG2 hepatocarcinoma cells). When assessing the ability of compound **71** to induce apoptosis in CCRF-CEM cells, Mbaveng and co-authors identified the caspases activation, MMP alteration, and increased ROS production (Mbaveng et al., 2019). The cytotoxicity of the triterpenoid trichadonic acid (**83**) isolated from methanol extract from the bark of *Hypericum roeperianum* reported on nine cancer cell lines, namely CCRF-CEM, CEM/ADR5000, HCT116 *p53*$^{+/+}$, HCT116 *p53*$^{-/-}$, U87. MG, U87. MGΔ*EGFR*, MDA-MB-231-pcDNA3, MDA-MB-231-*BCRP*, and HepG2 cell lines and a normal AML12 hepatocyte (Guefack et al., 2020). The reported IC_{50} values vary from 14.44 μM (against CCRF-CEM leukemia cells) to 44.20 μM (against the resistant HCT116 (p53$^{-/-}$) cells) (Guefack et al., 2020). Compound **83** induced apoptosis in CCRF-CEM cells *via* caspases activation and enhanced ROS production (Guefack et al., 2020).

Steroidal saponins. The spirostanol saponin, progenin III (**55**) displayed antiproliferative activity against CCRF-CEM, CEM/ADR5000, HCT116 *p53*$^{+/+}$, HCT116 *p53*$^{-/-}$, U87. MG, U87. MGΔ*EGFR*, MDA-MB-231-pcDNA3, MDA-MB-231-*BCRP*, HepG2, CC531, B16-F1, B16-F10, A2058, SK-Mel505, MaMel-80a, MV3, SkMel-28 and Mel-2A cells (Mbaveng, Chi, Nguenang, et al., 2020). The reported IC_{50} values varied from 1.59 μM (towards CCRF-CEM cells) to 31.61 μM (against the BRAF-V600E homozygous mutant SKMel-28 melanoma cells). Mbaveng and collaborators also found that normal sensitivity was achieved with CEM/ADR5000 cells and HCT116*p53*$^{-/-}$ cells respectively compared to their sensitive congeners CCRF-CEM cells and HCT116 *p53*$^{+/+}$ cells was achieved (Mbaveng, Chi, Nguenang, et al., 2020). They also demonstrated that compound **55** induces apoptosis in CCRF-CEM cells mediated by caspases 3/7 activation, MMP alteration, and increase ROS production.

Triterpene saponins. The triterpene saponin ardisiacrispin B (**99**) had antiproliferative potential against CCRF-CEM, CEM/ADR5000, HCT116 *p53*$^{+/+}$, HCT116 *p53*$^{-/-}$, U87. MG, U87. MGΔ*EGFR*, MDA-MB-231-pcDNA3, MDA-MB-231-*BCRP*, and HepG2 by the RRA (Mbaveng, Ndontsa, et al., 2018). The reported IC_{50} values were below 10 μM. Collateral sensitivity of resistant HCT116*p53*$^{-/-}$ *vs* HCT116*p53*$^{+/+}$ cells to compound **99** was observed. In addition to ferroptosis inhibition, compound **99** induced apoptosis in CCRF-CEM cells *via* activation of initiator caspases 8 and 9 and effector caspase 3/7, alteration of MMP and increase in ROS production (Mbaveng, Ndontsa, et al., 2018). The triterpene saponins, oleanan-12-en-3-*β*-*O*-D-glucopyranoside (**105**) also displayed prominent cytotoxic effects on a panel of nine human cancer cell lines, namely CCRF-CEM, CEM/ADR5000, HCT116 $p53^{+/+}$, HCT116 $p53^{-/-}$, U87. MG, U87. MGΔ*EGFR*, MDA-MB-231-pcDNA3, MDA-MB-231-*BCRP*, HepG2 (Mbaveng, Chi, Bonsou, et al., 2020). The recorded IC_{50} values ranged from 4.76 μM (against CCRF-CEM cells) to 12.92 μM (against HepG2 hepatocarcinoma cells). Compound **105** induces apoptosis in CCRF-CEM cells, mediated by caspases activation, MMP alteration, and increased ROS production (Mbaveng, Chi, Bonsou, et al., 2020).

5.2 Induction of ferroptotic, necroptotic, and autophagic cell death by terpenoids

Mbaveng and co-workers have demonstrated that saponin **55** (progenin III) is an excellent autophagy and necroptosis inducer (Mbaveng, Chi, Nguenang, et al., 2020). In effect, when applying the RRA to analyze the ability of compound **55** to interfere with ferrotoptic, necroptotic cell death or autophagy in CCRF-CEM cells using appropriate inhibitors, they found that the presence of the ferroptosis inhibitors ferrostatin-1 (FS-1) or deferoxamine (DFA) did not influence the cytotoxicity of compound **55**, suggesting that the compound is not involved in the ferroptotic cell death. By contrast, the presence of the necroptosis inhibitor, necrostatin-1, and the autophagy inhibitor, 3-MA decreased the cytotoxicity of compound **55** by 2.44-fold and 5.05-fold, respectively, clearly suggesting that compound **55** induced necroptotic cell death and autophagic in CCRF-CEM cells.

Mbaveng and collaborators have also applied the RRA to measure the effect of ferroptosis inhibitors (FS-1 and DFA) on the cytotoxicity of saponin **99** towards CCRF-CEM cells (Mbaveng, Ndontsa, et al., 2018). Cells were pre-incubated for 1 h in the presence of FS-1 (at 50 μM) or DFA (0.2 μM) to allow precipitation of cellular iron, then treated with various concentrations of compound **99**. The fluorescence was measured after 72 h incubation. The addition of DFA and FS-1 decreased the cytotoxicity of compound **99** (IC_{50}: 1.20 μM) by 1.31- and 3.81-fold, respectively, with IC_{50} values of 1.69 and 4.59 μM. These data indicate that compound **99** induced ferroptosis in CCRF-CEM cells.

6. Conclusion

In this Chapter, we have identified 111 terpenoids isolated from African medicinal plants with cytotoxic effects on various human cancer cell lines. They include five sesquiterpenoids twenty-six diterpenoids, twenty-nine steroids and sterol saponins, and fifty-one triterpenoids and triterpene saponins. The biosynthesis of various classes of terpenoids was also described. The mode of action of the most active terpenoids was also reported. The most potent sesquiterpenoids were 2*α*-hydroxyalantolactone (**1**), damsin (**2**), neoambrosin (**3**), vernomelitensin (**4**), and vernopicrin (**5**) while the most active diterpenes include *ent*-trachyloban-3*β*-ol (**16**), *ent*-trachyloban-3-one (**17**), and salvimulticanol (**28**). The best steroids and steroidal saponins were 16*β*-formyloxymelianthugenin (**31**), 16*β*-hydroxybersaldegenin 1,3,5-orthoacetate (**32**), 2*β*-acetoxy-3,5-di-*O*-acetylhellebrigenin (**33**), 2*β*-acetoxy-5*β*-*O*-acetylhellebrigenin (**34**), 2*β*-acetoxymelianthusigenin (**35**), 2*β*-hydroxy-3*β*,5*β*-di-*O*-acetylhellebrigenin (**36**), stigmasterol (**39**), (25*R*)-17α-hydroxyspirost-5-en-3*β*-yl *O*-α-L-rhamnopyranosyl-(1-2)-*O*-[*O*-*α*-L-rhamnopyranosyl-(1-4)-*α*-L-rhamnopyranosyl-(1-4)]-*β*-D-glucopyranoside (**42**), balanitin 4 (**46**), balanitin 6 (**47**), balanitin 7 (**48**), and progenin III (**55**). The most potent triterpenoids and triterpene saponins include 2*α*, 3*α*, 19α,20*β*, 23-pentahydroxyurs-12-en-28-oic acid (**65**), 3*β*-hydroxy-urs-11-en13(28)-olide (**67**), betulin (**70**), dichapetalin X (**73**), erythrodiol (**75**), ardisiacrispin B (**99**), and lebbeckosides A (**4**) and B (**103**). Terpenoids with the ability to combat cancer drug resistance were identified as sesquiterpenoids **1**, **2**, and **3**, diterpenoid **28**, sterols **31**, **33**, **34**, **35**, and **36**, and steroidal saponins **55**, triterpenoids **70** and **71**, and triterpene saponins alpha-hederin (**98**) and olean-12-en-3-*β*-*O*-D-glucopyranoside (**105**). Potent terpenoids such as compounds **99** and **105**, progenin III (**55**), betulinic acid (**71**), trichadonic acid (**83**) displayed apoptotic cell death, mainly mediated by caspase activation, alteration of MMP, and enhanced ROS production in cancer cells. The reported potent terpenoids should be further explored to develop novel cytotoxic products to fight cancer.

References

Abd El-Kader, A. M., Mahmoud, B. K., Hajjar, D., Mohamed, M. F. A., Hayallah, A. M., & Abdelmohsen, U. R. (2020). Antiproliferative activity of new pentacyclic triterpene and a saponin from *Gladiolus segetum* Ker-Gawl corms supported by molecular docking study. *RSC Advances, 10*(38), 22730–22741.

Ajikumar, P. K., Tyo, K., Carlsen, S., Mucha, O., Phon, T. H., & Stephanopoulos, G. (2008). Terpenoids: Opportunities for biosynthesis of natural product drugs using engineered microorganisms. *Molecular Pharmaceutics, 5*(2), 167–190.

Awouafack, M. D., Tane, P., Kuete, V., & Eloff, J. N. (2013). 2—Sesquiterpenes from the medicinal plants of Africa. In Kuete (Ed.). *Medicinal Plant Research in Africa* (pp. 33–103). Oxford: Elsevier.

Bedane, K. G., Brieger, L., Strohmann, C., Seo, E. J., Efferth, T., & Spiteller, M. (2020a). Cytotoxic bufadienolides from the leaves of a medicinal plant *Melianthus comosus* collected in South Africa. *Bioorganic Chemistry, 102*, 104102.

Bedane, K. G., Brieger, L., Strohmann, C., Seo, E. J., Efferth, T., & Spiteller, M. (2020b). Cytotoxic bufadienolides from the leaves of *Melianthus major. Journal of Natural Products, 83*(7), 2122–2128.

Block, S., Baccelli, C., Tinant, B., Van Meervelt, L., Rozenberg, R., Habib Jiwan, J. L., et al. (2004). Diterpenes from the leaves of *Croton zambesicus. Phytochemistry, 65*(8), 1165–1171.

Block, S., Stévigny, C., De Pauw-Gillet, M. C., de Hoffmann, E., Llabrès, G., Adjakidjé, V., et al. (2002). ent-trachyloban-3beta-ol, a new cytotoxic diterpene from *Croton zambesicus. Planta Medica, 68*(7), 647–649.

Chi, G. F., Sop, R. V. T., Mbaveng, A. T., Omollo Ombito, J., Fotso, G. W., Nguenang, G. S., et al. (2020). Steroidal saponins from *Raphia vinifera* and their cytotoxic activity. *Steroids,* 108724.

Cowan, M. M. (1999). Plant products as antimicrobial agents. *Clinical Microbiology Reviews, 12*(4), 564–582.

Efferth, T., Kadioglu, O., Saeed, M. E. M., Seo, E. J., Mbaveng, A. T., & Kuete, V. (2021). Medicinal plants and phytochemicals against multidrug-resistant tumor cells expressing ABCB1, ABCG2, or ABCB5: A synopsis of 2 decades. *Phytochemistry Reviews, 20*(1), 7–53.

Efferth, T., Saeed, M. E. M., Kadioglu, O., Seo, E. J., Shirooie, S., Mbaveng, A. T., et al. (2020). Collateral sensitivity of natural products in drug-resistant cancer cells. *Biotechnology Advances, 38*, 107342.

El-Baba, C., Baassiri, A., Kiriako, G., Dia, B., Fadlallah, S., Moodad, S., et al. (2021). Terpenoids' anti-cancer effects: Focus on autophagy. *Apoptosis: An International Journal on Programmed Cell Death, 26*(9-10), 491–511.

Fotso, W. G., Na-Iya, J., Mbaveng, T. A., Ango Yves, P., Demirtas, I., Kuete, V., et al. (2019). Polyacanthoside A, a new oleanane-type triterpenoid saponin with cytotoxic effects from the leaves of *Acacia polyacantha* (Fabaceae). *Natural Product Research, 33*(24), 3521–3526.

Fouedjou, T. R., Teponno, R. B., Quassinti, L., Bramucci, M., Petrelli, D., Vitali, L. A., et al. (2014). Steroidal saponins from the leaves of *Cordyline fruticosa* (L.) A. Chev. and their cytotoxic and antimicrobial activity. *Phytochemistry Letters, 7*, 62–68.

Gnoula, C., Mégalizzi, V., Nève, N. D. E., Sauvage, S., Ribaucour, F., Guissou, P., et al. (2008). Balanitin-6 and -7: Diosgenyl saponins isolated from Balanites aegyptiaca Del. display significant anti-tumor activity *in vitro* and *in vivo. International Journal of Oncology, 32*, 5–15.

Guan, Z., Xue, D., Abdallah, I. I., Dijkshoorn, L., Setroikromo, R., Lv, G., et al. (2015). Metabolic engineering of *Bacillus subtilis* for terpenoid production. *Applied Microbiology and Biotechnology, 99*(22), 9395–9406.

Guefack, M. G. F., Damen, F., Mbaveng, A. T., Tankeo, S. B., Bitchagno, G. T. M., Çelik, İ., et al. (2020). Cytotoxic constituents of the bark of *Hypericum roeperianum* towards multidrug-resistant cancer cells. *Evidence-Based Complementary and Alternative Medicine, 2020*, 4314807.

Hegazy, M. F., Dawood, M., Mahmoud, N., Elbadawi, M., Sugimoto, Y., Klauck, S. M., et al. (2021). 2α-Hydroxyalantolactone from *Pulicaria undulata*: Activity against multi-drug-resistant tumor cells and modes of action. *Phytomedicine, 81*, 153409.

Hegazy, M. F., Hamed, A. R., El-Halawany, A. M., Hussien, T. A., Abdelfatah, S., Ohta, S., et al. (2018). Cytotoxicity of abietane diterpenoids from *Salvia multicaulis* towards multidrug-resistant cancer cells. *Fitoterapia, 130*, 54–60.

Hegazy, M. F., Hamed, A. R., Ibrahim, M. A. A., Talat, Z., Reda, E. H., Abdel-Azim, N. S., et al. (2018). Euphosantianane A-D: Antiproliferative premyrsinane diterpenoids from the endemic Egyptian plant *Euphorbia Sanctae-Catharinae. Molecules (Basel, Switzerland), 23*(9), 2221.

Kang, A., & Lee, T. S. (2016). *Secondary metabolism for isoprenoid based biofuelds*. Elsevier.
Kashkooli, A. B., Krol, A. V. D., & Bouwmeester, H. J. (2018). Terpenoid biosynthesis in plants. In B. Siegmund, & E. Leitner (Eds.). *Flavour sciience* (pp. 1–9). Graz: Verlag der Technischen Universität Graz.
Kuete, V. (2010). Potential of Cameroonian plants and derived products against microbial infections: A review. *Planta Medica, 76*(14), 1479–1491.
Kuete, V. (2013). Medicinal plant research in Africa. In Kuete (Ed.). *Pharmacology and chemistry*. Oxford: Elsevier.
Kuete, V. (2023a). *African flora to fight bacterial resistance, part I: Standards for the activity of plant-derived products, 106*. London, Oxford, Cambridge, San Diego: Academic Press,.
Kuete, V. (2023b). African flora to fight bacterial resistance, Part II: The best source of herbal drugs and pharmaceuticals. In Advances in botanical research (107, pp. 1–660). London, Oxford, Cambridge, San Diego.
Kuete, V., & Efferth, T. (2015). African flora has the potential to fight multidrug resistance of cancer. *BioMed Research International, 2015*, 914813.
Kuete, V. (2025). Chapter Four—African medicinal plants and their derivative as the source of potent anti-leukemic products: Rationale classification of naturally occurring anticancer agents. *Advances in Botanical Research, 113*. https://doi.org/10.1016/bs.abr.2023.12.010.
Kuete, V., Ango, P. Y., Yeboah, S. O., Mbaveng, A. T., Mapitse, R., Kapche, G. D., et al. (2014). Cytotoxicity of four *Aframomum* species (*A. arundinaceum, A. alboviolaceum, A. kayserianum and A. polyanthum*) towards multi-factorial drug resistant cancer cell lines. *BMC Complementary and Alternative Medicine, 14*, 340.
Kuete, V., Fokou, F. W., Karaosmanoğlu, O., Beng, V. P., & Sivas, H. (2017). Cytotoxicity of the methanol extracts of *Elephantopus mollis, Kalanchoe crenata* and 4 other Cameroonian medicinal plants towards human carcinoma cells. *BMC Complementary and Alternative Medicine, 17*(1), 280.
Kuete, V., Mafodong, F. L. D., Celik, I., Fobofou, S. A. T., Ndontsa, B. L., Karaosmanoğlu, O., et al. (2017). *In vitro* cytotoxicity of compounds isolated from *Desbordesia glaucescens* against human carcinoma cell lines. *South African Journal of Botany, 111*, 37–43.
Kuete, V., Ngnintedo, D., Fotso, G. W., Karaosmanoğlu, O., Ngadjui, B. T., Keumedjio, F., et al. (2018). Cytotoxicity of seputhecarpan D, thonningiol and 12 other phytochemicals from African flora towards human carcinoma cells. *BMC Complementary and Alternative Medicine, 18*(1), 36.
Kuete, V., Omosa, L. K., Midiwo, J. O., Karaosmanoğlu, O., & Sivas, H. (2019). Cytotoxicity of naturally occurring phenolics and terpenoids from Kenyan flora towards human carcinoma cells. *Journal of Ayurveda and Integrative Medicine, 10*(3), 178–184.
Kuete, V., Sandjo, L., Seukep, J., Maen, Z., Ngadjui, B., & Efferth, T. (2015). Cytotoxic compounds from the fruits of Uapaca togoensis towards multi-factorial drug-resistant cancer cells. *Planta Medica, 81*, 32–38.
Kuete, V., Sandjo, L. P., Mbaveng, A. T., Zeino, M., & Efferth, T. (2015). Cytotoxicity of compounds from *Xylopia aethiopica* towards multi-factorial drug-resistant cancer cells. *Phytomedicine, 22*, 1247–1254.
Kuete, V., Tangmouo, J. G., Penlap Beng, V., Ngounou, F. N., & Lontsi, D. (2006). Antimicrobial activity of the methanolic extract from the stem bark of *Tridesmostemon omphalocarpoides* (Sapotaceae). *Journal of Ethnopharmacology, 104*(1–2), 5–11.
Kuete, V., Tankeo, S. B., Saeed, M. E., Wiench, B., Tane, P., & Efferth, T. (2014). Cytotoxicity and modes of action of five Cameroonian medicinal plants against multi-factorial drug resistance of tumor cells. *Journal of Ethnopharmacology, 153*(1), 207–219.
Lamarti, A., Badoc, A., Deffieux, G., & Carde, J. P. (1994). Biogénèse des monoterpènes II - la chaîne isoprénique. *Bulletin de la Societe Pharmaceutique de Bordeaux, 133*, 79–99.
Li, K., & Gustafson, K. R. (2021). Sesterterpenoids: Chemistry, biology, and biosynthesis. *Natural Product Reports, 38*(7), 1251–1281.

Ludwiczuk, A., Skalicka-Woźniak, K., & Georgiev, M. I. (2017). Chapter 11—Terpenoids. In S. Badal, & R. Delgoda (Eds.). *Pharmacognosy* (pp. 233–266). Boston: Academic Press.

Mbaveng, A. T., Chi, G. F., Bonsou, I. N., Ombito, J. O., Yeboah, S. O., Kuete, V., et al. (2020). Cytotoxic phytochemicals from the crude extract of *Tetrapleura tetraptera* fruits towards multi-factorial drug resistant cancer cells. *Journal of Ethnopharmacology, 267*, 113632.

Mbaveng, A. T., Chi, G. F., Nguenang, G. S., Abdelfatah, S., Tchangna Sop, R. V., Ngadjui, B. T., et al. (2020). Cytotoxicity of a naturally occuring spirostanol saponin, progenin III, towards a broad range of cancer cell lines by induction of apoptosis, autophagy and necroptosis. *Chemico-Biological Interactions, 326*, 109141.

Mbaveng, A. T., Damen, F., Simo Mpetga, J. D., Awouafack, M. D., Tane, P., Kuete, V., et al. (2019). Cytotoxicity of crude extract and isolated constituents of the *Dichrostachys cinerea* bark towards multifactorial drug-resistant cancer cells. *Evidence-Based Complementary and Alternative Medicine, 2019*, 8450158.

Mbaveng, A. T., Fotso, G. W., Ngnintedo, D., Kuete, V., Ngadjui, B. T., Keumedjio, F., et al. (2018). Cytotoxicity of epunctanone and four other phytochemicals isolated from the medicinal plants *Garcinia epunctata* and *Ptycholobium contortum* towards multi-factorial drug resistant cancer cells. *Phytomedicine, 48*, 112–119.

Mbaveng, A. T., Kuete, V., & Efferth, T. (2017). Potential of Central, Eastern and Western Africa medicinal plants for cancer therapy: Spotlight on resistant cells and molecular targets. *Frontiers in Pharmacology, 8*, 343.

Mbaveng, A. T., Ndontsa, B. L., Kuete, V., Nguekeu, Y. M. M., Celik, I., Mbouangouere, R., et al. (2018). A naturally occuring triterpene saponin ardisiacrispin B displayed cytotoxic effects in multi-factorial drug resistant cancer cells via ferroptotic and apoptotic cell death. *Phytomedicine, 43*, 78–85.

Mbaveng, A. T., Noulala, C. G. T., Samba, A. R. M., Tankeo, S. B., Fotso, G. W., Happi, E. N., et al. (2020). Cytotoxicity of botanicals and isolated phytochemicals from *Araliopsis soyauxii* Engl. (Rutaceae) towards a panel of human cancer cells. *Journal of Ethnopharmacology, 267*, 113535.

Mercer, E. I. (1984). The biosynthesis of ergosterol. *Pesticide Science, 15*(2), 133–155.

Miyoshi, N., Nakamura, Y., Ueda, Y., Abe, M., Ozawa, Y., Uchida, K., et al. (2003). Dietary ginger constituents, galanals A and B, are potent apoptosis inducers in Human T lymphoma Jurkat cells. *Cancer Letters, 199*(2), 113–119.

Mogue, L. D. K., Ango, P. Y., Fotso, G. W., Mapitse, R., Kapche, D. W. F. G., Karaosmanoğlu, O., et al. (2019). Two new polyhydroxylated pentacyclic triterpenes with cytotoxic activities from *Manilkara pellegriniana* (Sapotaceae). *Phytochemistry Letters, 31*, 161–165.

Mohamed, G. A., Ibrahim, S. R., Shaala, L. A., Alshali, K. Z., & Youssef, D. T. (2014). Urgineaglyceride A: A new monoacylglycerol from the Egyptian *Drimia maritima* bulbs. *Natural Product Research, 28*(19), 1583–1590.

Mosad, R. R., Ali, M. H., Ibrahim, M. T., Shaaban, H. M., Emara, M., & Wahba, A. E. (2017). New cytotoxic steroidal saponins from *Cestrum parqui*. *Phytochemistry Letters, 22*, 167–173.

Nagel, R., Bernholz, C., Vranová, E., Košuth, J., Bergau, N., Ludwig, S., et al. (2015). *Arabidopsis thaliana* isoprenyl diphosphate synthases produce the C25 intermediate geranylfarnesyl diphosphate. *The Plant Journal, 81*(5), 847–859

Noté, O. P., Azouaou, S. A., Simo, L., Antheaume, C., Guillaume, D., Pegnyemb, D. E., et al. (2016). Phenotype-specific apoptosis induced by three new triterpenoid saponins from *Albizia glaberrima* (Schumach. & Thonn.) Benth. *Fitoterapia, 109*, 80–86.

Noté, O. P., Jihu, D., Antheaume, C., Zeniou, M., Pegnyemb, D. E., Guillaume, D., et al. (2015). Triterpenoid saponins from *Albizia lebbeck* (L.) Benth and their inhibitory effect on the survival of high grade human brain tumor cells. *Carbohydrate Research, 404*, 26–33.

Noté, O. P., Simo, L., Mbing, J. N., Guillaume, D., Aouazou, S. A., Muller, C. D., et al. (2016). Two new triterpenoid saponins from the roots of *Albizia zygia* (DC.) J.F. Macbr. *Phytochemistry Letters, 18*, 128–135.
Nyaboke, H. O., Moraa, M., Omosa, L. K., Mbaveng, A. T., Vaderament-Alexe, N.-N., Masila, V., et al. (2018). Cytotoxicity of lupeol from the stem bark of *Zanthoxylum gilletii* against multi-factorial drug resistant cancer cell lines. *Investigational Medicinal Chemistry and Pharmacology, 1*(1), 10.
Nzowa, L. K., Barboni, L., Teponno, R. B., Ricciutelli, M., Lupidi, G., Quassinti, L., et al. (2010). Rheediinosides A and B, two antiproliferative and antioxidant triterpene saponins from *Entada rheedii*. *Phytochemistry, 71*(2), 254–261.
Omosa, L. K., Midiwo, J. O., Mbaveng, A. T., Tankeo, S. B., Seukep, J. A., Voukeng, I. K., et al. (2016). Antibacterial activity and structure-activity relationships of a panel of 48 compounds from kenyan plants against multidrug resistant phenotypes. *SpringerPlus, 5*, 901.
Osei-Safo, D., Dziwornu, G. A., Appiah-Opong, R., Chama, M. A., Tuffour, I., Waibel, R., et al. (2017). Constituents of the roots of *Dichapetalum pallidum* and their anti-proliferative activity. *Molecules (Basel, Switzerland), 22*(4), 532.
Pettit, G. R., Doubek, D. L., & Herald, D. L. (1991). Isolation and structure of cytostatic sterodial saponins from the african medicinal plant *Balanites aecgyptica*. *Journal of Natural Products, 54*(6), 1491–1502.
Ponou, K. B., Nono, N. R., Teponno, R. B., Lacaille-Dubois, M.-A., Quassinti, L., Bramucci, M., et al. (2014). Bafouoside C, a new triterpenoid saponin from the roots of *Cussonia bancoensis* Aubrev. & Pellegr. *Phytochemistry Letters, 10*, 255–259.
Quin, M. B., Flynn, C. M., & Dannert, S. C. (2014). Traversing the fungal terpenome. *Natural Product Report, 31*, 1449–1473.
Saeed, M., Jacob, S., Sandjo, L. P., Sugimoto, Y., Khalid, H. E., Opatz, T., et al. (2015). Cytotoxicity of the sesquiterpene lactones neoambrosin and damsin from *Ambrosia maritima* against multidrug-resistant cancer cells. *Frontiers in Pharmacology, 6*, 267.
Sandjo, L. P., & Kuete, V. (2013a). 3—Diterpenoids from the medicinal plants of Africa. In Kuete (Ed.). *Medicinal plant research in Africa* (pp. 105–133). Oxford: Elsevier.
Sandjo, L. P., & Kuete, V. (2013b). 4—Triterpenes and steroids from the medicinal plants of Africa. In Kuete (Ed.). *Medicinal plant research in Africa* (pp. 135–202). Oxford: Elsevier.
Sandjo, L. P., Fru, C. G., Kuete, V., Nana, F., Yeboah, S. O., Mapitse, R., et al. (2014). Elatumic acid: A new ursolic acid congener from *Omphalocarpum elatum* Miers (Sapotaceae). *Zeitschrift fur Naturforschung —Section C Journal of Biosciences, 69*(7-8), 276–282.
Sell, C. S. (2003). *A fragrant introduction to terpenoid chemistry*. London: The Royal Society of Chemistry.
Sell, C. S. (2007). *A fragrant introduction to terpenoid chemistry*. London: Royal Society of Chemistry.
Seukep, J. A., Sandjo, L. P., Ngadjui, B. T., & Kuete, V. (2016). Antibacterial activities of the methanol extracts and compounds from *Uapaca togoensis* against Gram-negative multi-drug resistant phenotypes. *South African Journal of Botany, 103*, 1–5.
Tabopda, T. K., Mitaine-offer, A.-C., Miyamoto, T., Tanaka, C., Mirjolet, J.-F., Duchamp, O., et al. (2012). Triterpenoid saponins from *Hydrocotyle bonariensis* Lam. *Phytochemistry, 73*, 142–147.
Tabopda, T. K., Mitaine-offer, A.-C., Tanaka, C., Miyamoto, T., Mirjolet, J.-F., & Lacaille-Dubois, M.-A. (2014). Steroidal saponins from *Dioscorea preussii*. *Fitoterapia, 97*, 198–203.
Tahsin, T., Wansi, J. D., Al-Groshi, A., Evans, A., Nahar, L., Martin, C., et al. (2017). Cytotoxic properties of the stem bark of *Citrus reticulata* Blanco (Rutaceae). *Phytotherapy Research, 31*(8), 1215–1219.

Talapatra, S. K., & Talapatra, B. (2015). *Chemistry of plant natural products. Stereochemistry, conformation, synthesis, biology, and medicine*Berlin, Heidelberg: Springer, 318–585.

Tapondjou, A. L., Jenett-siems, K., Böttger, S., & Melzig, M. F. (2013). Steroidal saponins from the flowers of *Dioscorea bulbifera* var. sativa. *Phytochemistry, 95*, 341–350.

Tapondjou, A. L., Nyaa, B. T. L., Tane, P., Ricciutelli, M., Quassinti, L., Bramucci, M., et al. (2011). Cytotoxic and antioxidant triterpene saponins from *Butyrospermum parkii* (Sapotaceae). *Carbohydrate Research, 346*(17), 2699–2704.

Tawila, A. M., Sun, S., Kim, M. J., Omar, A. M., Dibwe, D. F., Ueda, J. Y., et al. (2020). Chemical constituents of *Callistemon citrinus* from Egypt and their antiausterity activity against PANC-1 human pancreatic cancer cell line. *Bioorganic & Medicinal Chemistry Letters, 30*(16), 127352.

Tchimene, M. K., Okunji, C. O., Iwu, M. M., & Kuete, V. (2013). 1—Monoterpenes and related compounds from the medicinal plants of Africa. In Kuete (Ed.). *Medicinal plant research in Africa* (pp. 1–32). Oxford: Elsevier.

Thimmappa, R., Geisler, K., Louveau, T., O'Maille, P., & Osbourn, A. (2014). Triterpene biosynthesis in plants. *Annual Review of Plant Biology, 65*, 225–257.

Thoppil, R. J., & Bishayee, A. (2011). Terpenoids as potential chemopreventive and therapeutic agents in liver cancer. *World Journal of Hepatology, 3*(9), 228–249.

Toyang, N. J., Wabo, H. K., Ateh, E. N., Davis, H., Tane, P., Sondengam, L. B., et al. (2013). Cytotoxic sesquiterpene lactones from the leaves of *Vernonia guineensis* Benth. (Asteraceae). *Journal of Ethnopharmacology, 146*(2), 552–556.

Wang, G., Tang, W., & Bidigare, R. R. (2005). Terpenoids as therapeutic drugs and pharmaceutical agents. In L. Zhang, & A. L. Demain (Eds.). *Natural products: Drug discovery and therapeutic medicine* (pp. 197–227). Totowa, NJ: Humana Press.

Yan, Z., Xia, B., Qiu, M. H., Li Sheng, D., & Xu, H. X. (2013). Fast analysis of triterpenoids in *Ganoderma lucidum* spores by ultra-performance liquid chromatography coupled with triple quadrupole mass spectrometry. *Biomedical Chromatography, 27*(11), 1560–1567.

Yang, J., Xian, M., Su, S., Zhao, G., Nie, Q., Jiang, X., et al. (2012). Enhancing production of bio-isoprene using hybrid MVA pathway and isoprene synthase in *E. coli*. *PLoS One, 7*(4), e33509.

Youmbi, L. M., Makong, Y. S. D., Mbaveng, A. T., Tankeo, S. B., Fotso, G. W., Ndjakou, B. L., et al. (2023). Cytotoxicity of the methanol extracts and compounds of *Brucea antidysenterica* (Simaroubaceae) towards multifactorial drug-resistant human cancer cell lines. *BMC Complementary Medicine and Therapies, 23*, 48.

Zingue, S., Gbaweng Yaya, A. J., Michel, T., Ndinteh, D. T., Rutz, J., Auberon, F., et al. (2020). Bioguided identification of daucosterol, a compound that contributes to the cytotoxicity effects of *Crateva adansonii* DC (capparaceae) to prostate cancer cells. *Journal of Ethnopharmacology, 247*, 112251.

CHAPTER TWO

Coumarins from African medicinal plants: A review of their cytotoxic potential towards drug sensitive and multidrug-resistant cancer cell lines

Vaderament-A. Nchiozem-Ngnitedem[a,*], **Appolinaire Kene Dongmo**[b], **Leonel Donald Feugap Tsamo**[b], **and Victor Kuete**[c]
[a]Institute of Chemistry, University of Potsdam, Potsdam-Golm, Germany
[b]Department of Chemistry, Faculty of Science, University of Dschang, Dschang, Cameroon
[c]Department of Biochemistry, Faculty of Science, University of Dschang, Dschang, Cameroon
*Corresponding author. e-mail address: n.vaderamentalexe@gmail.com

Contents

Abstract

The chemical diversity and pharmacological potencies of coumarins are extremely broad. Coumarins, like many phenolic compounds, are widely distributed in nature, especially in plants families like Adoxaceae, Asclepidiaceae, Aspiaceae, Capparidaceae, Compositae, Ebenaceae, Fabaceae, Lauraceae, Meliaceae, Moraceae, Papilionaceae, Ptaeroxylaceae, Rutaceae, Euphorbiaceae, and Araliaceae. Their application has been documented in various research fields, including biology, cosmetic, agrochemical, molecular recognition, and imaging. In biology, natural coumarins, and their derivatives analogues have cytostatic/cytotoxicity effects against various cancer cell lines. In this book chapter,

Advances in Botanical Research, Volume 115
ISSN 0065-2296, https://doi.org/10.1016/bs.abr.2024.02.006

a compilation of various classes of coumarins purified from various African medicinal species with application in cancer treatment is described. The most potent ones include scopoletin, isoscopoletin, theraphin C, xanthoxyletin, clausarin, coladin, and isomesuol. They constitute useful tools for the development of novel cytotoxic drugs to combat cancer and its drug resistance.

Abbreviations

DNA deoxyribonucleic acid.
GC-MS gas chromatography mass spectrometry.
IC_{50} half inhibitory concentration.
IR infra-red.
MDR multidrug resistance.
NMR nuclear magnetic resonance.
Prep-HPLC preparative high pressure liquid chromatography.
SCs Sesquiterpene coumarins.

1. Introduction

Natural products (NPs) have historically proven their importance as a source of pharmacologically active molecules in traditional medicine for thousands of years, with the advantage of being effective and having few side effects (Dias, Urban, & Roessner, 2012; Newman & Cragg, 2016). Thus, an impressive number of allopathic medicines have been derived from natural sources (Butler, 2008; Newman, Cragg, & Snader, 2003). Most of these drugs have found much success in the pipeline of drug discovery, especially in the case of anticancer and anti-infective agents (Dias et al., 2012). For instance, within the orbital of cancer, >60% of the drugs approved by the federal agency (Food and Drug Administration, FDA) as anticancer drugs (doxorubicin, paclitaxel, bleomycin A2, calicheamicin) were NPs (Atanasov et al., 2021). In addition, several new secondary metabolites from the enormous biodiversity of the planet have been considered lead compounds. Subsequently, their structural modification yielded new agents with excellent pharmacological activity and extraordinary therapeutic possibilities against some refractory cell lines (Srivastava, Negi, Kumar, Gupta, & Khanuja, 2005). Coumarins constitute a large family of heterocyclic phenolic compounds containing a 1,2-benzopyrone or 2*H*-chromen-2-one motif, Fig. 1. This special group of secondary metabolites was first discovered ~200 years ago, from *Coumarouna odorata* (*Dipteryx odorata*, family Fabaceae) in 1820 (Sethena & Shah, 1945). The term coumarin derives from the French term of the tonka bean, "coumarou". Since their first isolation, coumarins (ubiquitous in the plant kingdom) and their

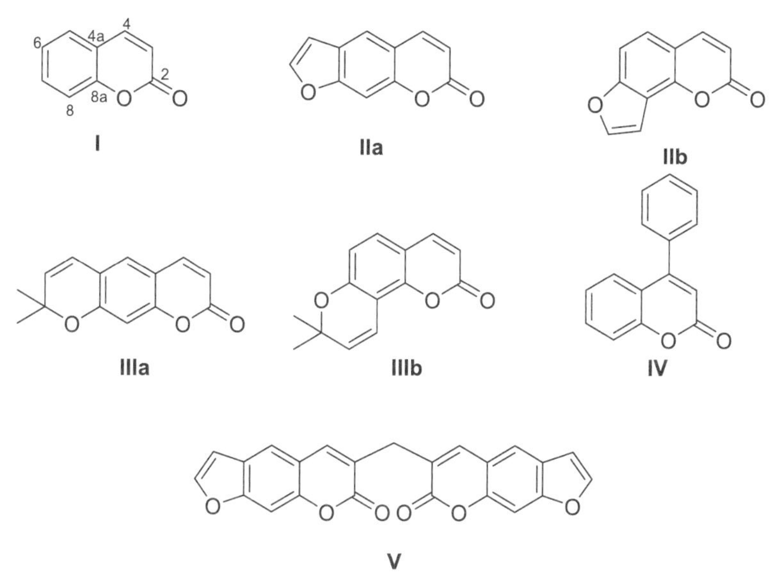

Fig. 1 Basic scaffold of natural coumarins: Simple coumarins (**I**); furanocoumarins (linear (**IIa**) and angular (**IIb**)); pyranocoumarins (linear (**IIIa**) and angular (**IIIb**)); phenyl-substituted coumarins (**IV**); biscoumarins (**V**).

derivatives have attracted a lot of attention from the scientific communities due to their medicinal, pharmacological, and industrial applications (Cao et al., 2019; Egan et al., 1990). From a pharmacological perspective, naturally occurring coumarins and their hemi/total synthesis analogues are well known for anticancer, cytotoxicity, anti-inflammatory, antimalarial, antioxidant, anti-HIV, anti-neurodegenerative, and antidiabetic activities (Akkol, Genç, Karpuz, Sobarzo-Sánchez, & Capasso, 2020; Bourgaud et al., 2006; Wu, Xu, Liu, Zeng, & Wu, 2020). Based on their biogenetic consideration, naturally occurring coumarins can be divided into several classes, including simple coumarins (**I**), furanocoumarins (**IIa** and **IIb**), pyranocoumarins (**IIIa** and **IIIb**), phenyl-substituted coumarins (**IV**), and biscoumarins (**V**) (Akkol et al., 2020; Bourgaud et al., 2006; Wu et al., 2020), Fig. 1. The following sections provide detailed information on the chemistry, biosynthesis, and phytochemical characterization of coumarins. Subsequently, a panel discussion on the cytotoxic activity of simple coumarins, furanocoumarins, pyranocoumarins, sesquiterpenes coumarins, and miscellaneous coumarins isolated from African medicinal plants towards drug sensitive and their multidrug-resistant (MDR) counterparts, is presented. The present book chapter is not exhaustive since myriads of coumarins have been isolated from African medicinal plants but were not investigated for their antineoplastic/cytotoxicity effects. Therefore, they could not be reported in this study.

2. Chemistry and biosynthesis

Coumarin, which constitutes a basic skeleton of various classes of coumarin analogues, is an aromatic compound with a molecular formula $C_9H_6O_2$ (seven double bond equivalents). Its planar structure which derives from an intramolecular condensation of *O*-coumaric acid, comprises a benzene unit with two vicinal hydrogen atoms substituted by an unsaturated lactone ring. The structural diversity of coumarins is based on various factors such as structure modification of the main scaffold; oxygenation of the aromatic carbons, *C*- or *O*-substitutions: methyl, prenyl, geranyl, farnesyl, epoxyprenyl groups, and the modification of substituents: oxygenation of side chains, and the cyclization leading to five- and six-membered rings. All these features in the benzene as well as in lactone rings, make coumarins architecturally diverse. Therefore, various subclasses of coumarins are known, including simple coumarins, furanocoumarins, pyranocoumarins, phenylcoumarins and biscoumarins (Fig. 2).

In the course of our investigation campaign, simple coumarins were found in various tissues of Araliopsis soyauxii, Cedrelopsis grevei, Citrus reticulata, Clausena anisata, Cleme viscosa, Ferula lutea, Murraya exotica, Olea Africana, Pedilanthus tithymaloides, Pituranthos chloranthus, Strychnos innocua, and Zanthoxylum zanthoxyloides. Structurally, positions C-5/6

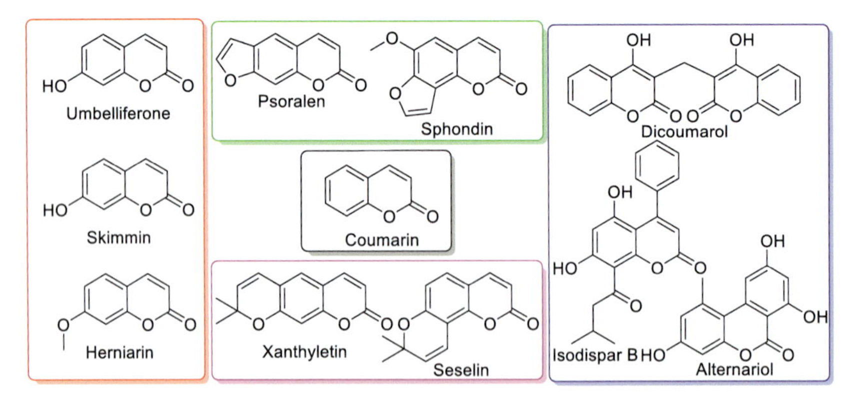

Fig. 2 Chemical structures of various coumarins.

and -7 were oxygenated in most molecules except for theraphin C, while C- and O-alkylations were mainly observed at C-3/6/8 and C-5/6/7, respectively. A representation of this subclass includes umbelliferone and skimmin (Ben, Jabrane, Harzallah-Skhiri, & Ben Jannet, 2013; Zhang, Yang et al., 2012). In addition, position 7 could be methylated, as exemplified by herniarin (Ang, Peteros, & Uy, 2019) (Fig. 2). The second important subclass of coumarins is furanocoumarins, which are obtained by fusing a furan ring with α-benzopyrone (coumarin). Depending on the junction, they can be divided into linear furanocoumarins (junction 6–7) (e.g. psoralen) and angular furanocoumarins (junction 7–8) (e.g. sphondin) (Abegaz et al., 2004; Yang et al., 2002). However, the existence of sixteen known furanocoumarins reported herein was linear mainly from eight genera such as Afraegle, Clausena, Diplolophium, Dorstenia, Ferula, Peucedanum, Pituranthos, and Thamnosma. Additionally, most furanocoumarins were O-substituted at the C-5 and/or C-8 positions, while C-prenylation occurs at the C-3 position. Unlike furanocoumarins, the six-membered pyran ring is made up pyranocoumarins such as xanthyletin (linear) and seselin (angular) (Mukandiwa, Ahmed, Eloff, & Naidoo, 2013; Tahsin et al., 2017). Pyranocoumarins were found in some plant species, including Clausena anisata, Citrus aurantium, and Citrus reticulata. Aryl-substituted coumarins consist of various classes of compounds containing a basic coumarins scaffold (Fig. 2).

As depicted in Scheme 1, the biosynthesis of coumarins, in general, is initiated from the shikimic acid pathway via cinnamic acid, which in turn originated from phenylalanine (Soine, 1964). Cinnamic acid then undergoes a series of reactions inter alia *ortho*-hydroxylation, glycosylation, isomerization, and cyclization, yielding coumarin. Stereochemical fidelity throughout this journey involves several enzymes; PAL: phenylalanine ammonia-lyase (PAL), cinnamate 2-hydroxylase, and β-glucosidase. Like in coumarins synthesis, those of simple coumarins, furanocoumarins, and pyranocoumarins follow a similar pathway. For the synthesis of various simple coumarins, PAL and tyrosine ammonia-lyase catalyzes phenylalanine and tyrosine into cinnamic- and *para*-coumaric acids, respectively. This is considered an intermediate step for the synthesis of esculetin, scopoletin, and umbelliferone (Dewick, 2002; Shimizu, 2014) (Scheme 1).

Based on the reaction mechanism, 6-hydroxycaffeoyl-CoA, 6-hydroxyferuloyl-CoA, and 6-hydroxy *p*-coumaroyl-CoA undergoes further intramolecular cyclization furnishing esculetin, scopoletin, and umbelliferone, respectively. Therefore, most simple coumarins are oxygenated at C-7 position. The furano and pyrano type coumarins which constitute a

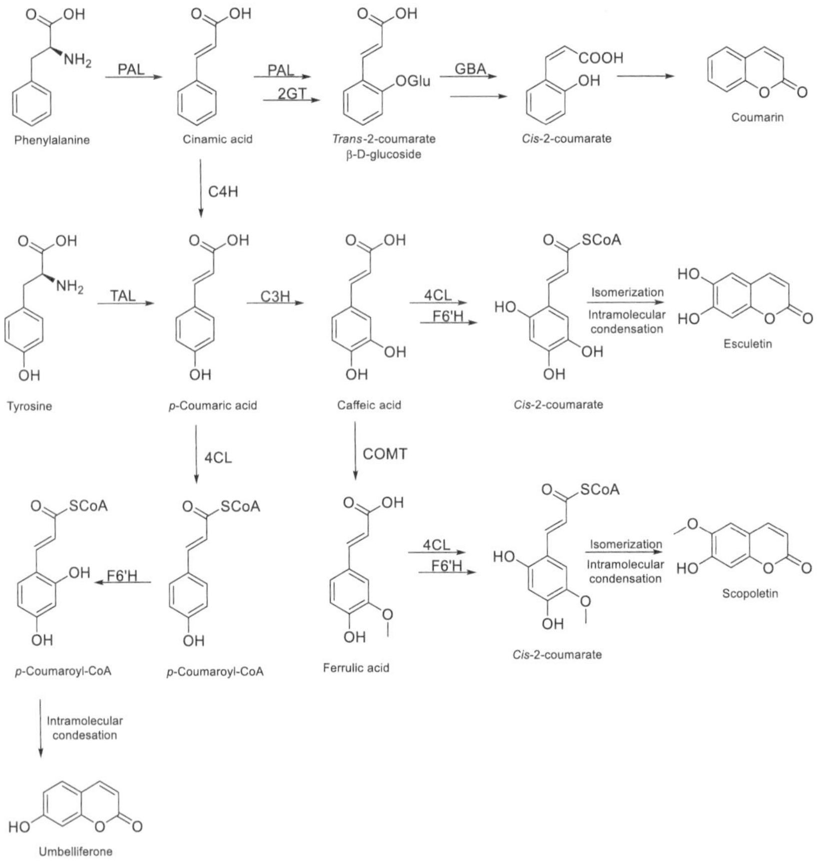

Scheme 1 Biosynthetic pathway of basic skeleton of coumarins and simple coumarins.

distinct cluster of simple coumarins use this simple class as a precursor (Scheme 2). For instance, umbelliferone is considered a key synthon for both furanocoumarins (linear and angular) and pyranocoumarins (linear and angular) (Floss & Mothes, 1964; Reddy et al., 2017). Various prenyltransferases catalyze the transfer of the dimethylallyl group from dimethylallyl pyrophosphate to positions 6 or 8, affording demethylsuberosin and osthenol, respectively (Bourgaud et al., 2006; Zhang, Li et al., 2012). Psoralen (linear five-membered ring) and xanthyletin (linear six-membered ring) are then synthesized following a series of reactions by ring formation of the demethylsuberosin which is catalyzed by marmesin synthase and cytochrome P450, respectively (Roselli et al., 2016; Zhao et al., 2015). On the other hand, columbianetin synthase and P-540 monooxygenase convert osthenol to

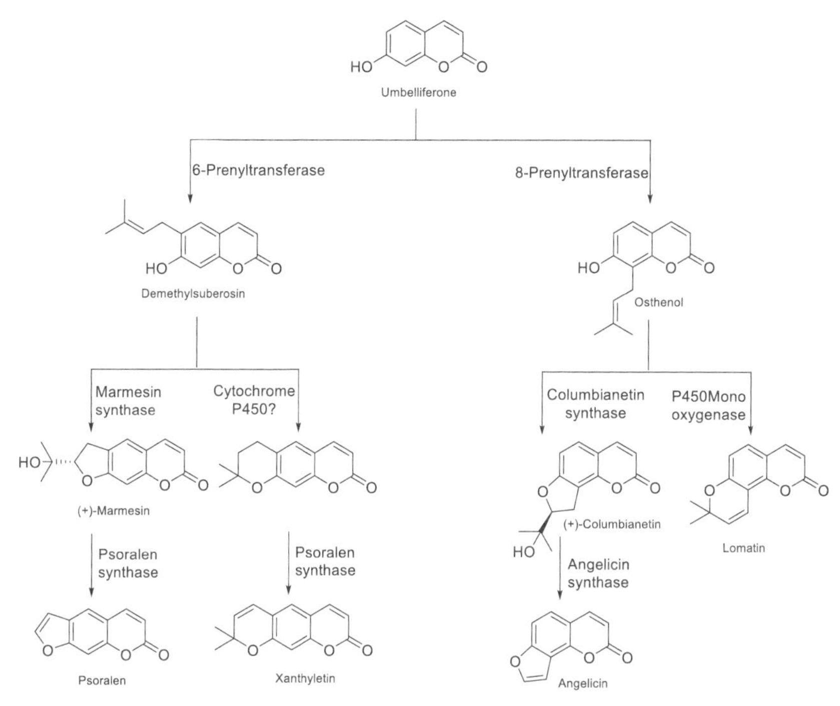

Scheme 2 Biosynthesis of linear and angular furano- and pyranocoumarins.

(+)-columbianetin and lomatin, respectively (Larbat et al., 2009; Radman et al., 2022). Moreover, (+)-columbianetin in the presence of angelicin synthase affords angelicin (Scheme 2).

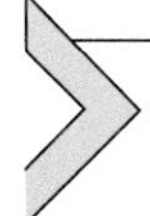

3. Phytochemical characterization of coumarins from a crude plant extract

The extraction of secondary metabolites from plants is one of the most delicate and important phases in phytochemical analysis. It is delicate in that, the use of optimum conditions should be prioritized to increase the extraction yield. Several extraction methods can be used to obtain coumarins from medicinal plants. Of these, the classical (or conventional) methods of extraction by maceration through the use of different solvent systems such as ethanol, methanol, benzene, chloroform, diethyl, and petroleum ethers (Bourgaud, Poutaraud, & Analysis, 1994; Lozhkin & Sakanyan, 2006). On the other hand, several other techniques can be used, such as ultrasonification, microwave-assisted solvent extraction in closed and open systems, accelerated solvent extraction, and supercritical fluid extraction, which allows better penetration of solvents into plant tissues (Waksmundzka-Hajnos et al., 2004). After the extraction phase, coumarins can be isolated and purified using various chromatographic techniques such as column chromatography, preparative thin layer chromatography, and preparative high-pressure liquid chromatography. The mobile phases are mainly composed of a mixture of both protic and aprotic solvents.

Several techniques are used to characterize or identify coumarins from various plant tissues. For instance, UV spectrophotometry is used to determine the characteristic chromophore groups of coumarins. Indeed, the UV spectra of coumarins show two absorption bands at 274 and 311 nm attributable to the benzene and pyrone rings, respectively. *O*-Methylation at positions C-5, C-7, and C-8 causes a bathochromic effect of the band at 274 nm, without affecting the 311 nm. Moreover, oxygenated coumarins at C-7 show strong absorption bands at 217, 240, 255, and between 315–330 nm. The subclass of linear furanocoumarins has four absorption zones at 205–255, 240–255, 260–270, and 298–316 nm, while angular furanocoumarins have their maxima located at 242–245 and 260–270 nm (Skalicka-Woźniak & Głowniak, 2006). Nuclear magnetic resonance (NMR) and infrared (IR) spectroscopy are also used. Infra-red is used to determine functional groups in coumarins, while NMR technique is employed in quality control and research for determining the content and purity of a sample as well as its molecular structure. Furthermore, the gas chromatography-mass spectrometry (GC-MS) technique is also used to identify coumarins. For example, GC-MS can quantitatively analyze mixtures containing known compounds. For unknown compounds, GC-MS can either be used to match against spectral libraries or to infer the basic structure directly.

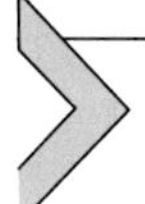

4. Cell lines used in the screening of the cytotoxicity of various coumarins from African medicinal plants

Various cancer cell lines have been used to evaluate the cytotoxicity of coumarins identified in African medicinal plants. They include murine leukemia (L1210, B16F10, P-388); human leukemia (HL-60; K562, K562/ADM, HT-29, CEM-SS), lung (A549, NCI-H322, SK-LU-1, Lu1), epidermoid carcinoma (A431), melanoma (A375, SK-MEL-2, MEL-8), prostate (PC-3, DU-145, LNCaP), colon (HCT-116, HT29, Col-1, Col2, COLO205, KM12), ovary (SKOV-3, A2780S, A2780), central nerve system (XF498), colon (HCT-15), mammary (BCAP), monocytes (U-937), breast

(BT-474, MDA-MB-231, HCC38, MDA-MB-468, SKBR3, MCF-7, BCA-1), oral (KB, KBv200), hepatocellular (HepG2), normal colon fibroblast cell (CCD-112CoN), normal lung fibroblast cell (MRC-5), man colon adenocarcinoma (CoLo 205), ewing sarcoma (A673, TC32), human oral squamous (CAL-27, FaDu, SCC-4, SCC-9, SCC-1, SCC-25), normal EBTr, nasopharyngeal (HK1, CNE1, TW01, SUNE1), adenocarcinoma (KKU-100), cervical (DoTc2, SiHa, HeLa, C33A), drug sensitive leukemia (CCRF-CEM) and its (CEM/ADR5000) MDR P-glycoprotein (P-gp)-over-expressing subline; breast (MDAMB-231-*pcDNA3*) and its resistant subline (MDA-MB-231-*BCRP* clone 23), colon (HCT116 ($p53^{+/+}$)) and its knockout clone (HCT116 ($p^{53-/-}$)), glioblastoma (U87MG) and its resistant subline (U87MG.Δ*EGFR*).

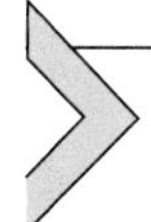

5. Cytotoxic coumarins from the African medicinal plants towards drug sensitive and MDR cancer cells

In the past three decades, coumarins found in various tissues in the plant kingdom, have attracted tremendous attention from the scientific communities due to their structural varieties and impressive biological effects. Among these properties, the cytotoxicity/antineoplastic activities of coumarins stand out. For instance, various coumarins which have been isolated so far from natural sources are well known for their extrinsic mechanism of action against different human cancer cell lines; induction of G2/M phase cell cycle arrest and upregulation of the p53/MDM2/p21 axis in SK-HEP-1 hepatoma cells, inhibiting the telomerase enzyme, inhibiting protein kinase activity, and downregulating oncogene expression by inducing the caspase-9 mediated apoptosis (P-gp-ATP-dependent membrane transport proteins) of the cancer cells (Fakai, Abd Malek, & Karsani, 2019; Guo, He, Bu, & Peng, 2019; Nasr, Bondock, & Youns, 2014; Park, Hong, Park, & Lee, 2020; Wen, Luo, Huang, Liao, & Yang, 2019). The cytotoxicity of various classes of coumarins isolated from African flora are shown in Table 1, and their chemical structures are in Fig. 3.

5.1 Simple coumarins

Umbelliferone/7-hydroxycoumarin (**1**) is the simplest coumarin largely found in various plants and is also considered the parent compound for the linear and furanocoumarins. From a pharmacological point of view, hydroxycoumarins, like umbelliferone (**1**), exert their antiproliferative potency by

Table 1 Cytotoxic of coumarins from African medicinal plants.

Compounds names	Source (plant part)	Country	Cancer cell lines, IC_{50} values (μM) and degree of resistance (D.R.) in bracket	References
Simple coumarins				
(–)-Deltoin (**34**)	*Ferula lutea* (Roots)	Tunisia	7.60 (HCT 116); 0.93 (HT-29)	Ben et al. (2013)
(–)-Prantschimgin (**33**)	*Ferula lutea* (Roots)	Tunisia	14.95 (HCT 116); 5.20 (HT-29); >10 (HTs 29); >10 (MCF-7); >10 (H-1299)	Ben et al. (2013); Sajjadi et al. (2015)
13-Hydroxyfeselol (**44**)	*Ferula tunetana* (Roots)	Tunisia	34.1 (HCT 116); 35.4 (HT-29)	Jabrane et al. (2010)
3-(1,1-Dimethyl allyl) xanthyletin (**35**)	*Clausena anisata* (Leaves)	Cameroon	3.8 (HeLa); 375.8 (Vero)	Tatsimo et al. (2015)
5,7-Dihydroxy-8-(2-methylbutyryl)-4-phenylcournarin (**64**)	*Pedilanthus tithymaloides*	Cameroon	40.63 (MDA-MB-231); 34.61 (MCF-7); 64.28 (HepG2); 35.63 (HeLa)	Chen et al. (2022); Sandjo et al. (2012)
6′-hydroxy-β-cycloauraptene (**18**)	*Cleme Viscosa* (Roots)	Sudan	57.3 (CEM-SS)	Almahy and Alagimi (2012)
6-Methylcoumarin (**2**)	*Zanthoxylum zanthoxyloides*	Senegal	>50 (A549); 41 (A431); >50 (NCI-H322); >50 (PC-3); >50 (A375); 37.0 (HCT-116)	Farooq et al. (2014); Tine et al. (2017)
7-[(*E*)-7-Hydroxy-3, 7-dimethylocta-2,5-dienyloxyl]-coumarin (**10**)	*Clausena anisata* (Leaves)	Cameroon	4.0 (HeLa); 935.8 (Vero)	Tatsimo et al. (2015)
Auraptene (**6**)	*Cleme Viscosa* (Roots)	Sudan	46.9 (CEM-SS)	Almahy and Alagimi (2012)

Badrakemin (**51**)	*Ferula assa-foetida* (Resins)	Egypt	>50 (COLO205); 9.1 (KM12); 20 (A498); 0.38 (UO31); >50 (A673); >50 (TC32)	Mohamed et al. (2001); Tosun, Beutler, Ransom, and Miski (2019)
Bergapten (**20**)	*Thamnosma rhodesica* (Roots); *Dorstenia kameruniana* (Roots)	Zimbabwe	>138.8 (L1210); >138.8 (HL-60); >138.8 (K562); >138.8 (B16F10); 41.6 (BCAP); 26.45 (MEL-8); 38.46 (U-937); 13.12 (DU-145); 20.56 (BT-474); 17.43 (MDA-MB-231); >231.4 (HL-60); 169.4 (P-388); 188.8 (CoLo 205), 114.3 (HeLa); 7.17 (CCRF-CEM) vs >185.02 (CEM/ADR5000) (D.R.: >25.8); 132.80 (MDA-MB231-*pcDNA*) vs 158.55 (MDA-MB231/*BCRP*) (D.R.: 1.19); 111.61 (HCT116 ($p53^{+/+}$)) vs >185.02 (HCT116 ($p53^{-/-}$)) (D.R.: >1.65); 163.61 (U87MG) vs 163.40 (U87. MGΔ*EGFR*) (D.R.: 0.99); 26.45 (MEL-8); 38.46 (U-937); 13.12 (DU-145); 20.56 (BT-474); 17.43 (MDA-MB-231); 0.06 (B16F10)	Adem et al. (2018); Ahua et al. (2004); Sumiyoshi et al. (2014); Thanh et al. (2004); Urbagarova et al. (2020); Yang et al. (2003)
Byakangelicin (**23**)	*Thamnosma rhodesica* (Roots)	Zimbabwe	42.8 (A549); 60.4 (SK-OV-3); 63.4 (SK-MEL-2); 85 (XF498); 54.1 (HCT15); 12.51 (B16F10)	Ahua et al. (2004); Kim, Kim, and Ryu (2007)
Cedrelopsin (**12**)	*Cedrelopsis grevei* (Trunk bark)	Madagascar	14.2 (A549)	Mahibalan et al. (2016); Um et al. (2003)

(continued)

Table 1 Cytotoxic of coumarins from African medicinal plants. (*cont'd*)

Compounds names	Source (plant part)	Country	Cancer cell lines, IC_{50} values (μM) and degree of resistance (D.R.) in bracket	References
Chalepin (**26**)	*Clausena anisata* (Stem bark)	Cameroon	17.8 (HT29); 22.3 (HCT116); 27.0 (MCF7); 63 (MDA-MB-231); >318 (MRC-5); 55.1 (HT29)	Fakai et al. (2017); Suhaimi et al. (2017); Tatsimo et al. (2015)
Clausarin (**37**)	*Citrus aurantium* (Stem bark)	Cameroon	17.6 (HepG2); 44.9 (HCT116); 6.9 (SK-LU-1); 78.2 (Vero); 7.8 (A549); 20.9 (MCF7); 4.2 (KB); 4.2 (KB-VIN)	Bissim et al. (2019); Jantamat et al. (2019); Su et al. (2009)
Cnidilin (**32**)	*Thamnosma rhodesica* (Roots)	Zimbabwe	27.7 (BCAP)	Ahua et al. (2004); You, An, Liang, and Wang (2013)
Coladin (**41**)	*Ferula tunetana* (Roots)	Tunisia	3.7 (HCT 116); 5.4 (HT-29)	Jabrane et al. (2010)
Colladonin (**42**)	*Ferula assa-foetida* (Roots); *Ferula tunetana* (Roots)	Egypt; Tunisia	19 (COLO205); 2.5 (KM12); 21 (A498); 0.75 (UO31); >50 (A673); 45 (TC32); 15.1 (HCT 116); 13.3 (HT-29)	Buddrus et al. (1985); Jabrane et al. (2010); Tosun, Aytar, Beutler, Wilson, and Miski (2021)
Diverse coumarins				
Excavatin D (**9**)	*Clausena anisata* (Leaves)	Cameroon	9.1 (HeLa)	Tatsimo et al. (2015)

Farnesiferol A (**50**)	*Ferula assa-foetida* (Resins)	Egypt	>130.7 (MCF-7); >130.7 (PC3); >261.4 (NIH)	Iranshahy et al. (2019); Mohamed et al. (2001)
Farnesiferol B (**45**)	*Ferula assa-foetida* (Resins)	Egypt	110.7 (MCF-7); 96.7 (PC3); >262.9 (NIH)	Iranshahy et al. (2019); Mohamed et al. (2001)
Farnesiferol C (**46**)	*Ferula assa-foetida* (Resins)	Egypt	109.0 (MCF-7); 112.7 (PC3); >261.4 (NIH); 78.88 (MCF-7)	Iranshahy et al. (2019); Kasaian et al. (2015); Mohamed et al. (2001)
Furanocoumarins				
Gravelliferone (**8**)	*Clausena anisata* (Stem bark)	Cameroon	6.1 (HeLa); 231.8 (Vero)	Tatsimo et al. (2015)
Gummosin (**49**)	*Ferula assa-foetida* (Resins)	Egypt	83.9 (MCF-7); 78.4 (PC3); >261 (NIH)	Iranshahy et al. (2019); Mohamed et al. (2001)
Imperatorin (**24**)	*Clausena anisata* (Roots); *Afraegle paniculate* (Stem bark)	Nigeria; Cameroon	>111 (L1210); 74.8 (HL-60); >111 (K562); >111 (B16F10); 60.7 (A549); 50.7 (SK-OV-3); 53.7 (SK-MEL-2); 45.2 (XF498); 71.8 (HCT15); 69.6 (HL-60); 74.8 (P-388); >185 (CoLo 205), >185 (HeLa)	Abe and Taylor (1971); Kim et al. (2007); Thanh et al. (2004); Tsassi et al. (2010); Yang et al. (2003)

(*continued*)

Table 1 Cytotoxic of coumarins from African medicinal plants. (*cont'd*)

Compounds names	Source (plant part)	Country	Cancer cell lines, IC_{50} values (μM) and degree of resistance (D.R.) in bracket	References
Isodispar B (**65**)	*Pedilanthus tithymaloides*	Cameroon	3.84 (SUNE1); 11.49 (TW01); 9.74 (CNE1); 5.58 (HK1); 56.73 (HCC38); 52.69 (MDA-MB-231); 52.75 (MDA-MB-468); 54.85 (SKBR3)	Lim et al. (2016); Sandjo et al. (2012)
Isoimperatorin (**31**)	*Pituranthos chloranthus* (Aerial parts)	Tunisia	45.2 (A549); 25.1 (SK-OV-3); 36.6 (SK-MEL-2); 39.6 (XF498); 20.7 (HCT15); 28.32 (MEL-8); 20.12 (U-937); 15.20 (DU-145); 27.82 (BT-474); 19.16 (MDA-MB-231); 24.5 (HeLa); 90.7 (L1210); 82.5 (HL-60); 74.4 (K562); 87.0 (B16F10)	Aloui, Kossentini, Geffroy-Rodier, Guillard, and Zouari (2015); Ameen (2014); Kim et al. (2007); Thanh et al. (2004); Urbagarova et al. (2020)
Isomarcandin (**43**)	*Ferula tunetana* (Roots)	Tunisia	>100 (HCT 116); >100 (HT-29)	Jabrane et al. (2010)
Isomesuol (**67**)	*Pedilanthus tithymaloides*	Cameroon	6.4 (P-388); 6.3 (KB); 8.7 (Col-2); 2.33 (Lu-1); 5.9 (BCA-1)	Reutrakul et al. (2003); Sandjo et al. (2012)
Isopimpinellin (**28**)	*Peucedanum zenkeri* (Seeds)	Cameroon	>203.1 (HL-60); >203.1 (P-388); 159.2 (CoLo 205), >203.1 (HeLa); >100 (B16F10)	Ngwendson et al. (2003); Sumiyoshi et al. (2014); Yang et al. (2003)

Isoscopoletin (**5**)	*Murraya exotica* (Branches); *Olea Africana* (Bark)	Egypt; South Africa	4.0 (CCRF-CEM) vs 1.6 (CEM/ADR5000) (D.R.:0.4)	Adams, Efferth, and Bauer (2006); Arango et al. (2010); Negi et al. (2015)
Mammea A/AA cyclo F (**62**)	*Ochrocarpos punctatus* (Bark)	Madagascar	15.2 (A2780)	Chaturvedula, Schilling, and Kingston (2002)
Mammea A/AB cyclo D (**63**)	*Ochrocarpos punctatus* (Bark)	Madagascar	11.1 (A2780); >49.5 (P-388); 37.3 (KB); >49.5 (Col-2); >49.5 (Lu-1); >49.5 (BCA-1)	Chaturvedula et al. (2002); Reutrakul et al. (2003)
Mammea A/AB cyclo F (**61**)	*Ochrocarpos punctatus* (Bark)	Madagascar	19.4 (A2780); 43.40 (MDA-MB-231); 34.69 (MCF-7); 64.69 (HepG2); 36.64 (HeLa)	Chaturvedula et al. (2002); Chen et al. (2022)
Mammea A/AC cyclo F (**59**)	*Ochrocarpos punctatus* (Bark)	Madagascar	10.8 (A2780); 52.56 (MDA-MB-231); 38.28 (MCF-7); 78.28 (HepG2); 28.86 (HeLa)	Chaturvedula et al. (2002); Chen et al. (2022)
Mammea A/AD cyclo D (**60**)	*Ochrocarpos punctatus* (Bark)	Madagascar	27.7 (A2780); >51.3 (P-388); >51.3 (KB); >51.3 (Col-2); >51.3 (Lu-1); >51.3 (BCA-1)	Chaturvedula et al. (2002); Reutrakul et al. (2003)
Mammea E/BB (**66**)	*Pedilanthus tithymaloides*	Cameroon	15.4 (SW-480); 11.6 (HT-29); 13.9 (HCT-116)	Sandjo et al. (2012); Yang et al. (2005)

(*continued*)

Table 1 Cytotoxic of coumarins from African medicinal plants. (*cont'd*)

Compounds names	Source (plant part)	Country	Cancer cell lines, IC_{50} values (µM) and degree of resistance (D.R.) in bracket	References
Marmesin (**27**)	*Thamnosma rhodesica* (Roots)	Zimbabwe	21.32 (MEL-8); 16.48 (U-937); 12.35 (DU-145); 29.34 (BT-474); 28.17 (MDA-MB-231)	Ahua et al. (2004); Urbagarova et al. (2020)
Microminutin (**14**)	*Murraya exotica* (Branches)	Egypt	6.2 (KKU-100)	Lekphrom, Kanokmedhakul, Kukongviriyapan, and Kanokmedhakul (2011); Negi et al. (2015)
Minumicrolin (**16**)	*Murraya exotica* (Branches)	Egypt	36.9 (KKU-100)	Lekphrom et al. (2011); Negi et al. (2015)
Murralongin (**13**)	*Murraya exotica* (Branches)	Egypt	34.9 (KKU-100); 470.2 (U-937)	Arango et al. (2010); Lekphrom et al. (2011); Negi et al. (2015)
Murrangatin (**15**)	*Murraya exotica* (Branches)	Egypt	10.5 (KKU-100)	Lekphrom et al. (2011); Negi et al. (2015)
Ochrocarpin A (**52**)	*Ochrocarpos punctatus* (Bark)	Madagascar	12.4 (A2780)	Chaturvedula et al. (2002)
Ochrocarpin B (**53**)	*Ochrocarpos punctatus* (Bark)	Madagascar	9.0 (A2780)	Chaturvedula et al. (2002)

Ochrocarpin C (**54**)	*Ochrocarpos punctatus* (Bark)	Madagascar	9.1 (A2780); 72.14 (MDA-MB-231); 52.55 (MCF-7); 62.55 (HepG2); 49.51 (HeLa)	Chaturvedula et al. (2002); Chen et al. (2022)
Ochrocarpin D (**55**)	*Ochrocarpos punctatus* (Bark)	Madagascar	14.5 (A2780)	Chaturvedula et al. (2002)
Ochrocarpin E (**56**)	*Ochrocarpos punctatus* (Bark)	Madagascar	12.0 (A2780)	Chaturvedula et al. (2002)
Ochrocarpin F (**57**)	*Ochrocarpos punctatus* (Bark)	Madagascar	22.0 (A2780)	Chaturvedula et al. (2002)
Ochrocarpin G (**58**)	*Ochrocarpos punctatus* (Bark)	Madagascar	21.3 (A2780)	Chaturvedula et al. (2002)
Osthol (**7**)	*Pituranthos chloranthus* (Aerial parts)	Tunisia	380 (A2780S); 108.1 (L1210); 63.5 (HL-60); 103.2 (K562); 115.5 (B16F10); 61.0 (HL-60); 38.1 (P-388); 122.4 (CoLo 205), 129.8 (HeLa)	Aloui et al. (2015); Shokoohinia, Sajjadi, Gholamzadeh, Fattahi, and Behbahani (2014); Thanh et al. (2004); Yang et al. (2003)

(*continued*)

Table 1 Cytotoxic of coumarins from African medicinal plants. *(cont'd)*

Compounds names	Source (plant part)	Country	Cancer cell lines, IC_{50} values (μM) and degree of resistance (D.R.) in bracket	References
Oxypeucedanin (**29**)	*Diplolophium buchanani* (Leaves)	Malawi	>105 (L1210); 96 (HL-60); 105 (K562); >105 (B16F10); 25.98 (L5178Y) vs 28.89 (MDR-L5178Y) (D.R.: 1.11); 39.4 (HeLa); 33 (A549); 67 (SK-OV-3); 57.7 (SK-MEL-2); 56.3 (XF498); 11.9 (HCT15)	Ameen (2014); Kim et al. (2007); Marston, Hostettmann, and Msonthi1 (1995); Mottaghipisheh et al. (2018); Thanh et al. (2004)
Oxypeucedanin hydrate (**30**)	*Diplolophium buchanani* (Leaves)	Malawi	>98.6 (L1210); >98.6 (HL-60); >98.6 (K562); >98.6 (B16F10); 138 (L5178Y) vs 199 (MDR-L5178Y) (D.R.: 1.44); 16.7 (HeLa); 21.7 (A549); 28.5 (A431); 32.0 (NCI-H322); >50 (PC-3); 32.8 (A375); 34.5 (HCT-116); 31.2 (A549); 63.4 (SK-OV-3); 54.2 (SK-MEL-2); 52.9 (XF498); 11.1 (HCT15); 25.16 (MEL-8); 22.45 (U-937); 11.56 (DU-145); 16.56 (BT-474); 24.62 (MDA-MB-231)	Ameen (2014); Farooq et al. (2014); Kim et al. (2007); Marston et al. (1995); Mottaghipisheh et al. (2018); Thanh et al. (2004); Urbagarova et al. (2020)
Phebalosin (**17**)	*Murraya exotica* (Branches)	Egypt	80.2 (U-937)	Arango et al. (2010); Negi et al. (2015)
Phellopterin (**22**)	*Clausena anisata* (Stem bark)	Cameroon	64.8 (HeLa); 0.52 (B16F10)	Sumiyoshi et al. (2014); Tatsimo et al. (2015)

Psoralen (**19**)	*Dorstenia elliptica* (Twigs); *Dorstenia turbinata* (Twigs)	Cameroon	>161.2 (L1210); 155.3 (HL-60); 103.7 (K562); >161.2 (B16F10); 473.4 (KB) vs 465.3 (KBv200) (D.R.: 0.98); 131.1 (K562) vs 336.4 (K562/ADM) (D.R.: 2.57); 45.60 (HCT 116); 41.70 (HT-29); 235.9 (HT-29); 176.8 (MCF-7); 242.9 (A549); 240.7 (HepG2); 0.11 (B16F10)	Abegaz et al. (2004); Ben et al. (2013); Mar, Je, and Seo (2001); Ngameni et al. (2006); Sumiyoshi et al. (2014); Thanh et al. (2004); Wang, Hong, Zhou, Xu, and Qu (2011)
Pyranocoumarins				
Rutamarin (**25**)	*Toona ciliata*	Uganda	7.3 (HT29); 8.7 (HCT116); 12.1 (MCF7); >280.6 (MDA-MB-231); >280.6 (MRC-5); 5.6 (HT29)	Fakai et al. (2017); Okot et al. (2023); Suhaimi et al. (2017)
Samarcandin (**47**)	*Ferula assa-foetida* (Resins)	Egypt	11 (AGS); 13 (WEHI-164); 86.63 (MCF-7)	Ghoran et al. (2016); Kasaian et al. (2015); Mohamed et al. (2001)
Scoparone (**3**)	*Citrus reticulata* (Bark); *Araliopsis soyauxii* (Bark)	Cameroon	88.3 (A549); 98.5 (MCF7); 90.7 (PC3); 92.89 (CCRF-CEM) vs >100 (CEM/ADR5000) (D.R.: >1.08); >100 (MDA-MB231-*pcDNA*) vs >100 (MDA-MB231/*BCRP*); >100 (HCT116 (*p53*$^{+/+}$)) vs >100 (HCT116 (*p53*$^{-/-}$)); >100 (U87MG) vs >100 (U87.MGΔ*EGFR*)	Mbaveng et al. (2021); Tahsin et al. (2017)

(continued)

Table 1 Cytotoxic of coumarins from African medicinal plants. (*cont'd*)

Compounds names	Source (plant part)	Country	Cancer cell lines, IC_{50} values (μM) and degree of resistance (D.R.) in bracket	References
Scopoletin (**4**)	*Olea Africana* (Bark)	South Africa	99.9 (KKU-100); 189.5 (KB); 1.6 (P-388); 25 (DoTc2); 15 (SiHa); 7.5 (HeLa); 25 (C33A); 90 (HCvEpC); 2.6 (CCRF-CEM) vs 1.6 (CEM/ADR5000) (D.R.:0.6)	Adams et al. (2006); Darmawan, Kosela, Kardono, and Syah (2012); Lekphrom et al. (2011); Tian et al. (2019); Tsukamoto, Hisada, Nishibe, Roux, and Rourke (1984)
Sesquiterpene coumarins				
Theraphin C (**11**)	*Pedilanthus tithymaloides*	Cameroon	42.8 (Lu1); 13.1 (Col2); 6.2 (KB); 6.4 (LNCaP); 1.6 (DLD-1); 4.7 (HeLa); 3.48 (MCF-7); 5.67 (NCI-H460)	Lee et al. (2003); Ngo et al. (2010); Sandjo et al. (2012)
Tunetacoumarin A (**40**)	*Ferula tunetana* (Roots)	Tunisia	>100 (HCT 116); >100 (HT-29)	Jabrane et al. (2010)
Umbelliferone (**1**)	*Strychnos innocua* (Root bark); *Ferula lutea* (Roots)	Nigeria; Tunisia	>50 (A549); 11.3 (A431); >50 (NCI-H322); >50 (PC-3); >50 (A375); >50 (HCT-116); 497.2 (HL-60); 4.35 (HT-29)	Ben et al. (2013); Farooq et al. (2014); Uttu et al. (2022); Wagh et al. (2021);
Umbelliprenin (**39**)	*Ferula tunetana* (Roots)	Tunisia	>100 (HCT 116); >100 (HT-29)	Jabrane et al. (2010)

Xanthotoxin (**21**)	*Peucedanum galbanum* (Leaves); *Ferula lutea* (Roots)	South Africa; Tunisia	133.7 (L1210); 77.3 (HL-60); 113.4 (K562); 118.5 (B16F10); 32.4 (HeLa); >231.4 (HL-60); 231.4 (P-388); 163.4 (CoLo 205): >231.4 (HeLa); 0.14 (B16F10); >100 (HCT 116); >100 (HT-29)	Ameen (2014); Ben et al. (2013); Finkelstein, Albrecht, and van Jaarsveld (1993); Sumiyoshi et al. (2014); Thanh et al. (2004); Yang et al. (2003)
Xanthoxyletin (**36**)	*Clausena anisata* (Leaves)	Cameroon	78.2 (HepG2); 79.8 (HCT116); 94.4 (SK-LU-1); >100 (Vero); 10 (SCC-1); 15 (SCC-4); 10 (SCC-9); 25 (SCC-25); 15 (CAL-27); 30 (FaDu); 95 (EBTr)	Jantamat et al. (2019); Tatsimo et al. (2015); Wen et al. (2019)
Xanthyletin (**38**)	*Citrus reticulata* (Bark)	Cameroon	351.0 (A549); 84.6 (MCF7); 425.9 (PC3); 85.5 (HL-60); 85.5 (MCF-7); 117.4 (HT-29); 111.7 (HeLa)	Sharif et al. (2013); Tahsin et al. (2017)

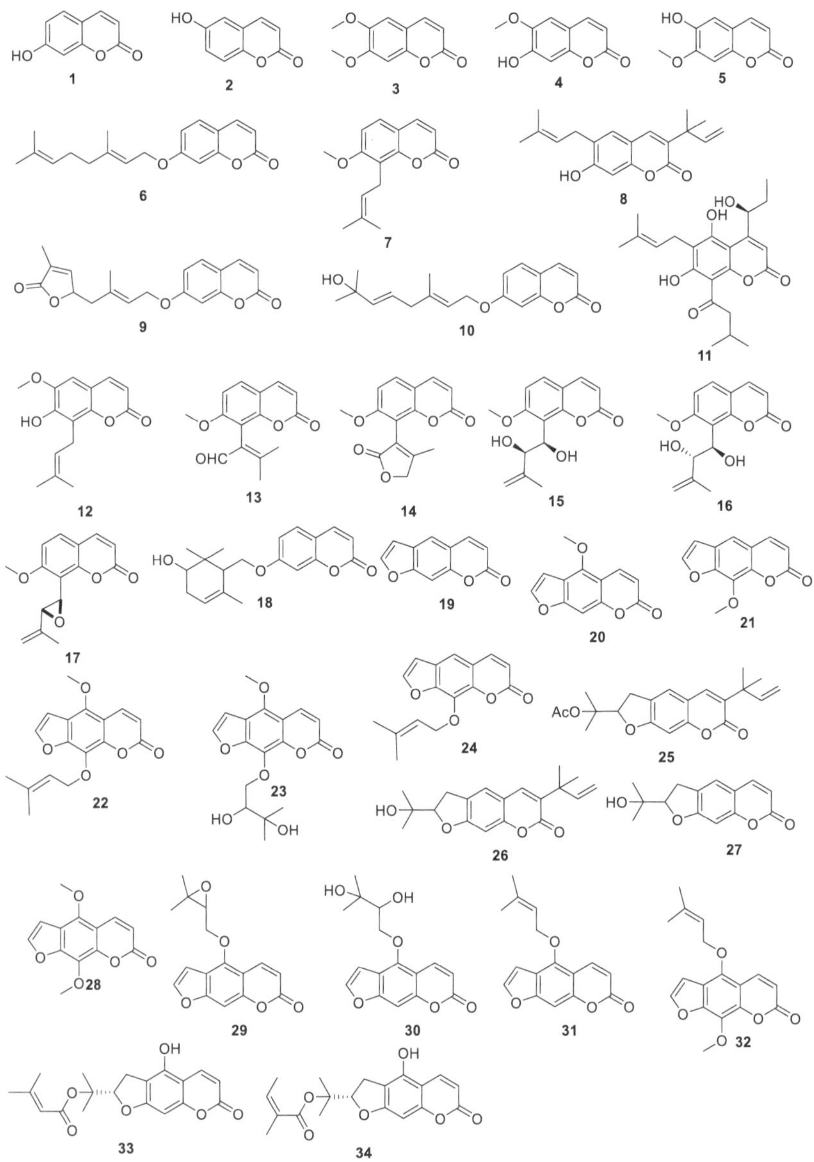

Fig. 3 Chemical structures of cytotoxic coumarins from African medicinal plants. Simple coumarins: **1**: umbelliferone; **2**: 6-methylcoumarin; **3**: scoparone; **4**: scopoletin; **5**: isoscopoletin; **6**: auraptene; **7**: osthol; **8**: gravelliferone; **9**: excavatin D; **10**: 7-[(*E*)-7-hydroxy-3,7-dimethylocta-2,5-dienyloxyl]-coumarin; **11**: theraphin C; **12**: cedrelopsin; **13**: murralongin; **14**: microminutin; **15**: murrangatin; **16**: minumicrolin; **17**: phebalosin; **18**: 6′-hydroxy-β-cycloauraptene. **Furanocoumarins**: **19**: psoralen; **20**: bergapten; **21**: xanthotoxin; **22**: phellopterin; **23**: byakangelicin; **24**: imperatorin; **25**: rutamarin; **26**: chalepin; **27**: marmesin; **28**: isopimpinellin; **29**: oxypeucedanin; **30**: oxypeucedanin hydrate; **31**: isoimperatorin; **32**: cnidilin; **33**: (−)-prantschimgin; **34**: (−)-deltoin.

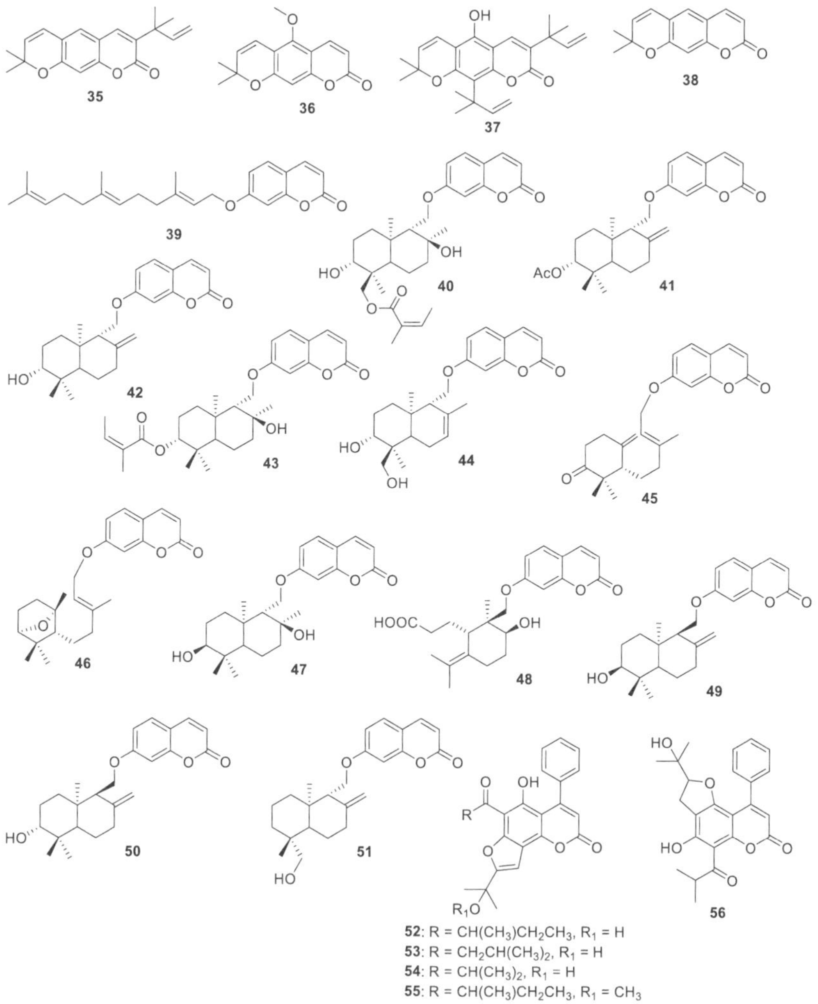
35
36
37
38
39
40
41
42
43
44
45
46
47
48
49
50
51
56
52: R = CH(CH3)CH2CH3, R1 = H
53: R = CH2CH(CH3)2, R1 = H
54: R = CH(CH3)2, R1 = H
55: R = CH(CH3)CH2CH3, R1 = CH3

Fig. 3 (*Continued*)

Pyranocoumarins: **35**: 3-(1,1-dimethyl allyl) xanthyletin; **36**: xanthoxyletin; **37**: clausarin; **38**: xanthyletin. **Sesquiterpene coumarins**: **39**: umbelliprenin; **40**: tunetacoumarin A; **41**: coladin; **42**: colladonin; **43**: isomarcandin; **44**: 13-hydroxyfeselol; **45**: farnesiferol B; **46**: farnesiferol C; **47**: samarcandin; **48**: galbanic acid; **49**: gummosin; **50**: farnesiferol A; **51**: badrakemin. **Diverse coumarins**: **52**: ochrocarpin A; **53**: ochrocarpin B; **54**: ochrocarpin C; **55**: ochrocarpin D; **56**: ochrocarpin E; **57**: ochrocarpin F; **58**: ochrocarpin G; **59**: mammea A/AC cyclo F; **60**: mammea A/AD cyclo D; **61**: mammea A/AB cyclo F; **62**: mammea A/AA cyclo F; **63**: mammea A/AB cyclo D; **64**: 5,7-dihydroxy-8-(2-methylbutyryl)-4-phenylcournarin; **65**: isodispar B; **66**: mammea E/BB; **67**: isomesuol.

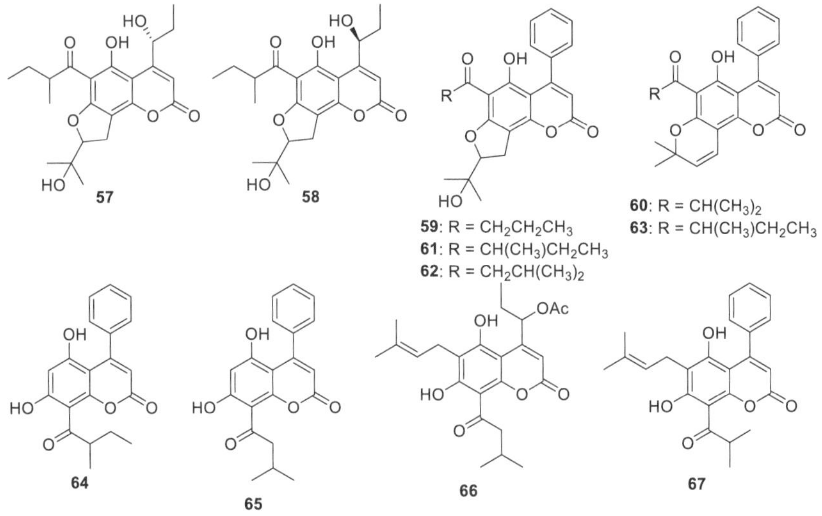

57 58

59: R = $CH_2CH_2CH_3$
61: R = $CH(CH_3)CH_2CH_3$
62: R = $CH_2CH(CH_3)_2$

60: R = $CH(CH_3)_2$
63: R = $CH(CH_3)CH_2CH_3$

64 65 66 67

Fig. 3 (*Continued*)

generating free radical species in cancer cells, thus producing oxidative stress leading to cell death. For instance, umbelliferone found in various African medicinal spices; *Strychnos innocua* (root bark), and *Ferula lutea* (roots) showed moderate to low cytotoxicity ($IC_{50} > 11$ μM) against several human cancer cell lines, including lung (A549 and NCI-H322), epidermoid carcinoma (A431), melanoma (A375), prostate (PC-3) and colon (HCT-116) cell lines using MTT assay (Ben et al., 2013; Farooq et al., 2014; Uttu, Sallau, Iyun, & Ibrahim, 2022; Wagh, Butle, & Raut, 2021), Table 1. Its positional isomer, 6-hydroxycoumarin (**2**), also showed similar activity against the tested cell lines (Farooq et al., 2014; Tine, Renucci, Costa, Wélé, & Paolini, 2017). The mechanistic study further demonstrated that umbelliferone (**1**) was cytotoxic against liver hepatocellular carcinoma (HCC) and human oral carcinoma KB cells in a concentration-dependent manner. The activity was established to be via the induction of apoptosis, cell cycle arrest, and DNA fragmentation (Vijayalakshmi & Sindhu, 2017; Yu, Hu, & Zhang, 2015). Furthermore, treatment of human renal cell carcinoma cells (786-O, OS-RC-2, and ACHN) with umbelliferone (**1**) reduced cell proliferation in a concentration-dependent manner and induced dose-dependent apoptotic events (Wang et al., 2019). Scoparone (**3**), a natural compound isolated from *Citrus reticulata* (Bark) and *Araliopsis soyauxii* (Bark), has been used in Chinese herbal medicine to treat neonatal jaundice. Scoparone (**3**) inhibited the proliferation of DU145 prostate cancer cells via cell cycle arrest in the G1 phase. This effect was also mediated by the inhibition of signal transducer and activator of transcription 3, which has been recognized as a promising therapeutic target in metastatic prostate cancer (Kim et al., 2013). Regioisomer such as scopoletin (**4**) and isoscopoletin (**5**) are potent cytotoxic agents with various mechanisms of action, Table 1 and Fig. 3. Scopoletin (**4**) induces apoptotic cell death in HeLa cervical cancer, enhancement of the Bax, Caspase 3, 8, and 9 expression and decline in the Bcl-2 expression, and causes DNA damage in the HeLa cells (Tian et al., 2019). Another study suggested that simple coumarins with prenyl side chain exhibited potent cytotoxicity. As illustrated by auraptene (**6**), which is a natural bioactive monoterpene coumarin ether. First purified from the genus *Citrus*. Since then, their occurrence has been extended to other plant species, including *Cleme viscosa* (Almahy & Alagimi, 2012). Auraptene (**6**) acts as a potent inhibitor of glioblastoma (U87GBM) cell line by inducing apoptosis through Bax/Bcl-2 regulation, blocking cell cycle progression (Afshari et al., 2019). In addition, auraptene (**6**) notably induces apoptosis in PC3 and DU145 prostate cancer cells via Mcl-1-mediated activation of caspases (Lee et al., 2017).

5.2 Furanocoumarins

Psoralen (**19**) is the simplest member of furanocoumarin present in *Psoralea corylifolia* as well as in many other botanicals, including *Dorstenia* species. Psoralen (**19**) and some biosynthetic derivatives have been reported as potent anticancer agents against various human carcinoma, Table 1 and Fig. 3. A mechanistic study revealed that psoralen (**19**) inhibited the proliferation of human breast (MCF-7/ADR) cell lines as shown by G0/G1 phase arrest rather than encouraging apoptosis (Wang et al., 2016). It was also observed that psoralen reversed multidrug resistance (MDR) by inhibiting ATPase activity rather than reducing P-gp expression (Wang et al., 2016). On the other hand, psoralen (**19**) acts as MDR-reversing activity of docetaxel against human lung cancer (A549/D16) cell line (Hsieh, Chen, Yu, Sheu, & Chiou, 2014). Furanocoumarin having a methoxy group, such as bergapten (**20**), xanthotoxin (**21**), phellopterin (**22**), and byakangelicin (**23**) showed anti-proliferative activity against B16F10 melanoma cells and caused G2/M arrest in concentration-dependent manner 0.05–15.0 μM (Sumiyoshi, Sakanaka, Taniguchi, Baba, & Kimura, 2014). In another study, bergapten (**20**), isolated from *Thamnosma rhodesica* (roots) and *Dorstenia kameruniana* (roots), exhibited cytotoxicity against some refractory cell lines inter alia drug sensitive leukemia (CCRF-CEM) and its MDR P-gp-over-expressing subline (CEM/ADR5000), breast cancer cells (MDAMB-231-pcDNA3) and its resistant subline (MDA-MB-231-BCRP clone 23), colon cancer cells (HCT116 (*p53*$^{+/+}$)), and its knockout clone (HCT116 (*p53*$^{-/-}$)), glioblastoma cells (U87MG) and its resistant subline (U87MG.ΔEGFR), (Adem et al., 2018; Ahua et al., 2004; Sumiyoshi et al., 2014; Thanh, Jin, Song, Bae, & Sam, 2004; Urbagarova et al., 2020; Yang, Wang, Chen, & Wang, 2003), Table 1 and Fig. 3. The cytotoxic effect of bergapten (**20**) on human NSCLC cell lines A549 and NCI-H460 was further investigated by an MTT assay. After 24 h of incubation, bergapten (**20**) decreased the viability of A549 and NCI-H460 cells to 90.2% and 87.3%, respectively. This compound also induces G0/G1 arrest and increases the percentage of cells in the sub-G1 phase of non-small cell lung cancer A549 and NCI-H460 cells (Chiang, Lin, Lin, Shieh, & Kao, 2019). Imperatorin (**24**), a demethoxy derivative of phellopterin, has been reported to have many pharmacological properties, including anticancer activity with various extrinsic and intrinsic mechanisms of action, including autophagy induction, necrosis, and inhibition of the formation of DNA adducts (Appendino et al., 2004; Ju, Gong, Su, & Mou, 2017; Li et al., 2014; Li, Chen, Tsai, & Yeh, 2019). Imperatorin (**24**) inhibited the proliferation of human hepatoma (HepG2) cells through apoptosis induction in a time- and dose-dependent manner, as evidenced by nuclear morphology, DNA fragmentation, loss of mitochondrial membrane potential, and activation of caspase-3, -8, -9, and poly(ADP-ribose) polymerase cleavage (Luo et al., 2011). Furthermore, an equal mixture (50 μM) of quercetin and imperatorin (**24**) showed a synergism effect, which led to increased activity against human (HeLa and Hep-2) cell lines (Bądziul, Jakubowicz-Gil, Paduch, Głowniak, & Gawron, 2014). Rutamarin (**25**) from *Toona ciliata* showed significant cytotoxicity against HT29, HCT116, MCF7, and HT29 cancer cell lines (Fakai, Karsani, & Malek, 2017; Okot et al., 2023; Suhaimi, Hong, & Abdul Malek, 2017). The morphological and biochemical hallmarks of apoptosis, including activation of caspases 3, 8, and 9, were observed in rutamarin-treated HT29 colon cancer cells (Suhaimi et al., 2017). Chalepin (**26**) from a Cameroonian medicinal plant, *Clausena anisata* (stem bark) exhibited cytotoxic potency against HT29, HCT116, MCF7, MDA-MB-231, MRC-5 cell lines (Fakai et al., 2017; Suhaimi et al., 2017; Tatsimo et al., 2015), Table 1 and Fig. 3. Induction of apoptosis by chalepin (**26**) through phosphatidylserine externalizations and DNA fragmentation in MCF7 cells was observed by Fakai and collaborators (Fakai et al., 2019). Parallel to this, chalepin (**26**) additionally exhibited moderate cytotoxicity (IC_{50} = 28 μM) against the A549 cell line (Richardson, Sethi, Lee, & Malek, 2016). Furthermore, chalepin (**26**) causes S phase cell cycle arrest, nuclear factor-kappa B pathway inhibition, and STAT-3 inhibition induces extrinsic apoptotic pathway in non-small cell lung cancer carcinoma (A549) (Richardson, Aminudin, & Malek, 2017).

5.3 Pyranocoumarins

Unlike simple and furanocoumarins, the cytotoxicity potency of pyranocoumarin is scarce. Nevertheless, naturally occurring pyranocoumarins are reputed for their cytotoxic and antitumor-promoting activities. For instance, some Cameroonian plants belonging to the Rutaceae family; *Clausena anisata*, *Citrus aurantium*, and *Citrus reticulata* are known as sources of pyranocoumarin namely 3-(1,1-dimethyl allyl)xanthyletin (**35**), xanthoxyletin (**36**), clausarin (**37**), and xanthyletin (**38**), Table 1 and Fig. 3. The antiproliferative effect of xanthoxyletin (**36**) against human gastric adenocarcinoma (SGC-7901) cells and its ability to induce apoptosis was reported (Rasul et al., 2011). In parallel to this, xanthoxyletin (**36**) showed cytotoxicity against various oral cancer cells (IC_{50} = 10–30 μM) without obvious cytotoxicity in normal embryonic bovine tracheal epithelial (EBTr) cells (IC_{50} = 95 μM) (Wen et al., 2019). Clausarin (**37**) acts as a potent antineoplastic agent against human HCC (HepG2) (IC_{50} = 17.6 μM); human colon carcinoma cell line (HCT116) (IC_{50} = 44.9 μM); human lung adenocarcinoma cell line (SK-LU-1) (IC_{50} = 6.9 μM); non-cancerous African green monkey kidney epithelial (Vero) (IC_{50} = 78.2 μM); lung (A549) (IC_{50} = 7.8 μM); breast (MCF7) (IC_{50} = 20.9 μM); nasopharynx (KB) (IC_{50} = 4.2 μM); nasopharynx MDR (KB-VIN) (IC_{50} = 4.2 μM) (Bissim et al., 2019; Jantamat, Weerapreeyakul, & Puthongking, 2019; Su et al., 2009), while selectivity was observed for 3-(1,1-dimethyl allyl)xanthyletin (**35**) against HeLa (IC_{50} = 3.8 μM) (Tatsimo et al., 2015). Moreover, xanthyletin (**38**) acts as a DNA-damaging agent (Gunatilaka et al., 1994).

5.4 Sesquiterpene coumarins

Sesquiterpene coumarins (SCs) are also found in some tissues (roots, resins) of African medicinal plants, including those of *Ferula assa-foetida* and *Ferula tunetana*, Table 1 and Fig. 3. Their common feature includes a coumarin moiety (umbelliferone) linked (*O*-prenylation) to an aliphatic side chain made up of three isoprenyl groups. Umbelliprenin (**39**), which is considered a key synthon for the synthesis of various sesquiterpene umbelliferyl (7-hydroxycoumarin) ethers, represents the simplest SCs found in some genera (Gliszczyńska & Brodelius, 2012). In a panel of several human cell lines, umbelliprenin (**39**) (IC_{50} = 12.4 μM) was more potent than cisplatin

(IC_{50} = 23.1 μM) against human metastatic pigmented malignant melanoma cells (M4Beu). After 48 h of incubation, umbelliprenin (**39**) also induced cell death disturbance and apoptosis in M4Beu cells (Barthomeuf, Lim, Iranshahi, & Chollet, 2008). Umbelliprenin (**39**), tunetacoumarin A (**40**) coladin (**41**), colladonin (**42**), isomarcandin (**43**), and 13-hydroxyfeselol (**44**), were isolated from the roots of *Ferula tunetana*. The cytotoxic activity of the isolated compounds was tested on the human colorectal cancer cell lines HCT 116 and HT-29. Out of the tested compounds, coladin (**41**) showed significant cytotoxicity ($IC_{50} < 10$ μM) against both cell lines, while moderate activity was recorded for colladonin (**42**) and 13-hydroxyfeselol (**44**) ($IC_{50} > 10$ μM) (Jabrane et al., 2010). MDR mediated by P-gp or ATP-binding cassette sub-family B member 1 is an important cell membrane protein that pumps most chemotherapeutic drugs, thereby decreasing their bioavailability. This phenomenon is known as one of the major reasons for the failure of cancer therapy. It was found that several SCs increased the intracellular accumulation of ABCB1 substrate doxorubicin, which demonstrated that SCs could reverse ABCB1-mediated MDR. For instance, umbelliprenin (**39**), farnesiferol B (**45**), farnesiferol C (**46**) and samarcandin (**47**) at 50 μM showed potentiation effect on doxorubicin against MCF-7/Adr cell line (IC_{50} = 21.41 μM, tested alone), with IC_{50} values of 14.7, 10.7, 6.7, and 17.8, respectively (Kasaian et al., 2015), Table 1 and Fig. 3. Galbanic acid (**48**) from the Egyptian plant *Ferula assa-foetida* is well known to have cytotoxic, anti-angiogenic, and apoptotic effects in prostate cancer and murine Lewis lung cancer cells (Kim et al., 2011; Kim et al., 2019; Mohamed, El-Razek, Ohta, Ahmed, & Hirata, 2001).

The cytotoxicity of phytochemicals will be appreciated according to the classification criteria established as follows: outstanding activity ($IC_{50} \leq 0.5$ μM), excellent activity ($0.5 < IC_{50} \leq 2$ μM), very good activity ($2 < IC_{50} \leq 5$ μM), good activity ($5 < IC_{50} \leq 10$ μM), average activity ($10 < IC_{50} \leq 20$ μM), weak activity ($20 < IC_{50} \leq 60$ μM), very weak activity ($60 < IC_{50} \leq 150$ μM), and not active ($IC_{50} > 150$ μM) (Kuete, 2025). This basis of classification will be used to discuss the cytotoxicity of coumarins isolated from African medicinal plants. From the data summarized in Table 1, it appears that outstanding cytotoxic effects were obtained for compounds **19**, **20**, **21**, **22**, and **51** against B16F10 cells; B16F10 cells; B16F10 cells; B16F10 cells; and UO31 cells, respectively; excellent cytotoxic were obtained with compounds **4**, **5**, **11**, **34**, and **42** against CEM/ADR5000 cells; CEM/ADR5000 cells; DLD-1 cells; HT-29 cells; and UO31 cells, respectively; very good cytotoxic effects were obtained with compounds **1**, **4**, **5**, **10**, **11**, **35**, **37**, **41**, **42**, **65**, and **67** against HT-29 cells;

CCRF-CEM cells; CCRF-CEM cells; HeLa cells; HeLa, MCF-7 cells; HeLa cells; KB and KB-VIN cells; HCT 116 cells; KM12 cells; SUNE1 cells; and Lu-1 cells, respectively; good cytotoxic effects were obtained with compounds **4**, **8**, **9**, **11**, **14**, **20**, **25**, **36**, **37**, **41**, **51**, **53**, **54**, **65**, and **67** against HeLa cells; HeLa cells; HeLa cells; KB, LNCaP, NCI-H460 cells; KKU-100 cells; CCRF-CEM cells; HCT116 and HT29 cells; SCC-1 and SCC-9 cells; SK-LU-1, A549 cells; HT-29 cells; KM12 cells; A2780 cells; A2780 cells; CNE1 and HK1 cells; P-388, KB, Col-2 and BCA-1 cells, respectively.

Collateral sensitivity (D.R.: < 1) resistant CEM/ADR5000 cells vs sensitive CCRF-CEM cells to compounds **4** (D.R.: 0.6), **5** (D.R.: 0.4), CEM/ADR5000 cells vs CCRF-CEM cells, **19** (D.R.: 0.98), KBv200 cells vs KB cells, **20** (D.R: 0.99), U87. MGΔEGFR cells vs U87MG cells were also recorded (Table 1). This is an indication that compounds these compounds (**4**, **5**, **19**, and **20**) are potent cytotoxic coumarins that could be used to fight cancer.

6. Conclusion

The present chapter provides an overview cytotoxic potential of natural coumarins isolated from African medicinal plants toward drug sensitive and MDR cell lines. Due to the increasing incidence of cancer diseases and the emergence of cases annually, sustainable solutions should be taken to address this menace. Thus, African pharmacopoeia appears as a potential source of secondary metabolites, particularly coumarins. These compounds have various biological properties, and they can be easily synthesized and derivatized in order to produce new analogues with better pharmacokinetic and pharmacodynamic profiles for future anticancer drugs. Even though the reduction of cell viability in some coumarins was associated with the induction of apoptosis, the molecular targets and mechanisms of action of other potent compounds remain unclear, and significant work and room for improvement are warranted.

References

Abe, M. O., & Taylor, D. A. H. (1971). A quinolone alkaloid from *Oricia suaveolens*. *Phytochemistry,* 1167–1169.

Abegaz, B. M., Ngadjui, B. T., Folefoc, G. N., Fotso, S., Ambassa, P., Bezabih, M., ... Petersen, D. (2004). Prenylated flavonoids, monoterpenoid furanocoumarins and other constituents from the twigs of *Dorstenia elliptica* (Moraceae). *Phytochemistry, 65*(2), 221–226.

Adams, M., Efferth, T., & Bauer, R. (2006). Activity-guided isolation of scopoletin and isoscopoletin, the inhibitory active principles towards CCRF-CEM leukaemia cells and multidrug resistant CEM/ADR5000 cells, from *Artemisia argyi*. *Planta Medica, 72*(9), 862–864.

Adem, F. A., Kuete, V., Mbaveng, A. T., Heydenreich, M., Ndakala, A., Irungu, B., ... Yenesew, A. (2018). Cytotoxic benzylbenzofuran derivatives from *Dorstenia kameruniana*. *Fitoterapia, 128*, 26–30.

Afshari, A. R., Karimi Roshan, M., Soukhtanloo, M., Ghorbani, A., Rahmani, F., Jalili-Nik, M., ... Roshan, K. M. (2019). Cytotoxic effects of auraptene against a human malignant glioblastoma cell line. *Avicenna Journal of Phytomedicine, 9*(4), 334–346.

Ahua, K. M., Ioset, J. R., Ransijn, A., Mauël, J., Mavi, S., & Hostettmann, K. (2004). Antileishmanial and antifungal acridone derivatives from the roots of *Thamnosma rhodesica*. *Phytochemistry, 65*(7), 963–968.

Akkol, E. K., Genç, Y., Karpuz, B., Sobarzo-Sánchez, E., & Capasso, R. (2020). Coumarins and coumarin-related compounds in pharmacotherapy of cancer. *Cancers, 12*(7), 1–25.

Almahy, H. A., & Alagimi, A. A. (2012). Coumarins from the roots of *Cleme Viscosa* (L.) antimicrobial and cytotoxic studies. *Arabian Journal of Chemistry, 5*(2), 241–244.

Aloui, L., Kossentini, M., Geffroy-Rodier, C., Guillard, J., & Zouari, S. (2015). Phytochemical investigation, isolation and characterization of coumarins from aerial parts and roots of Tunisian *Pituranthos chloranthus* (Apiaceae). *Pharmacognosy Communications, 5*(4), 237–243.

Ameen, B. A. H. (2014). Phytochemical study and cytotoxic activity of *Ferulago angulata* (Schlecht) Boiss, from Kurdistan-region of Iraq. *International Journal of Innovative Research in Advanced Engineering, 1*(9), 2349 –2163.

Ang, A. M. G., Peteros, N. P., & Uy, M. M. (2019). Antioxidant and toxicity assay-guided isolation of herniarin from *Equisetum debile* Roxb. (Equisetaceae). *Asian Journal of Biological and Life Sciences, 8*, 30–35.

Appendino, G., Bianchi, F., Bader, A., Campagnuolo, C., Fattorusso, E., Taglialatela-Scafati, O., ... Muñoz, E. (2004). Coumarins from *Opopanax chironium*. New dihydrofuranocoumarins and differential induction of apoptosis by imperatorin and heraclenin. *Journal of Natural Products, 67*(4), 532–536.

Arango, V., Robledo, S., Séon-Méniel, B., Figadère, B., Cardona, W., Sáez, J., & Otálvaro, F. (2010). Coumarins from *Galipea panamensis* and their activity against *Leishmania panamensis*. *Journal of Natural Products, 73*(5), 1012–1014.

Atanasov, A. G., Zotchev, S. B., Dirsch, V. M., Orhan, I. E., Banach, M., Rollinger, J. M., ... Supuran, C. T. (2021). Natural products in drug discovery: Advances and opportunities. *Nature Reviews. Drug Discovery, 20*(3), 200–216.

Bądziul, D., Jakubowicz-Gil, J., Paduch, R., Głowniak, K., & Gawron, A. (2014). Combined treatment with quercetin and imperatorin as a potent strategy for killing HeLa and Hep-2 cells. *Molecular and Cellular Biochemistry, 392*(1–2), 213–227.

Barthomeuf, C., Lim, S., Iranshahi, M., & Chollet, P. (2008). Umbelliprenin from *Ferula szowitsiana* inhibits the growth of human M4Beu metastatic pigmented malignant melanoma cells through cell-cycle arrest in G1 and induction of caspase-dependent apoptosis. *Phytomedicine, 15*(1–2), 103–111.

Ben, S., Jabrane, A., Harzallah-Skhiri, F., & Ben Jannet, H. (2013). New bioactive dihydrofuranocoumarins from the roots of the Tunisian *Ferula lutea* (Poir.) Maire. *Bioorganic and Medicinal Chemistry Letters, 23*(14), 4248–4252.

Bissim, S., Kenmogne, S. B., Tcho, A. T., Lateef, M., Ahmed, A., Ngeufa Happi, E., ... Kamdem Waffo, A. F. (2019). Bioactive acridone alkaloids and their derivatives from *Citrus aurantium* (Rutaceae). *Phytochemistry Letters, 29*, 148–153.

Bourgaud, F., Hehn, A., Larbat, R., Doerper, S., Gontier, E., Kellner, S., & Matern, U. (2006). Biosynthesis of coumarins in plants: A major pathway still to be unravelled for cytochrome P450 enzymes. *Phytochemistry Reviews, 5*(2–3), 293–308.

Bourgaud, F., Poutaraud, A., & Analysis, A. G.-P. (1994). Extraction of coumarins from plant material (Leguminosae). & 1994, undefined *Phytochemical Analysis, 5*(3), 127–132.

Buddrus, J., Bauer, H., Abu-mustafa, E., Khattab, A., Mishaal, S., M El-khrisy, E. A., & Linscheid, M. (1985). Foetidin, a sesquiterpenoid coumarin from *Ferula assa-foetida. Phytochemistry, 24*(4), 869–870.

Butler, M. S. (2008). Natural products to drugs: Natural product-derived compounds in clinical trials. *Natural Product Reports, 25*(3), 475–516.

Cao, D., Liu, Z., Verwilst, P., Koo, S., Jangjili, P., Kim, J. S., & Lin, W. (2019). Coumarin-based small-molecule fluorescent chemosensors. *Chemical Reviews, 119*(18), 10403–10519.

Chaturvedula, V. S. P., Schilling, J. K., & Kingston, D. G. I. (2002). New cytotoxic coumarins and prenylated benzophenone derivatives from the bark of *Ochrocarpos punctatus* from the Madagascar rainforest. *Journal of Natural Products, 65*(7), 965–972.

Chen, X., Liu, Z., Zhong, B., Zhu, M., Yao, H., Chen, X., ... Guo, Y. (2022). Cytotoxic 4-phenylcoumarins from the flowering buds of *Mesua ferrea. Natural Product Research,* 1–10.

Chiang, S. R., Lin, C. S., Lin, H. H., Shieh, P. C., & Kao, S. H. (2019). Bergapten induces G1 arrest of non-small cell lung cancer cells, associated with the p53-mediated cascade. *Molecular Medicine Reports, 19*(3), 1972–1978.

Darmawan, A., Kosela, S., Kardono, L. B. S., & Syah, Y. M. (2012). Scopoletin, a coumarin derivative compound isolated from *Macaranga gigantifolia* Merr. *Journal of Applied Pharmaceutical Science, 2*(12), 175–177.

Dewick, P. (2002). *Medicinal natural products: A biosynthetic approach.* John Wiley & Sons,.

Dias, D. A., Urban, S., & Roessner, U. (2012). A historical overview of natural products in drug discovery. *Metabolites, 2*(2), 303–336.

Egan, D., O'kennedy, R., Moran, E., Cox, D., Prosser, E., & Thornes, R. D. (1990). The pharmacology, metabolism, analysis, and applications of coumarin and coumarin-related compounds. *Drug Metabolism Reviews, 22*(5), 503–529.

Fakai, M. I., Abd Malek, S. N., & Karsani, S. A. (2019). Induction of apoptosis by chalepin through phosphatidylserine externalisations and DNA fragmentation in breast cancer cells (MCF7). *Life Sciences, 220,* 186–193.

Fakai, M. I., Karsani, S. A., & Malek, S. N. A. (2017). Chalepin and rutamarin isolated from *Ruta angustifolia* inhibit cell growth in selected cancer cell lines (MCF7, MDA-MB-231, HT29, and HCT116). *Journal of Information System and Technology Management, 2*(5), 8–17.

Farooq, S., Shakeel-U-Rehman, Dangroo, N. A., Priya, D., Banday, J. A., Sangwan, P. L., ... Saxena, A. K. (2014). Isolation, cytotoxicity evaluation and hplc-quantification of the chemical constituents from *Prangos pabularia. PLoS One, 9*(10), e108713.

Finkelstein, N., Albrecht, C. F., & van Jaarsveld, P. P. (1993). Isolation and structure elucidation of xanthotoxin, a phototoxic furanocoumarin, from *Peucedanum galbanum. South African Journal of Botany, 59*(1), 81–84.

Floss, H., & Mothes, U. (1964). On the biochemistry of furocoumarin in *Pimpinella magna. Zeitschrift Fur Naturforschung B, 19,* 770–771.

Ghoran, S. H., Atabaki, V., Babaei, E., Olfatkhah, S. R., Dusek, M., Eigner, V., ... Khalaji, A. D. (2016). Isolation, spectroscopic characterization, x-ray, theoretical studies as well as in vitro cytotoxicity of Samarcandin. *Bioorganic Chemistry, 66,* 27–32.

Gliszczyńska, A., & Brodelius, P. E. (2012). Sesquiterpene coumarins. *Phytochemistry Reviews, 11*(1), 77–96.

Gunatilaka, A. A. L., Kingston, D. G. I., Wijeratne, E. M. K., Bandara, B. M. R., Hofmann, G. A., & Johnson, R. K. (1994). Biological activity of some coumarins from Sri Lankan rutaceae. *Journal of Natural Products, 57*(4), 518–520.

Guo, H., He, Y., Bu, C., & Peng, Z. (2019). Antitumor and apoptotic effects of 5-methoxypsoralen in U87MG human glioma cells and its effect on cell cycle, autophagy and PI3K/Akt signaling pathway. *Archives of Medical Science, 15*(6), 1530–1538.

Hsieh, M. J., Chen, M. K., Yu, Y. Y., Sheu, G. T., & Chiou, H. L. (2014). Psoralen reverses docetaxel-induced multidrug resistance in A549/D16 human lung cancer cells lines. *Phytomedicine, 21*(7), 970–977.

Iranshahy, M., Farhadi, F., Paknejad, B., Zareian, P., Iranshahi, M., Karami, M., & Abtahi, S. R. (2019). Gummosin, a sesquiterpene coumarin from *Ferula assa-foetida* is preferentially cytotoxic to human breast and prostate cancer cell lines. *Avicenna Journal of Phytomedicine, 9*(5), 446–453.

Jabrane, A., Jannet, H., Ben, Mighri, Z., Mirjolet, J. F., Duchamp, O., ... Lacaille-Dubois, M. A. (2010). Two new sesquiterpene derivatives from the Tunisian endemic *Ferula tunetana* POM. *Chemistry and Biodiversity,* 7(2), 392–399.

Jantamat, P., Weerapreeyakul, N., & Puthongking, P. (2019). Cytotoxicity and apoptosis induction of coumarins and carbazole alkaloids from *Clausena harmandiana. Molecules (Basel, Switzerland), 24*(18), 1–14.

Ju, A.-H., Gong, W.-J., Su, Y., & Mou, Z.-B. (2017). Imperatorin shows selective antitumor effects in SGC-7901 human gastric adenocarcinoma cells by inducing apoptosis, cell cycle arrest and targeting PI3K/Akt/m-TOR signalling pathway. *JBUON, 22*(6), 1471–1476.

Kasaian, J., Mosaffa, F., Behravan, J., Masullo, M., Piacente, S., Ghandadi, M., & Iranshahi, M. (2015). Reversal of P-glycoprotein-mediated multidrug resistance in MCF-7/Adr cancer cells by sesquiterpene coumarins. *Fitoterapia, 103*, 149–154.

Kim, J. K., Kim, J. Y., Kim, H. J., Park, K. G., Harris, R. A., Cho, W. J., ... Lee, I. K. (2013). Scoparone exerts anti-tumor activity against DU145 prostate cancer cells via inhibition of STAT3 activity. *PLoS One, 8*(11), e80391.

Kim, K. H., Lee, H. J., Jeong, S. J., Lee, H. J., Lee, E. O., Kim, H. S., ... Kim, S. H. (2011). Galbanic acid isolated from *Ferula assafoetida* exerts in vivo anti-tumor activity in association with anti-angiogenesis and anti-proliferation. *Pharmaceutical Research, 28*(3), 597–609.

Kim, Y. H., Shin, E. A., Jung, J. H., Park, J. E., Ku, J. S., Koo, J., ... Kim, S. H. (2019). Galbanic acid potentiates TRAIL induced apoptosis in resistant non-small cell lung cancer cells via inhibition of MDR1 and activation of caspases and DR5. *European Journal of Pharmacology, 847*, 91–96.

Kim, Y.-K., Kim, Y. S., & Ryu, S. Y. (2007). Antiproliferative effect of furanocoumarins from the root of *Angelica dahurica* on cultured human tumor cell lines. *Phytotherapy Research, 21*(4), 288–290.

Kuete, V. (2025). Chapter Four-African medicinal plants and their derivative as the source of potent anti-leukemic products: Rationale classification of naturally occurring anticancer agents. *Advances in Botanical research,* 113. https://doi.org/10.1016/bs.abr.2023.12.010.

Larbat, R., Hehn, A., Hans, J., Schneider, S., Jugde, H., Schneider, B., ... Bourgaud, F. (2009). Isolation and functional characterization of CYP71AJ4 encoding for the first P450 monooxygenase of angular furanocoumarin biosynthesis. *Journal of Biological Chemistry, 284*(8), 4776–4785.

Lee, J. C., Shin, E. A., Kim, B., Kim, B. I., Chitsazian-Yazdi, M., Iranshahi, M., & Kim, S. H. (2017). Auraptene induces apoptosis via myeloid cell leukemia 1-mediated activation of caspases in PC3 and DU145 prostate cancer cells. *Phytotherapy Research, 31*(6), 891–898.

Lee, K. H., Chai, H. B., Tamez, P. A., Pezzuto, J. M., Cordel, G. A., Win, K. K., & Tin-Wa, M. (2003). Biologically active alkylated coumarins from *Kayea assamica. Phytochemistry, 64*(2), 535–541.

Lekphrom, R., Kanokmedhakul, S., Kukongviriyapan, V., & Kanokmedhakul, K. (2011). C-7 oxygenated coumarins from the fruits of *micromelum minutum. Archives of Pharmacal Research, 34*(4), 527–531.

Li, X., Zeng, X., Sun, J., Li, H., Wu, P., Fung, K. P., & Liu, F. (2014). Imperatorin induces Mcl-1 degradation to cooperatively trigger Bax translocation and Bak activation to suppress drug-resistant human hepatoma. *Cancer Letters, 348*(1–2), 146–155.

Li, Y. Z., Chen, J. H., Tsai, C. F., & Yeh, W. L. (2019). Anti-inflammatory property of imperatorin on alveolar macrophages and inflammatory lung injury. *Journal of Natural Products, 82*(4), 1002–1008.

Lim, C. K., Hemaroopini, S., Gan, S. Y., Loo, S. M., Low, J. R., Jong, V. Y. M., ... Chee, C. F. (2016). *In vitro* cytotoxic activity of isolated compounds from Malaysian *Calophyllum* species. *Medicinal Chemistry Research, 25*(8), 1686–1694.

Lozhkin, A. V., & Sakanyan, E. I. (2006). Natural coumarins: Methods of isolation and analysis. *Pharmaceutical Chemistry Journal, 40*(6), 337–346.

Luo, K. W., Sun, J. G., Chan, J. Y. W., Yang, L., Wu, S. H., Fung, K. P., & Liu, F. Y. (2011). anticancer effects of imperatorin isolated from *Angelica dahurica*: Induction of Apoptosis in HepG2 cells through both death-receptor- and mitochondria-mediated pathways. *Chemotherapy, 57*(6), 449–459.

Mahibalan, S., Rao, P. C., Khan, R., Basha, A., Siddareddy, R., Masubuti, H., ... Begum, A. S. (2016). Cytotoxic constituents of *Oldenlandia umbellata* and isolation of a new symmetrical coumarin dimer. *Medicinal Chemistry Research, 25*(3), 466–472.

Mar, W., Je, K.-H., & Seo, E.-K. (2001). Cytotoxic constituents of *Psoralea corylifolia. Archives of Pharmacal Research, 24*(3), 211–213.

Marston, A., Hostettmann, K., & Msonthi1, J. D. (1995). Isolation of antifungal and larvicidal constituents of *Diplolophium buchanani* by centrifugal partition chromatography. *Journal of Natural Products, 58*(1), 128–130.

Mbaveng, A. T., Noulala, C. G. T., Samba, A. R. M., Tankeo, S. B., Fotso, G. W., Happi, E. N., ... Efferth, T. (2021). Cytotoxicity of botanicals and isolated phytochemicals from *Araliopsis soyauxii* Engl. (Rutaceae) towards a panel of human cancer cells. *Journal of Ethnopharmacology, 267*, 113535.

Mohamed, H., El-Razek, A., Ohta, S., Ahmed, A. A., & Hirata, T. (2001). Sesquiterpene coumarins from the roots of *Ferula assa-foetida. Phytochemistry, 58*, 1289–1295.

Mottaghipisheh, J., Nové, M., Spengler, G., Kúsz, N., Hohmann, J., & Csupora, D. (2018). Antiproliferative and cytotoxic activities of furocoumarins of *Ducrosia anethifolia. Pharmaceutical Biology, 56*(1), 658–664.

Mukandiwa, L., Ahmed, A., Eloff, J., & Naidoo, V. (2013). Isolation of seselin from *Clausena anisata* (Rutaceae) leaves and its effects on the feeding and development of *Lucilia cuprina* larvae may explain its use in ethnoveterinary medicine. *Journal of Ethnopharmacology, 150*(3), 886–891.

Nasr, T., Bondock, S., & Youns, M. (2014). Anticancer activity of new coumarin substituted hydrazide–hydrazone derivatives. *European Journal of Medicinal Chemistry, 76*, 539–548.

Negi, N., Abou-Dough, A. M., Kurosawa, M., Kitaji, Y., Saito, K., Ochi, A., ... Ito, C. (2015). Coumarins from *Murraya exotica* collected in Egypt. *NPC. Natural Product Communications, 10*(4), 617–620.

Newman, D. J., & Cragg, G. M. (2016). Natural products as sources of new drugs from 1981 to 2014. *Journal of Natural Products, 79*(3), 629–661.

Newman, D. J., Cragg, G. M., & Snader, K. M. (2003). Natural products as sources of new drugs over the period 1981-2002. *Journal of Natural Products, 66*(7), 1022–1037.

Ngameni, B., Touaibia, M., Patnam, R., Belkaid, A., Sonna, P., Ngadjui, B. T., ... Roy, R. (2006). Inhibition of MMP-2 secretion from brain tumor cells suggests chemopreventive properties of a furanocoumarin glycoside and of chalcones isolated from the twigs of *Dorstenia turbinata. Phytochemistry, 67*(23), 2573–2579.

Ngo, N., Nguyen, V., Vo, H., Van, Vang, O., Duus, F., ... Nguyen, L. (2010). Cytotoxic coumarins from the bark of *Mammea siamensis. Chemical and Pharmaceutical Bulletin, 58*(11), 1487–1491.

Ngwendson, J. N., Bedir, E., Efange, S. M. N., Okunji, C. O., Iwu, M. M., Schuster, B. G., ... Khan, I. A. (2003). Constituents of *Peucedanum zenkeri* seeds and their anti-microbial effects. *Die Pharmazie-An. International Journal of Pharmaceutical Sciences, 58*(8), 587–589.

Okot, D. F., Namukobe, J., Vudriko, P., Anywar, G., Heydenreich, M., Omowumi, O. A., & Byamukama, R. (2023). *In vitro* anti-venom potentials of aqueous extract and oils of *Toona ciliata* M. Roem against cobra venom and chemical constituents of oils. *Molecules (Basel, Switzerland), 28*(7), 1–28.

Park, S. H., Hong, J. Y., Park, H. J., & Lee, S. K. (2020). The antiproliferative activity of oxypeucedanin via induction of G2/M phase cell cycle arrest and p53-dependent MDM2/p21 expression in human hepatoma cells. *Molecules (Basel, Switzerland), 25*(3), 1–15.

Radman, S., Babojelić, M. S., Khandy, M. T., Sofronova, A. K., Gorpenchenko, T. Y., & Chirikova, N. K. (2022). Plant pyranocoumarins: Description, biosynthesis, application. *Plants, 11*, 1–32.

Rasul, A., Khan, M., Ma, T., Yu, B., Ma, T., & Yang, H. (2011). Xanthoxyletin, a coumarin induces S Phase arrest and apoptosis in human gastric adenocarcinoma SGC-7901 cells. *Asian Pacific Journal of Cancer Prevention, 12*(5), 1219–1223.

Reddy, C. H., Kim, S. C., Hur, M., Kim, Y. B., Park, C. G., Lee, W. M., ... Koo, S. C. (2017). Natural Korean medicine Dang-Gui: Biosynthesis, effective extraction, and formulations of major active pyranocoumarins, their molecular action mechanism in cancer. *Molecules (Basel, Switzerland), 22*, 1–16.

Reutrakul, V., Leewanich, P., Tuchinda, P., Pohmakotr, M., Jaipetch, T., Sophasan, S., ... Tuchindá, P. (2003). Cytotoxic coumarins from *Mammea harmandii. Planta Medica, 69*(11), 1048–1051.

Richardson, J. S. M., Aminudin, N., & Malek, S. N. A. (2017). Chalepin: A compound from Ruta angustifolia L. pers exhibits cell cycle arrest at S phase, suppresses nuclear factor-kappa B (NF-κB) pathway, signal transducer and activation of transcription 3 (STAT3) phosphorylation and extrinsic apoptotic pathway in non-small cell lung cancer carcinoma (A549). *Pharmacognosy Magazine, 13*(51), S489–S498.

Richardson, J. S. M., Sethi, G., Lee, G. S., & Malek, S. N. A. (2016). Chalepin: Isolated from *Ruta angustifolia* L. Pers induces mitochondrial mediated apoptosis in lung carcinoma cells. *BMC Complementary and Alternative Medicine, 16*(1).

Roselli, S., Olry, A., Vautrin, S., Coriton, O., Ritchie, D., Galati, G., ... El Ene Berg Es, H. (2016). In F. Eric Bourgaud, & A. Hehn (Vol. Eds.), *A bacterial artificial chromosome (BAC) genomic approach reveals partial clustering of the furanocoumarin pathway genes in parsnip: 89*, (pp. 1119–1132). (pp. 1119), Wiley Online Library.

Sajjadi, S. E., Jamali, M., Shokoohinia, Y., Abdi, G., Shahbazi, B., & Fattahi, A. (2015). Antiproliferative evaluation of terpenoids and terpenoid coumarins from *Ferulago macrocarpa* (Fenzl) Boiss. fruits. *Pharmacognosy Research*, 7(4), 322–328.

Sandjo, L. P., Foster, A. J., Rheinheimer, J., Anke, H., Opatz, T., & Thines, E. (2012). Coumarin derivatives from *Pedilanthus tithymaloides* as inhibitors of conidial germination in *Magnaporthe oryzae. Tetrahedron Letters, 53*(17), 2153–2156.

Sethena, S., & Shah, M. (1945). The chemistry of coumarins. *Chemical Reviews, 36*(1), 1–62.

Sharif, N., Mustahil, N., Mohd, N. H. S., Sukari, M. A., Rahmani, M., Taufiq-Yap, Y. H., & Ee, G. C. L. (2013). Cytotoxic constituents of *Clausena excavata. African Journal of Biotechnology, 10*(72), 16337–16341.

Shimizu, B. I. (2014). 2-Oxoglutarate-dependent dioxygenases in the biosynthesis of simple coumarins. *Frontiers in Plant Science, 53*, 1–7.

Shokoohinia, Y., Sajjadi, S. E., Gholamzadeh, S., Fattahi, A., & Behbahani, M. (2014). Antiviral and cytotoxic evaluation of coumarins from *Prangos ferulacea. Pharmaceutical Biology, 52*(12), 1543–1549.

Skalicka-Woźniak, K., & Głowniak, K. (2006). Coumarins: Analytical and preparative techniques. encyclopedia of analytical chemistry: Applications, theory and instrumentation. *In Encyclopedia of analytical chemistry: Applications, theory and instrumentation*, 1–26.

Soine, T. O. (1964). Naturally occurring coumarins and related physiological activities. *Journal of Pharmaceutical Sciences, 53*(3), 231–264.

Srivastava, V., Negi, A. S., Kumar, J. K., Gupta, M. M., & Khanuja, S. P. S. (2005). Plant-based anticancer molecules: A chemical and biological profile of some important leads. *Bioorganic and Medicinal Chemistry, 13*(21), 5892–5908.

Su, C. R., Yeh, S. F., Liu, C. M., Damu, A. G., Kuo, T. H., Chiang, P. C., ... Wu, T. S. (2009). Anti-HBV and cytotoxic activities of pyranocoumarin derivatives. *Bioorganic and Medicinal Chemistry, 17*(16), 6137–6143.

Suhaimi, S. A., Hong, S. L., & Abdul Malek, S. N. (2017). Rutamarin, an active constituent from *Ruta angustifolia* Pers., induced apoptotic cell death in the HT29 colon adenocarcinoma cell line. *Pharmacognosy Magazine, 13*(50), 179–188.

Sumiyoshi, M., Sakanaka, M., Taniguchi, M., Baba, K., & Kimura, Y. (2014). Anti-tumor effects of various furocoumarins isolated from the roots, seeds and fruits of *Angelica* and *Cnidium* species under ultraviolet A irradiation. *Journal of Natural Medicines, 68*(1), 83–94.

Tahsin, T., Wansi, J. D., Al-Groshi, A., Evans, A., Nahar, L., Martin, C., & Sarker, S. D. (2017). Cytotoxic properties of the stem bark of *Citrus reticulata* blanco (Rutaceae). *Phytotherapy Research, 31*(8), 1215–1219.

Tatsimo, S. J. N., Tamokou, J. D. D., Lamshöft, M., Mouafo, F. T., Lannang, A. M., Sarkar, P., ... Spiteller, M. (2015). LC-MS guided isolation of antibacterial and cytotoxic constituents from *Clausena anisata. Medicinal Chemistry Research, 24*(4), 1468–1479.

Thanh, P. N., Jin, W. Y., Song, G. Y., Bae, K. H., & Sam, S. K. (2004). Cytotoxic coumarins from the root of *Angelica dahurica. Archives of Pharmacal Research, 27*(12), 1211–1215.

Tian, Q., Wang, L., Sun, X., Zeng, F., Pan, Q., & Xue, M. (2019). Scopoletin exerts anticancer effects on human cervical cancer cell lines by triggering apoptosis, cell cycle arrest, inhibition of cell invasion and PI3K/AKT signalling pathway. *JBUON, 24*(3), 997–1002.

Tine, Y., Renucci, F., Costa, J., Wélé, A., & Paolini, J. (2017). A method for LC-MS/MS profiling of coumarins in *Zanthoxylum zanthoxyloides* (Lam.) B. zepernich and timler extracts and essential oils. *Molecules (Basel, Switzerland), 22*(1), 1–13.

Tosun, F., Aytar, E. C., Beutler, J. A., Wilson, J. A., & Miski, M. (2021). Cytotoxic sesquiterpene coumarins from the roots of *heptaptera cilicica. Records of Natural Products, 15*(6), 529–536.

Tosun, F., Beutler, J. A., Ransom, T. T., & Miski, M. (2019). Anatolicin, a highly potent and selective cytotoxic sesquiterpene coumarin from the root extract of *Heptaptera anatolica. Molecules (Basel, Switzerland), 24*(6), 1–8.

Tsassi, V. B., Hussain, H., Meffo, Y., Kouam, S. F., Dongo, E., Schulz, B., ... Krohn, K. (2010). Antimicrobial coumarins from the stem bark of *Afraegle paniculata. Natural Product Communications, 5*(4), 559–561.

Tsukamoto, H., Hisada, S., Nishibe, S., Roux, D. G., & Rourke, J. P. (1984). Coumarins from *Olea africana* and *Olea capensis. Phytochemistry, 23*(3), 699–700.

Um, B. H., Lobstein, A., Weniger, B., Spiegel, C., Yice, F., Rakotoarison, O., ... Anton, R. (2003). New coumarins from *Cedrelopsis grevei. Fitoterapia, 74*(7–8), 638–642.

Urbagarova, B. M., Shults, E. E., Taraskin, V. V., Radnaeva, L. D., Petrova, T. N., Rybalova, T. V., ... Ganbaatar, J. (2020). Chromones and coumarins from *Saposhnikovia divaricata* (Turcz.) Schischk. Growing in Buryatia and Mongolia and their cytotoxicity. *Journal of Ethnopharmacology, 261*, 112517.

Uttu, A. J., Sallau, M. S., Iyun, O. R. A., & Ibrahim, H. (2022). Coumarin and fatty alcohol from root bark of *Strychnos innocua* (delile): Isolation, characterization and *in silico* molecular docking studies. *Bulletin of the National Research Centre, 46*(1), 1–12.

Vijayalakshmi, A., & Sindhu, G. (2017). Umbelliferone arrest cell cycle at G0/G1 phase and induces apoptosis in human oral carcinoma (KB) cells possibly via oxidative DNA damage. *Biomedicine and Pharmacotherapy, 92*, 661–671.

Wagh, A., Butle, S., & Raut, D. (2021). Isolation, identification, and cytotoxicity evaluation of phytochemicals from chloroform extract of *Spathodea campanulata*. *Future Journal of Pharmaceutical Sciences, 7*(1), 1–8.

Waksmundzka-Hajnos, M., Petruczynik, A., Dragan, A., Wianowska, D., Dawidowicz, A. L., & Sowa, I. (2004). Influence of the extraction mode on the yield of some furanocoumarins from *Pastinaca sativa* fruits. *Journal of Chromatography B, 800*, 181–187.

Wang, X., Cheng, K., Han, Y., Zhang, G., Dong, J., Cui, Y., & Yang, Z. (2016). Effects of psoralen as an anti-tumor agent in human breast cancer MCF-7/ADR Cells. *Biological & Pharmaceutical Bulletin, 39*(5), 815–822.

Wang, X., Huang, S., Xin, X., Ren, Y., Weng, G., & Wang, P. (2019). The antitumor activity of umbelliferone in human renal cell carcinoma via regulation of the p110γ catalytic subunit of PI3Kγ. *Acta Pharmaceutica, 69*(1), 111–119.

Wang, Y., Hong, C., Zhou, C., Xu, D., & Qu, H.-B. (2011). Screening antitumor compounds psoralen and isopsoralen from *Psoralea corylifolia* L. seeds. *Evidence-Based Complementary and Alternative Medicine*, 1–7.

Wen, Q., Luo, K., Huang, H., Liao, W., & Yang, H. (2019). Xanthoxyletin inhibits proliferation of human oral squamous carcinoma cells and induces apoptosis, autophagy, and cell cycle arrest by modulation of the. *International Medical Journal of Experimental and Clinical Research, 25*, 8025–8033.

Wu, Y., Xu, J., Liu, Y., Zeng, Y., & Wu, G. (2020). A review on anti-tumor mechanisms of coumarins. *Frontiers in Oncology, 10*, 592853.

Yang, H., Protiva, P., Gil, R. R., Jiang, B., Baggett, S., Basile, M. J., ... Kennelly, E. J. (2005). Antioxidant and cytotoxic isoprenylated coumarins from *Mammea americana*. *Planta Medica, 71*(9), 852–860.

Yang, L. L., Wang, M. C., Chen, L. G., & Wang, C. C. (2003). Cytotoxic activity of coumarins from the fruits of *Cnidium monnieri* on leukemia cell lines. *Planta Medica, 69*(12), 1091–1095.

Yang, L.-L., Liang, Y.-C., Chang, C.-W., Lee, W.-S., Kuo, C.-T., Wang, C.-C., ... Lin, C.-H. (2002). Effects of sphondin, isolated from *Heracleum laciniatum*, on IL–1h-induced cyclooxygenase-2 expression in human pulmonary epithelial cells. *Life Sciences, 72*, 199–213.

You, L., An, R., Liang, K., & Wang, X. (2013). Anti-breast cancer agents from Chinese herbal medicines. *Mini-Reviews in Medicinal Chemistry, 13*, 101–105.

Yu, S. M., Hu, D. H., & Zhang, J. J. (2015). Umbelliferone exhibits anticancer activity via the induction of apoptosis and cell cycle arrest in HepG2 hepatocellular carcinoma cells. *Molecular Medicine Reports, 12*(3), 3869–3873.

Zhang, S., Yang, J., Li, H., Li, Y., Liu, Y., Zhang, D., ... Chen, X. (2012). Skimmin, a coumarin, suppresses the streptozotocin-induced diabetic nephropathy in wistar rats. *European Journal of Pharmacology, 692*(1–3), 78–83.

Zhang, J., Li, L., Jiang, C., Xing, C., Kim, S.-H., & Lu, J. (2012). Anticancer and other bioactivities of Korean *Angelica gigas* Nakai (AGN) and its major pyranocoumarin compounds. *Anticancer Agents in Medicinal Chemistry, 12*(10), 1239–1254.

Zhao, Y., Liu, T., Luo, J., Zhang, Q., Xu, S., Han, C., ... ong, L. (2015). Integration of a decrescent transcriptome and metabolomics dataset of *Peucedanum praeruptorum* to investigate the CYP450 and MDR genes involved in coumarins biosynthesis and transport. *Frontiers in Plant Science, 6*, 1–16.

CHAPTER THREE

Quinones from African medicinal plants as potential anticancer pharmaceuticals

Jenifer R.N. Kuete[a], Leonidah K. Omosa[b], and Victor Kuete[c,*]
[a]Department of Chemistry, Faculty of Science, University of Dschang, Dschang, Cameroon
[b]Department of Chemistry, Faculty of Science and Technology, University of Nairobi, Nairobi, Kenya
[c]Department of Biochemistry, Faculty of Science, University of Dschang, Dschang, Cameroon
*Corresponding author. e-mail address: kuetevictor@yahoo.fr

Contents

Abstract

Quinones are secondary metabolites isolated mostly from plants and having aromatic di-one or di-ketone systems. Several quinones are significant roles in biology and display pharmacological effects such as anticancer, antimicrobial and antiparasitic, or antidiabetic activities. In this chapter, we have identified 33 quinones isolated from African medicinal plants with cytotoxic effects on various human cancer cell lines. They include benzoquinones, naphthoquinones, and anthraquinones. The most active benzoquinones were 1,2,4,5-tetraacetate-3-methyl-6-(14-nonadecenyl)-cyclohexadi-2,5-diene (**1**), 2,5-dihydroxy-3-pentadecyl-2,5-cyclohexadiene-1,4-dione (**4**), ardisiaquinone B (**6**), rapanone (**7**), and ardisiaquinones J (**8**), K (**9**), and N (**10**). The most potent cytotoxic naphthoquinones include 2-acetyl-7-methoxynaphtho[23-*b*]furan-4,9-quinone (**11**), 2-acetylfuro-1,4-naphthoquinone (**12**), 7-methyljuglone (**13**), diospyrin (**14**), isodiospyrin (**15**), and plumbagin (**17**). The best cytotoxic anthraquinones were 6,7-dimethoxy xanthopurpurin (**20**), damnacanthal (**24**), emodin (**26**),

Advances in Botanical Research, Volume 115
ISSN 0065-2296, https://doi.org/10.1016/bs.abr.2024.02.005

schimperiquinone B (**30**), and scutianthraquinones A (**31**), B (**32**), and D (**33**). Quinones having the ability to combat the multidrug resistance of cancer cell lines were **11** and **24**. Some of the active quinones such as **7, 11, 12, 17,** and **24** displayed apoptotic cell death, mainly mediated by caspase activation, alteration of MMP, and enhanced ROS production in cancer cells. The most potent cytotoxic quinones reported herein should be further explored to develop novel cytotoxic products to fight cancer.

Abbreviations

D.R. degree of resistance
HSP70 70 kd heat shock proteins
IC_{50} Inhibitiry concentration 50
MDR multidrug-resistant
MMP mitochondrial membrane potential
PGE2 prostaglandin E2
P-GSK3β phosphorylated-glycogen synthase kinase
ROS reactive oxygen species

1. Introduction

Quinones are secondary metabolites isolated mostly from plants and having aromatic di-one or di-ketone systems and are generally derived from the oxidation of hydroquinone (Eyong, Kuete, & Efferth, 2013). Naturally occurring quinones are widely distributed, including benzoquinones, naphthoquinones, and anthraquinones (Fig. 1). Benzoquinones contain two carbonyl groups on a saturated hexacyclic aromatic ring system usually at ortho or para positions meanwhile naphthoquinones contain the naphthalene nucleus with two carbonyl groups on one nucleus usually at the ortho or para position. Anthraquinones contain the anthracene nucleus with two carbonyl groups usually on ring B at para positions (Eyong et al., 2013). Naphthoquinones and anthraquinones are fungi metabolites and are extremely common in higher plants (Eyong et al., 2013). Several quinones are significant roles in biology such as vitamin K, which is involved in the coagulation of blood, ubiquinone-10 which is a naturally occurring 1,4-benzoquinone involved in respiration apparatus, or plastoquinone, a redox relay involved in photosynthesis. The pharmacological importance of several quinones has been documented. In effect, they are anticancer cytotoxins such as daunorubicin, which is antileukemic (O'Brien, 1991), purgatives such as sennosides, antimicrobial and antiparasitic agents such as rhein and saprorthoquinone, and atovaquone, antitumor phytochemicals

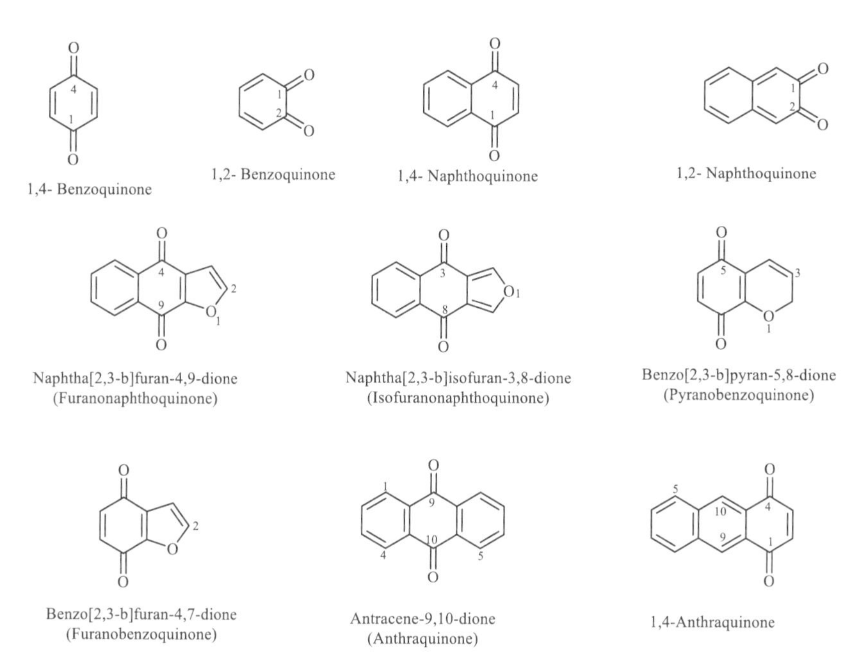

Fig. 1 Basic skeleton of benzoquinones, naphthoquinones, and anthraquinones.

such as emodin or juglone, inhibitors of prostaglandin E2 (PGE2) biosynthesis such as arnebinone and arnebifuranone, and drugs for cardiovascular disease such as tanshinone (Liu, 2011). Examples of established quinone-based drugs are Mecarbinate (dimecarbine) for the treatment of patients with hypertensive disease (Savenkov & Sabirova, 1967) and the antineoplastic, apaziquone (Caramés Masana & de Reijke, 2017). Several quinones isolated from African medicinal plants had antimicrobial, antiparasitic, antidiabetic, anti-inflammatory, as well as cytotoxic activities. Some of them include diospyrone and crassiflorone (Kuete, Tangmouo, Marion Meyer, & Lall, 2009), febrifuquinone and adamabianthrone (Tsaffack et al., 2009), emodin, plumbagin, and rapanone (Omosa et al., 2016), newbouldiaquinone, 2-acetylfuro-1,4-naphthoquinone, and lapachol (Eyong et al., 2005; Kuete et al., 2007), pycnanthuquinone C, 2, 3-dimethoxy-1,4-benzoquinone and 3-hydroxy-2-methyl-5-(3-methyl 2 butenyl)benzo-1,4-quinone (McGaw, Lall, Meyer, & Eloff, 2008), and 7-methyljuglone (Gu et al., 2004) with the antimicrobial activity, damnacanthal (Sandjo et al., 2016) and newbouldiaquinone A (Eyong et al., 2006) with antiparasitic activity, or emodin with antidiabetic activity (Eyong et al., 2013). In this chapter, the overview of the cytotoxic potential of benzoquinones, naphthoquinones, and anthraquinones isolated from African medicinal plants against human cancer cell lines including the multidrug-resistant (MDR) phenotypes is discussed.

2. Biosynthesis of quinones

Two pathways are known for the biosynthesis of natural quinones in higher plants and other organisms: the acetate-malonate pathway and the shikimate or *o*-succinoyl benzoic acid (OSB) pathway (Schemes 1 and 2) (Eyong et al. 2013). The biosynthetic pathways of illustrative benzoquinone coenzyme Q (Schemes 3), 2-geranyl-1,4-naphthoquinone and its related anthraquinone (Scheme 4), as well as other anthraquinones *via* acetic-mevolonate and shikimate pathways, are also shown (Scheme 5). The detailed biosynthesis of quinone was previously described by Eyong and collaborators (Eyong et al., 2013).

3. Cell lines used in the screenings of the cytotoxicity of quinones isolated from African medicinal plants

Several cancer cell lines were used to assess the cytotoxicity of quinones isolated from African medicinal plants. They include cell lines from breast cancer (MCF-7, MCF-7/Adr, MDA-MB-231-*pcDNA*, MDA-MB-231-*BCRP*, and SKBR3), cervical cancer (Caski and HeLa),

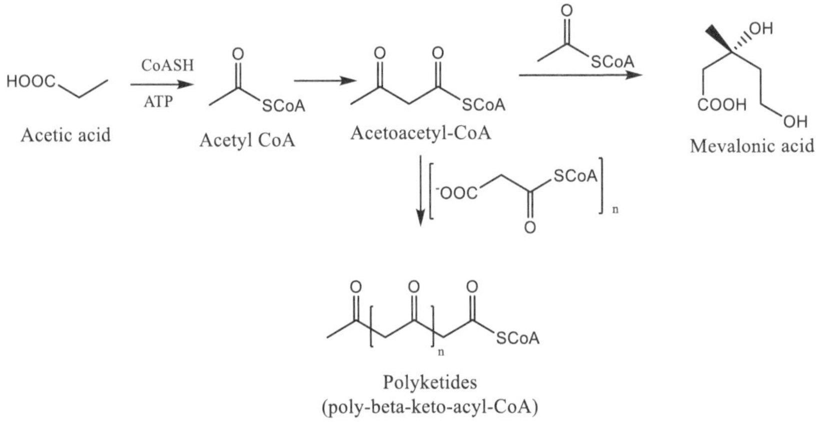

Scheme 1 Acetate-malonate pathway for the synthesis of polyketides.

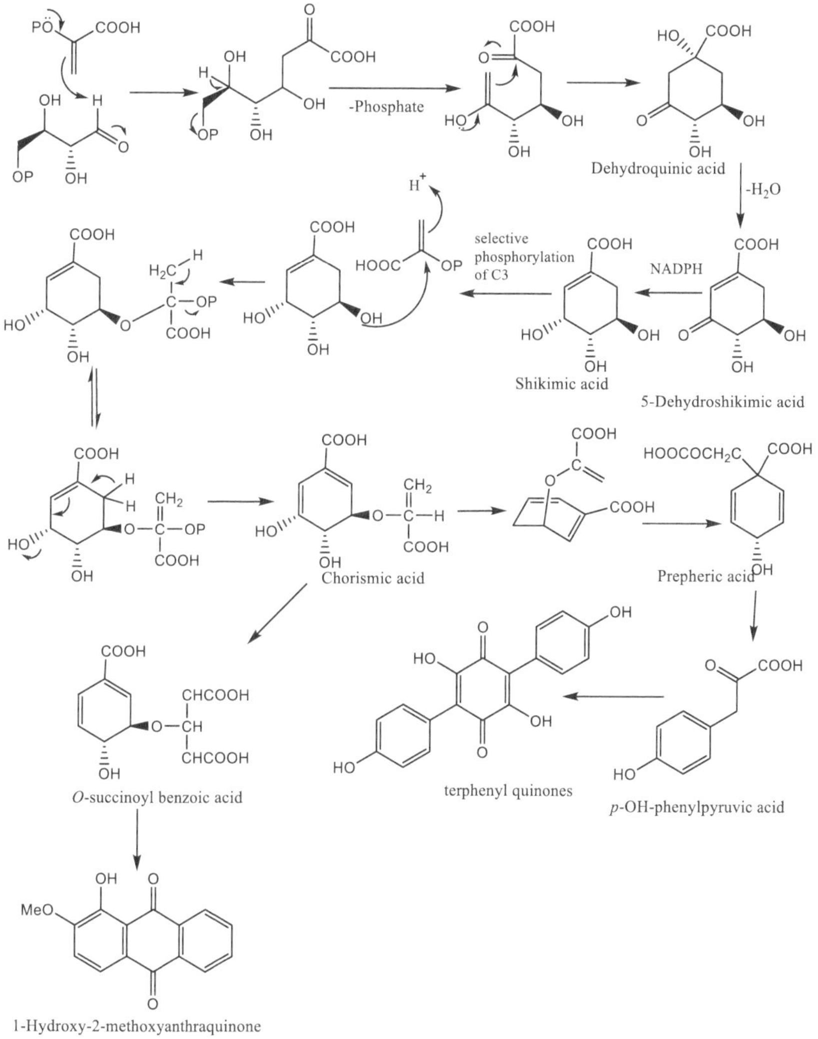

Scheme 2 The shikimic acid pathway for biosynthesis of quinone.

cholangiocarcinoma (CL-6), colon cancer (Caco-2, colo-38, Colo-205, DLD-1, HCT116 *p53*$^{+/+}$, HCT116 *p53*$^{-/-}$, SW480, and SW-680), endometrial cancer (Ishikawa), epidermoid cancer (A431), Ehrlich ascites carcinoma cells (EAC (mouse)), gastric cancer (MKN45), central nervous system cancer (U87MG and U87MG.Δ*EGFR*, XF498, and C6 (rat)), hepatocarcinoma (Hep-3B, HepG2, HepG2/C3A, PLC/PRF/5, and SK-HEP-1), laryngeal cancer (Hep2), leukemia (CCRF-CEM, CEM/ADR5000, HEL, HL60, HL60AR, Jurkat, K-562, PF-382, THP-1, U937,

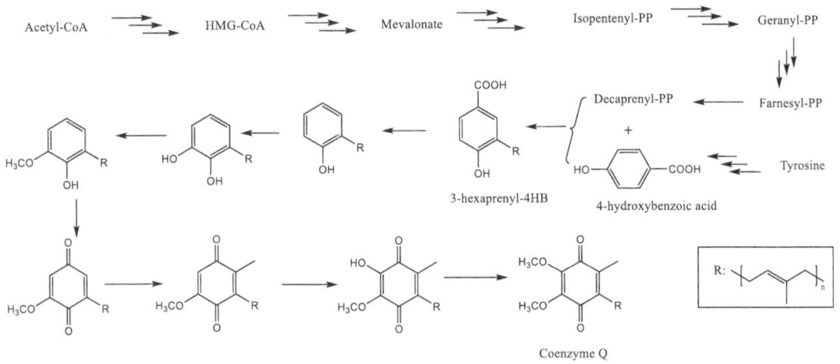

Scheme 3 Simplified scheme of the biosynthesis of coenzyme Q (Lenaz and Genova, 2013). *HMG-CoA: 3-hydroxy-3-methyl-glutaryl coenzyme A.*

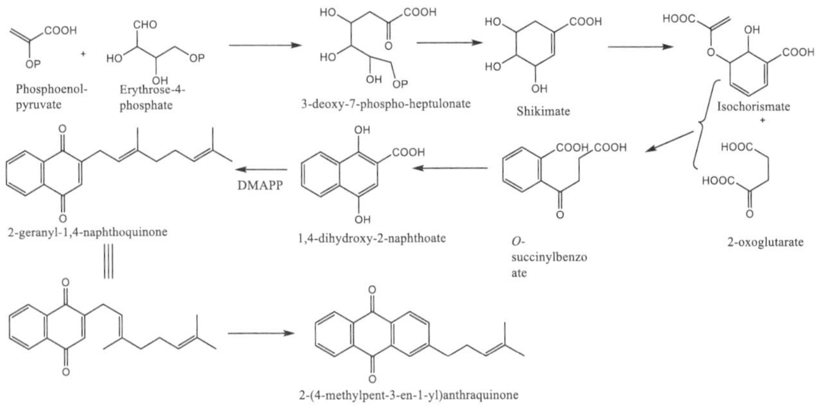

Scheme 4 Biosynthesis of 2-geranyl-1,4-naphthoquinone and its related anthraquinone (Furumoto & Hoshikuma, 2011).

and murine leukemia CH27, P388 and Wehi-3B), lung cancer (A549, H460, SPC212, and SK-LU1), melanoma (A375), ovarian cancer (A2780 and HO-8910PM), pancreatic cancer (Capan-1, MiaPaCa-2, Panc-1, PC-3, and PSN-1), prostate cancer (DU-145, LNCaP, and PC-3), renal cancer (786−0), and squamous cell cancer (KB) cell lines (Table 1).

4. Cytotoxic quinones from African medicinal plants towards drug sensitive and MDR cancer cells

A total of 33 cytotoxic quinones including benzoquinones, naphthoquinones, and anthraquinones have been isolated from African medicinal plants.

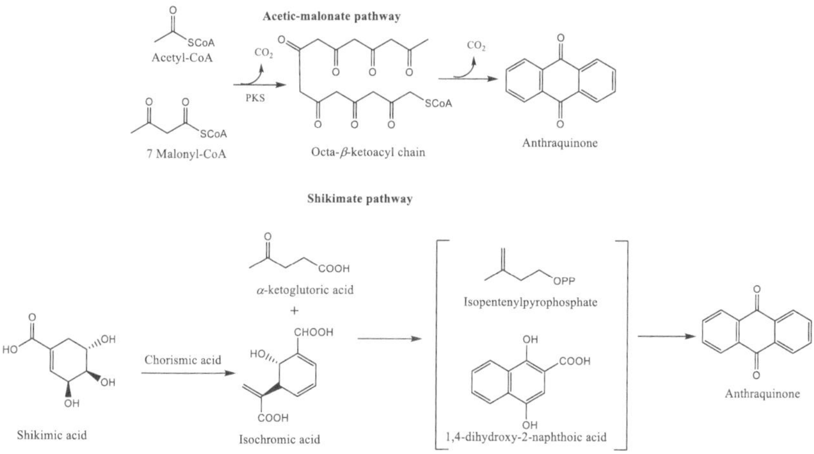

Scheme 5 Natural anthraquinone biosynthesis pathways (Dulo, Phan, Githaiga, Raes, & De Meester, 2021).

Their effects on various human cancer cell lines are summarized in Table 1. The degree of resistance (D.R.) is determined as the ratio of the IC_{50} value in resistant cells *versus* the IC_{50} values in the corresponding sensitive cell line: CEM/ADR5000 cells *vs* CCRF-CEM cells, MDA-MB-231-*BCRP* *vs* MDA-MB-231-pcDNA3, HCT116 $p53^{-/-}$ cells *vs* HCT116 $p53^{+/+}$ cells, and U87MG.Δ*EGFR* *vs* U87MG, is also depicted in Table 1 (Mbaveng et al., 2020). Collateral sensitivity is achieved when the D.R., defined as the ratio of the IC_{50} values of the cytotoxic agent towards the resistant cell line *versus* that in its sensitive counterpart, is below 1 whilst the D.R. above 1 defines normal sensitivity. It has been established that the D.R. < 0.9 defines the collateral sensitivity, whilst D.R. between 0.9 and 1.2 defines normal sensitivity. The cross-resistance is noted if the cytotoxic agent is more active in the sensitive cell line than its resistant subline, with D.R above 1.2 (Efferth et al., 2020; Efferth et al., 2021; Mbaveng, Kuete, & Efferth, 2017). Collateral or normal sensitivities should be achieved for samples with the ability to combat drug resistant cancer cells (Efferth et al., 2020; Kuete & Efferth, 2015). In this section, the activity of these quinones will be discussed according to the cut-off points of cytotoxic phytochemicals defined as follows: outstanding activity ($IC_{50} \leq 0.5\,\mu M$), excellent activity ($0.5 < IC_{50} \leq 2\,\mu M$), very good activity ($2 < IC_{50} \leq 5\,\mu M$), good activity ($5 < IC_{50} \leq 10\,\mu M$), average activity ($10 < IC_{50} \leq 20\,\mu M$), weak activity ($20 < IC_{50} \leq 60\,\mu M$), very weak activity ($60 < IC_{50} \leq 150\,\mu M$), and not active ($IC_{50} > 150\,\mu M$) (Kuete, 2025). Herein, we have focused only on active quinones, meanwhile, not active ones were not considered.

Table 1 Cytotoxic quinones from African medicinal plants and their effects on sensitive and drug-resistant cancer cell lines.

Compounds names	Source	Country	Cancer Cell Lines and IC_{50} values (μM) and degree of resistance in bracket	References
Benzoquinones				
1,2,4,5-Tetraacetate-3-methyl-6-(14-nonadecenyl)-cyclohexadi-2,5-diene (1)	*Maesa lanceolata*	Kenya	9.25 (MCF-7); 24.63 (Caco-2); 25.09 (DLD-1); 18.17 (HepG2); 47.53 (A549); 36.21 (SPC212)	Kuete et al. (2016), Omosa et al. (2016)
2,5-Dihydroxy-3-butyl-2,5-cyclohexadiene-1,4-dione (2)	*Maesa lanceolata*	Kenya	64.59 (MCF-7); 89.72 (Caco-2); 176.17 (HepG2); 68.62 (A549); 87.91 (SPC212)	Kuete et al. (2016), Omosa et al. (2016)
2,5-Dihydroxy-3-heptyl-2,5-cyclohexadiene-1,4-dione (3)	*Maesa lanceolata*	Kenya	388 (MCF-7); 63.93 (Caco-2); 94.22 (HepG2); 107.52 (A549); 8.05 (SPC212)	Kuete et al. (2016), Omosa et al. (2016)
2,5-Dihydroxy-3-pentadecyl-2,5-cyclohexadiene-1,4-dione (4)	*Maesa lanceolata*	Kenya	30.37 (MCF-7); 27.81 (Caco-2); 51.21 (DLD-1); 48.35 (HepG2); 43.32 (A549); 8.39 (SPC212)	Kuete et al. (2016), Omosa et al. (2016)
5-O-methylembelin (5)	*Mvrsine africana*	Kenya	22.56 (MCF-7); 38.89 (Caco-2); 45.37 (DLD-1); 61.28 (HepG2); 50.26 (A549); 38.28 (SPC212)	Kuete et al. (2016), Omosa et al. (2016)
Ardisiaquinone B (6)	*Myrsine Africana*	Kenya	114.60 (HepG2); 21.68 (A549); 3.14 (SPC212)	Kuete et al. (2016), Omosa et al. (2016)

Ardisiaquinone N (10)	Ardisia kivuensis	Cameroon	3.90 (THP-1); 40.82 (HeLa); 87.20 (A549); 53.80 (MCF-7)	Paul et al. (2014)
Ardisiaquinone K (9)	Ardisia kivuensis	Cameroon	3.87 (THP-1); 85.11 (A549); 90.91 (PC-3); 9.61 (HeLa); 10.42 (A431); 10.27 (MCF-7); 6.64 (Ishikawa)	Ndonsta et al. (2011), Paul et al. (2014)
Ardisiaquinone J (8)	Ardisia kivuensis	Cameroon	3.77 (THP-1); 56.50 (A549); 15.40 (HeLa); 9.74 (A431); 6.88 (MCF-7); 10.27 (Ishikawa)	Ndonsta et al. (2011), Paul et al. (2014)
Rapanone (7)	*Maesa lanceolata*	Kenya	16.94 (MCF-7); 22.95 (Caco-2); 46.62 (DLD-1); 32.69 (HepG2); 27.35 (A549); 2.27 (SPC212)	Kuete et al. (2016), Omosa et al. (2016)
Naphthoquinones				
2-Acetyl-7-methoxynaphtho[23-*b*] furan-4,9-quinone (11)	*Milletia versicolor*	Cameroon	0.57 (CCRF-CEM) *vs* 1.02 (CEM/ADR5000) [1.79]; 2.14 (MDA-MB-231-*pcDNA*) *vs* 3.26 (MDA-MB-231-*BCRP*) [1.53]; 1.00 (HCT116 *p53*$^{+/+}$) *vs* 2.25 (HCT116 *p53*$^{-/-}$) [2.26]; 1.00 (U87MG) *vs* 0.96 (U87MG.Δ*EGFR*) [0.96]; 0.79 (HepG2)	Kuete et al. (2017)

(*continued*)

Table 1 Cytotoxic quinones from African medicinal plants and their effects on sensitive and drug-resistant cancer cell lines. *(cont'd)*

Compounds names	Source	Country	Cancer Cell Lines and IC_{50} values (μM) and degree of resistance in bracket	References
2-Acetylfuro-1,4-naphthoquinone (12)	*Newbouldia laevis*	Cameroon	35.88 (CCRF-CEM) *vs* 69.21 (CEM/ADR5000) [1.92]; 19.88 (HL60); 2.36 (PF-382); 8.02 (U87MG); 15.74 (Capan-1); 7.48 (MiaPaCa-2); 22.69 (A549); 6.86 (MCF-7); 14.67 (SW-680); 28.14 (786–0); 2.77 (Colo-38); 1.65 (HeLa); 0.70 (Caski)	Eyong et al. (2005), Kuete et al. (2007), Kuete, Wabo, et al. (2011)
7-Methyljuglone (13)	*Euclea natalensis*	South Africa	4.1 (KB); 13.2 (SK-LU1); 3.7 (LNCaP)	Bapela, Lall, Fourie, Franzblau, and Van Rensburg (2006), Gu et al. (2004), 80
Diospyrin (14)	*Euclea natalensis*	South Africa	0.8 (A375); 3.6 (Hep2); 0.8 (EAC)	Das Sarma, Ghosh, Patra, and Hazra (2007), McGaw et al. (2008), Van der Kooy, Meyer, and Lall (2006)
Isodiospyrin (15)	*Euclea natalensis*	South Africa	0.67 (Hep-3B); 4.84 (KB); 0.35 (Colo-205); 0.72 (HeLa)	Kuo et al. (1997), McGaw et al. (2008), Van der Kooy et al. (2006)

Lapachol (16)	*Newbouldia laevis*	Cameroon	64.59 (DU-145)	Eyong et al. (2008), Kuete et al. (2007)
Plumbagin (17)	*Diospyros spp*	Cameroon	0.06 (MCF-7); 0.07 (Caco-2); 0.98 (DLD-1); 1.01 (HepG2); 1.14 (A549); 0.27 (SPC212); 24 (CL-6); 7.7 (C6)	Kuete, Alibert-Franco, et al. (2011), Kuete et al. (2016,2009),Majiene et al. (2019), Omosa et al. (2016), Panrit, Plengsuriyakarn, Martviset, and Na-Bangchang (2018)
Anthraquinones				
3-Hydroxy-2-hydroxymethylanthraquinone (18)	*Pentas schimperi*	Cameroon	37.29(CCRF-CEM) *vs* 55.29 (CEM/ADR5000) [1.48]; 88.00 (HCT116 *p53*$^{+/+}$) *vs* 121.41 (HCT116 *p53*$^{-/-}$) [1.39]; 117.33 (U87MG); 56.82 (HepG2)	Kuete et al. (2015)
5-Hydroxy-6-hydroxymethyl anthragallol 1,3-dimethyl ether (19)	*Galium sinaicum*	Egypt	30.82 (P388)	el-Gamal et al. (1996)
6,7-Dimethoxy xanthopurpurin (20)	*Galium sinaicum*	Egypt	3.33 (P388)	el-Gamal et al. (1996)
6-Hydroxy-7-methoxy rubiadin (21)	*Galium sinaicum*	Egypt	34.55 (P388)	el-Gamal et al. (1996)

(continued)

Table 1 Cytotoxic quinones from African medicinal plants and their effects on sensitive and drug-resistant cancer cell lines. (*cont'd*)

Compounds names	Source	Country	Cancer Cell Lines and IC_{50} values (μM) and degree of resistance in bracket	References
7-Carboxy-anthragallol 1,3-dimethyl ether (22)	*Galium sinaicum*	Egypt	10.38 (P388)	el-Gamal et al. (1996)
Chrysophanol (23)	*Rumex abyssinicus*	Kenya	52.24 (A549); 145.63 (SPC212)	Kuete et al. (2016)
Damnacanthal (24)	*Garcinia huillensis; Pentas schimperi*	Cameroon; Congo	3.12 (CCRF-CEM) *vs* 28.72 (CEM/ADR5000) [9.20]; 14.89 (HL60); 23.87 (MDA-MB-231-*pcDNA*) *vs* 7.02 (MDA-MB-231-*BCRP*) [0.29]; 22.27 (HCT116 *p53*$^{+/+}$) *vs* 23.48 (HCT116 *p53*$^{-/-}$) [1.05]; 22.06 (U87MG) *vs* 30.32 (U87MG.Δ*EGFR*) [1.37]; 17.62 (HepG2); 11.70 (Wehi-3B); 23.05 (KB); 102.84 (SW480); 4.47 (Panc-1); 3.79 (PSN-1)	Abu et al. (2014), Dibwe et al. (2012), Kuete et al. (2015), Thani, Vallisuta, Siripong, and Ruangwises (2010)

Damnacanthol (25)	*Pentas schimperi*	Cameroon	12.18 (CCRF-CEM) *vs* 21.55 (CEM/ADR5000) [1.77]; 66.65 (MDA-MB-231-*pcDNA*) *vs* 68.63 (MDA-MB-231-*BCRP*) [1.09]; 77.46 (HCT116 $p53^{+/+}$) *vs* 59.40 (HCT116 $p53^{-/-}$) [0.77]; 72.82 (U87MG) *vs* 80.11 (U87MG.Δ*EGFR*) [1.10]; 54.40 (HepG2)	Kuete et al. (2015)
Emodin (26)	*Rumex abyssinicus*	Kenya	37.57 (MCF-7); 73.63 (Caco-2); 77.28 (DLD-1); 71.7 (HepG2); 66.30 (A549); 99.31 (SPC212); 20 (PC-3); 37 (MCF-7/Adr); 26 (MCF-7); 25 (SKBR3); 35.30 (HO-8910PM); 20 (CH27); 60 (H460); 42.5 (HepG2/C3A); 46.6 (PLC/PRF/5); 53.1 (SK-HEP-1); 47.5 (MKN45); 30 (U937); 38.25 (K562); 4.19 (HEL); 20 (Jurkat); 5.79 (HL-60/ADR); 13.5 (XF498)	Chen et al. (2007), He and Hu (2011), Kuete et al. (2016), Omosa et al. (2016), Semwal, Semwal, Combrinck, and Viljoen (2021)
Physcion (27)	*Vismia laurentii*	Cameroon	203.1 (MCF-7)	Kuete, Wabo, et al. (2011), Wabo et al. (2007), Zhang et al. (2021)

(continued)

Table 1 Cytotoxic quinones from African medicinal plants and their effects on sensitive and drug-resistant cancer cell lines. (*cont'd*)

Compounds names	Source	Country	Cancer Cell Lines and IC_{50} values (μM) and degree of resistance in bracket	References
Ruberythric acid 1-methyl ether (28)	*Galium sinaicum*	Egypt	72.29 (P388)	el-Gamal et al. (1996)
Rubiadin (29)	*Pentas longiflora; Pentas lanceolata*	Kenya	11.81 (CEM-SS); 39.37 (MCF-7); 222.91 (HeLa)	Ali et al. (2000), Endale et al. (2012), Padmaa, Shruthi, Sujan Ganapathy, & Vedamurthy, 2016
Schimperiquinone B (30)	*Pentas schimperi*	Cameroon	4.41 (CCRF-CEM) *vs* 74.33 (CEM/ADR5000) [16.85]	Kuete et al. (2015)
Scutianthraquinone A (31)	*Scutia myrtina*	Madagascar	7.6 (A2780)	Hou et al. (2009)
Scutianthraquinone B (32)	*Scutia myrtina*	Madagascar	5.8 (A2780)	Hou et al. (2009)
Scutianthraquinones D (33)	*Scutia myrtina*	Madagascar	4.3 (A2780)	Hou et al. (2009)

4.1 Cytotoxic benzoquinones

Ten potent cytotoxic benzoquinones were identified in African medicinal plants. They include 1,2,4,5-tetraacetate-3-methyl-6-(14-nonadecenyl)-cyclohexadi-2,5-diene (**1**), 2,5-dihydroxy-3-butyl-2,5-cyclohexadiene-1,4-dione (**2**), 2,5-dihydroxy-3-heptyl-2,5-cyclohexadiene-1,4-dione (**3**), 2,5-dihydroxy-3-pentadecyl-2,5-cyclohexadiene-1,4-dione (**4**), 5-*O*-methylembelin (**5**), ardisiaquinone B (**6**), rapanone (**7**), and ardisiaquinones J (**8**), K (**9**), and N (**10**). Their chemical structures are depicted in Fig. 2 while the cytotoxic activities in human cancer cell lines are shown in Table 1. Amongst them, very good antiproliferative activities ($2 < IC_{50} \leq 5\ \mu M$) were obtained with benzoquinones **6** and **7** against SPC212 cells, and **8, 9,** and **10** against THP-1 cells meanwhile good antiproliferative activities ($5 < IC_{50} \leq 10\ \mu M$) were obtained with benzoquinones **1** against MCF-7 cells, **4** and **8** against SPC212 cells, **8** against A431 cells and MCF-7 cells, and **9** against HeLa cells and Ishikawa cells.

4.2 Cytotoxic naphthoquinones

Potent cytotoxic naphthoquinones isolated from African medicinal plants include 2-acetyl-7-methoxynaphtho[23-*b*]furan-4,9-quinone (**11**), 2-acetylfuro-1,4-naphthoquinone (**12**), 7-methyljuglone (**13**), diospyrin (**14**),

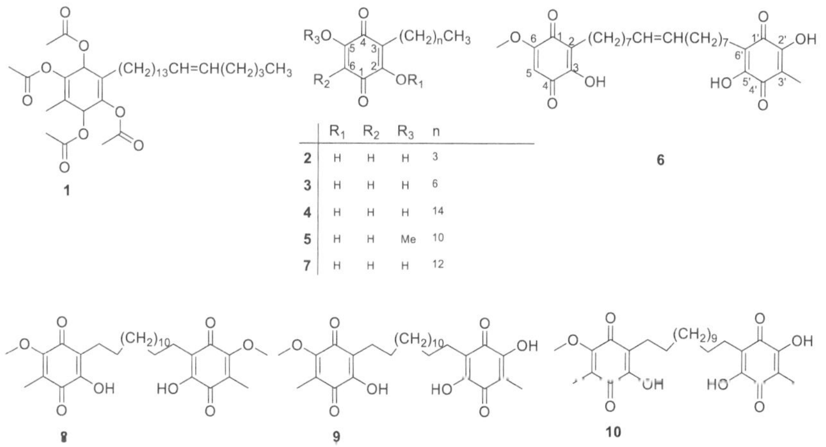

	R_1	R_2	R_3	n
2	H	H	H	3
3	H	H	H	6
4	H	H	H	14
5	H	H	Me	10
7	H	H	H	12

Fig. 2 Chemical structures of potent benzoquinones isolated from African medicinal plants. 1: 1,2,4,5-tetraacetate-3-methyl-6-(14-nonadecenyl)-cyclohexadi-2,5-diene; 2: 2,5-dihydroxy-3-butyl-2,5-cyclohexadiene-1,4-dione; 3: 2,5-dihydroxy-3-heptyl-2,5-cyclohexadiene-1,4-dione; 4: 2,5-dihydroxy-3-pentadecyl-2,5-cyclohexadiene-1,4-dione; 5: 5-*O*-methylembelin; 6: ardisiaquinone B; 7: rapanone; 8: ardisiaquinone J; 9: ardisiaquinone K; 10: ardisiaquinone N.

isodiospyrin (**15**), lapachol (**16**), and plumbagin (**17**). Their chemical structures are depicted in Fig. 3 while the cytotoxic activities in human cancer cell lines are shown in Table 1. Amongst them, outstanding cytotoxic activities ($IC_{50} \leq 0.5\ \mu M$) were obtained with naphthoquinones **15** towards Colo-205 cells, and **17** towards MCF-7 cells, Caco-2 cells, and SPC212 cells meanwhile excellent cytotoxic activities ($0.5 < IC_{50} \leq 2\ \mu M$) were recorded with naphthoquinones with **11** towards CCRF-CEM cells, CEM/ADR5000 cells, HCT116 $p53^{+/+}$ cells, HCT116 $p53^{-/-}$ cells, U87MG cells, and U87MG.Δ*EGFR* cells, **11** and **17** towards HepG2 cells, **12** and **15** towards HeLa cells, **12** towards Caski cells, **14** towards A375 cells and EAC cells, **15** towards Hep-3B cells, and **17** towards DLD-1 cells and A549 cells. Very good antiproliferative activities were obtained with naphthoquinones **11** towards MDA-MB-231-*pcDNA* cells, MDA-MB-231-*BCRP* cells, and HCT116 $p53^{-/-}$ cells, **12** towards PF-382 cells and Colo-38 cells, **13** towards KB cells and LNCaP cells, and **14** towards Hep2 cells meanwhile good antiproliferative activities were obtained with naphthoquinones **12** against U87MG cells, MiaPaCa-2 cells, and MCF-7 cells, and **17** towards C6 cells. Importantly, a normal sensitivity of U87MG.Δ*EGFR* cells *vs* U87MG cells (D.R. 0.96) to **11** was also achieved.

4.3 Cytotoxic anthraquinones

Sixteen potent cytotoxic anthraquinones were isolated from African medicinal plants. They were 3-hydroxy-2-hydroxymethylanthraquinone

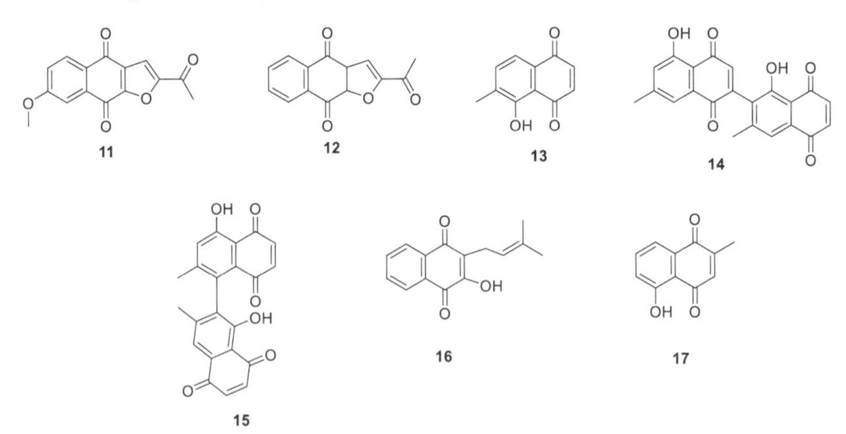

Fig. 3 Chemical structures of potent naphthoquinones isolated from African medicinal plants. 11: 2-acetyl-7-methoxynaphtho[23-b]furan-4,9-quinone; 12: 2-acetylfuro-1,4-naphthoquinone; 13: 7-methyljuglone; 14: diospyrin; 15: isodiospyrin; 16: lapachol; 17: plumbagin.

(**18**), 5-hydroxy-6-hydroxymethyl anthragallol 1,3-dimethyl ether (**19**), 6,7-dimethoxy xanthopurpurin (**20**), 6-hydroxy-7-methoxy rubiadin (**21**), 7-carboxy-anthragallol 1,3-dimethyl ether (**22**), chrysophanol (**23**), damnacanthal (**24**), damnacanthol (**25**), emodin (**26**), physcion (**27**), ruberythric acid l-methyl ether (**28**), rubiadin (**29**), schimperiquinone B (**30**), and scutianthraquinones A (**31**), B (**32**), and D (**33**). Their chemical structures are depicted in Fig. 4 while the cytotoxic activities in human cancer cell lines are shown in Table 1. Amongst them, very good antiproliferative activities were obtained with anthraquinones **20** towards P388 cells, **24** and **30** towards CCRF-CEM cells, **24** towards Panc-1 cells and PSN-1 cells, **26** towards HEL cells, and **33** towards A2780 cells meanwhile good antiproliferative activities were obtained with anthraquinones **24** towards MDA-MB-231-*BCRP* cells, **26** towards HL-60/ADR cells, **31** and **32** towards A2780 cells. Interestingly, a collateral sensitivity of MDA-MB-231-*BCRP* cells *vs* MDA-MB-231 cells to **24** (D.R. 0.29) was also obtained.

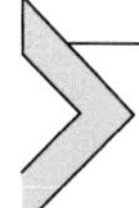

5. Modes of action of cytotoxic quinones isolated from African medicinal plants

The cytotoxicity of the naphthoquinone plumbagin (**17**) and benzoquinones rapanone (**7**) was determined in 6 human carcinoma cell lines and normal CRL2120 fibroblasts (Kuete et al., 2016). The studied cell lines were A549, SPC212, DLD-1, Caco-2, MCF7, and HepG2. The two quinones displayed interesting activities with IC_{50} values below 50 μM in the six tested cancer cell lines. The IC_{50} values ranged from 0.06 μM (MCF-7 cells) to 1.14 μM (A549 cells) for compound **17**, and from 2.27 μM (mesothelioma SPC212 cells) to 46.62 μM (colorectal adenocarcinoma DLD-1 cells) for compound **7**. Compounds **7** and **17** induced apoptosis in MCF-7 cells mediated by increased reactive oxygen species (ROS) production and mitochondrial membrane potential (MMP) loss, respectively (Kuete et al., 2016). The naphthoquinone plumbagin (**17**) also displayed highly cytotoxic activity on C6 cells and caused necrotic cell death. It also increased the amount of intracellular ROS, and significantly uncoupled mitochondrial oxidation from phosphorylation impairing energy production in C6 cells (Majiene, Kuseliauskyte, Stimbirys, & Jekabsone, 2019).

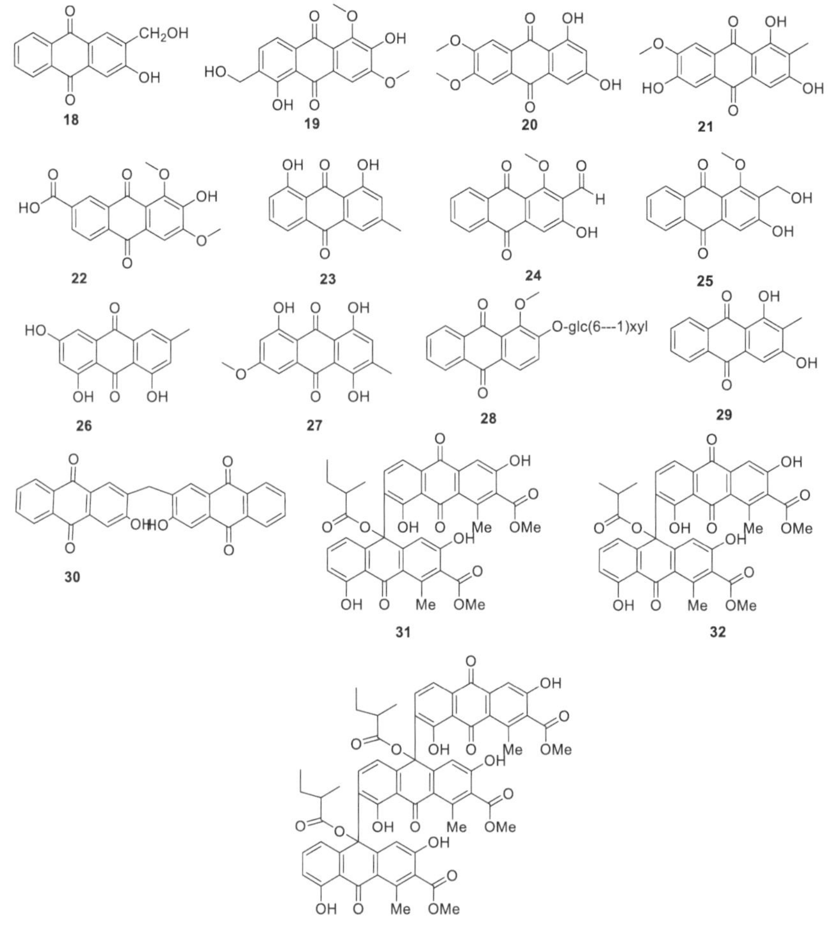

Fig. 4 Chemical structures of potent anthraquinones isolated from African medicinal plants. 18: 3-hydroxy-2-hydroxymethylanthraquinone; 19: 5-hydroxy-6-hydroxymethyl anthragallol 1,3-dimethyl ether; 20: 6,7-dimethoxy xanthopurpurin; 21: 6-hydroxy-7-methoxy rubiadin; 22: 7-carboxy-anthragallol 1,3-dimethyl ether; 23: chrysophanol; 24: damnacanthal; 25: damnacanthol; 26: emodin; 27: physcion; 28: ruberythric acid l-methyl ether; 29: rubiadin; 30: schimperiquinone B; 31: scutianthraquinone A; 32: scutianthraquinone B; 33: scutianthraquinones D.

The anthraquinones, damnacanthal (**24**) and damnacanthol (**25**) had cytotoxic effects against several cancer cell lines including drug sensitive and MDR phenotypes; they induced apoptosis in CCRF-CEM leukemia cells, mediated by the disruption of the MMP and increase in ROS production (Kuete et al., 2015). It was reported that caspase activation is slightly involved in damnacanthal-induced apoptosis in CCRF-CEM cells. However, it was also shown that damnacanthal-induced apoptosis in MCF-7 cells included activation of p21 and caspase-7 (Aziz et al., 2014). Compound **24** also caused cell growth arrest as well as induction of caspase activity in colon HCT116, SW480, and LoVo cells (Dibwe, Awale, Kadota, & Tezuka, 2012).

The cytotoxicity of 2-acetyl-7-methoxynaphtho[2,3-b]furan-4,9-quinone, a naphthoquinone (**11**) was determined against 9 drug sensitive and MDR cancer cell lines by Kuete and collaborators (Kuete, Mbaveng, Sandjo, Zeino, & Efferth, 2017). The investigated cancer cell lines were CCRF-CEM, CEM/ADR5000, HCT116 $p53^{+/+}$, HCT116 $p53^{-/-}$, U87. MG, U87. MGΔ*EGFR*, MDA-MB-231-*pcDNA3*, MDA-MB-231-*BCRP*, and HepG2. They found that the compound had impressive antiproliferative effects with the IC_{50} values ranging from 0.79 μM (against HepG2 cells) to 3.26 μM (against MDA-MB231/*BCRP* cells) on the 9 tested cancer cell lines. Interestingly, the IC_{50} values below 1 μM were recorded with compound **11** towards CCRF-CEM cells (0.57 μM), U87MG.Δ*EGFR* cells (0.96 μM cells), and HepG2 cells (0.76 μM). They finally demonstrated that the alteration of MMP was the main mode of induction of apoptosis in CCRF-CEM cells by the tested compound.

The antiproliferative effect of 2-acetylfuro-1,4-naphthoquinone (**12**) was evaluated on a panel of 14 cancer cell lines by the team of Kuete and co-workers (Kuete, Wabo, et al., 2011). The investigated cancer cell lines included CCRF-CEM, CEM/ADR5000, *PF-382,* HL-60, MiaPaCa-2, Capan-1, MCF-7, SW-680, *786–0,* U87MG, A549, Caski, HeLa, Colo-38, as well as the normal AML12 hepatocytes. IC_{50} values around or below 10 μM were obtained on 78.57% of the tested cancer cell lines. Upon treatment with two-fold IC_{50} and after 72 h, compound **12** induced cell cycle arrest in S-phase, and significant apoptosis in CCRF-CEM cells (Kuete, Wabo, et al., 2011).

Emodin (**26**) efficiently inhibited the proliferation of HEL cells, inducing apoptosis, probably mediated by the downregulation of P-Akt, phosphorylated-glycogen synthase kinase (P-GSK3β), and 70 kd heat shock proteins (HSP70) proteins expression (He & Hu, 2011). Anthraquinone **26** also efficiently inhibits growth and induces apoptosis on HL-60/ADR cells, probably through the downregulation of mitochondrial transmembrane potential and expressions of bcl-2 and c-myc, as well as up-regulation of caspase-3 activity (Chen et al., 2007).

6. Conclusion

In this chapter, we have identified 33 quinones isolated from African medicinal plants with cytotoxic effects on various human cancer cell lines. They include benzoquinones, naphthoquinones, and anthraquinones. The most active benzoquinones were 1,2,4,5-tetraacetate-3-methyl-6-(14-nonadecenyl)-cyclohexadi-2,5-diene (**1**), 2,5-dihydroxy-3-pentadecyl-2,5-cyclohexadiene-1,4-dione (**4**), ardisiaquinone B (**6**), rapanone (**7**), and ardisiaquinones J (**8**), K (**9**), and N (**10**). The most potent cytotoxic naphthoquinones include 2-acetyl-7-methoxynaphtho[23-*b*]furan-4,9-quinone (**11**), 2-acetylfuro-1,4-naphthoquinone (**12**), 7-methyljuglone (**13**), diospyrin (**14**), isodiospyrin (**15**), and plumbagin (**17**). The best cytotoxic anthraquinones were 6,7-dimethoxy xanthopurpurin (**20**), damnacanthal (**24**), emodin (**26**), schimperiquinone B (**30**), and scutianthraquinones A (**31**), B (**32**), and D (**33**). Quinones having the ability to combat the multidrug resistance of cancer cell lines were **11** and **24**. Some of the active quinones such as **7, 11, 12, 17,** and **24** displayed apoptotic cell death, mainly mediated by caspase activation, alteration of MMP, and enhanced ROS production in cancer cells. The most potent cytotoxic quinones herein should be further explored to develop novel cytotoxic products to fight cancer.

References

Abu, N., Ali, N. M., Ho, W. Y., Yeap, S. K., Aziz, M. Y., & Alitheen, N. B. (2014). Damnacanthal: a promising compound as a medicinal anthraquinone. *Anti-cancer Agents in Medicinal Chemistry, 14*(5), 750–755.

Ali, A. M., Ismail, N. H., Mackeen, M. M., Yazan, L. S., Mohamed, S. M., Ho, A. S., et al. (2000). Antiviral, cyototoxic and antimicrobial activities of anthraquinones isolated from the roots of *Morinda elliptica. Pharmaceutical Biology, 38*(4), 298–301.

Aziz, M. Y., Omar, A. R., Subramani, T., Yeap, S. K., Ho, W. Y., Ismail, N. H., et al. (2014). Damnacanthal is a potent inducer of apoptosis with anticancer activity by stimulating p53 and p21 genes in MCF-7 breast cancer cells. *Oncology Letters,* 7(5), 1479–1484.

Bapela, N. B., Lall, N., Fourie, P. B., Franzblau, S. G., & Van Rensburg, C. E. (2006). Activity of 7-methyljuglone in combination with antituberculous drugs against *Mycobacterium tuberculosis. Phytomedicine, 13*(9–10), 630–635.

Caramés Masana, F., & de Reijke, T. M. (2017). The efficacy of Apaziquone in the treatment of bladder cancer. *Expert Opinion on Pharmacotherapy, 18*(16), 1781–1788.

Chen, Y. Y., Zheng, H. Y., Hu, J. D., Zheng, Z. H., Zheng, J., Lian, X. L., et al. (2007). Inhibitory effects of emodin on drug-resistant HL-60/ADR cell proliferation and its induction of apoptosis. *Zhongguo Shi Yan Xue Ye xue Za Zhi/Zhongguo Bing li Sheng li xue hui = Journal of Experimental Hematology/Chinese Association of Pathophysiology, 15*(5), 955–960.

Das Sarma, M., Ghosh, R., Patra, A., & Hazra, B. (2007). Synthesis and antiproliferative activity of some novel derivatives of diospyrin, a plant-derived naphthoquinonoid. *Bioorganic & Medicinal Chemistry, 15*(11), 3672–3677.

Dibwe, D. F., Awale, S., Kadota, S., & Tezuka, Y. (2012). Damnacanthal from the Congolese medicinal plant *Garcinia huillensis* has a potent preferential cytotoxicity against human pancreatic cancer PANC-1 cells. *Phytotherapy Research, 26*(12), 1920–1926.

Dulo, B., Phan, K., Githaiga, J., Raes, K., & De Meester, S. (2021). Natural quinone dyes: a review on structure, extraction techniques, analysis and application potential. *Waste and Biomass Valorization, 12*(12), 6339–6374.

Efferth, T., Kadioglu, O., Saeed, M. E. M., Seo, E. J., Mbaveng, A. T., & Kuete, V. (2021). Medicinal plants and phytochemicals against multidrug-resistant tumor cells expressing ABCB1. *ABCG2, or ABCB5: a synopsis of 2 decades. Phytochemistry Reviews, 20*(1), 7–53.

Efferth, T., Saeed, M. E. M., Kadioglu, O., Seo, E. J., Shirooie, S., Mbaveng, A. T., et al. (2020). Collateral sensitivity of natural products in drug-resistant cancer cells. *Biotechnology Advances, 38*, 107342.

el-Gamal, A. A., Takeya, K., Itokawa, H., Halim, A. F., Amer, M. M., Saad, H. E., et al. (1996). Anthraquinones from the polar fractions of *Galium sinaicum. Phytochemistry, 42*(4), 1149–1155.

Endale, M., Alao, J. P., Akala, H. M., Rono, N. K., Eyase, F. L., Derese, S., et al. (2012). Antiplasmodial quinones from *Pentas longiflora* and *Pentas lanceolata. Planta Medica, 78*(1), 31–35.

Eyong, K. O., Folefoc, G. N., Kuete, V., Beng, V. P., Krohn, K., Hussain, H., et al. (2006). Newbouldiaquinone A: A naphthoquinone–anthraquinone ether coupled pigment, as a potential antimicrobial and antimalarial agent from Newbouldia laevis. *Phytochemistry, 67*(6), 605–609.

Eyong, K. O., Krohn, K., Hussain, H., Folefoc, G. N., Nkengfack, A. E., Schulz, B., et al. (2005). Newbouldiaquinone and newbouldiamide: a new naphthoquinone-anthraquinone coupled pigment and a new ceramide from *Newbouldia laevis. Chemical and Pharmaceutical Bulletin, 53*(6), 616–619.

Eyong, K. O., Kuete, V., & Efferth, T. (2013). 10 - Quinones and benzophenones from the medicinal plants of Africa. In V. Kuete (Ed.). *Medicinal Plant Research in Africa: Pharmacology and Chemistry* (pp. 351–391)Oxford: Elsevier.

Eyong, K. O., Kumar, P. S., Kuete, V., Folefoc, G. N., Nkengfack, E. A., & Baskaran, S. (2008). Semisynthesis and antitumoral activity of 2-acetylfuranonaphthoquinone and other naphthoquinone derivatives from lapachol. *Bioorganic and Medicinal Chemistry Letters, 18*(20), 5387–5390.

Furumoto, T., & Hoshikuma, A. (2011). Biosynthetic origin of 2-geranyl-1,4-naphthoquinone and its related anthraquinone in a *Sesamum indicum* hairy root culture. *Phytochemistry, 72*(9), 871–874.

Gu, J. Q., Graf, T. N., Lee, D., Chai, H. B., Mi, Q., Kardono, L. B., et al. (2004). Cytotoxic and antimicrobial constituents of the bark of *Diospyros maritima collected* in two geographical locations in Indonesia. *Journal of Natural Products, 67*(7), 1156–1161.

He, X. C., & Hu, J. D. (2011). Effects of emodin on human erythroleukemia cell line HEL. *Zhongguo Shi Yan Xue Ye xue Za Zhi/Zhongguo Bing li Sheng li xue hui = Journal of Experimental Hematology/Chinese Association of Pathophysiology, 19*(5), 1121–1124.

Hou, Y., Cao, S., Brodie, P. J., Callmander, M. W., Ratovoson, F., Rakotobe, E. A., et al. (2009). Antiproliferative and antimalarial anthraquinones of *Scutia myrtina* from the Madagascar forest. *Bioorganic &. Medicinal Chemistry, 17*(7), 2871–2876.

Kuete, V. (2025). Chapter Four-African medicinal plants and their derivative as the source of potent anti-leukemic products: rationale classification of naturally occurring anticancer agents. *Advances in Botanical Research; a Journal of Science and its Applications, 113*. https://doi.org/10.1016/bs.abr.2023.12.010.

Kuete, V., Alibert-Franco, S., Eyong, K. O., Ngameni, B., Folefoc, G. N., Nguemeving, J. R., et al. (2011). Antibacterial activity of some natural products against bacteria expressing a multidrug-resistant phenotype. *International Journal of Antimicrobial Agents, 37*(2), 156–161.

Kuete, V., Donfack, A. R., Mbaveng, A. T., Zeino, M., Tane, P., & Efferth, T. (2015). Cytotoxicity of anthraquinones from the roots of *Pentas schimperi* towards multi-factorial drug-resistant cancer cells. *Investigational New Drugs, 33*(4), 861–869.

Kuete, V., & Efferth, T. (2015). African flora has the potential to fight multidrug resistance of cancer. *BioMed Res Int, 2015*, 914813.

Kuete, V., Eyong, K. O., Folefoc, G. N., Beng, V. P., Hussain, H., Krohn, K., et al. (2007). Antimicrobial activity of the methanolic extract and of the chemical constituents isolated from *Newbouldia laevis*. *Die Pharmazie, 62*(7), 552–556.

Kuete, V., Mbaveng, A. T., Sandjo, L. P., Zeino, M., & Efferth, T. (2017). Cytotoxicity and mode of action of a naturally occurring naphthoquinone, 2-acetyl-7-methoxynaphtho[2,3-b]furan-4,9-quinone towards multi-factorial drug-resistant cancer cells. *Phytomedicine, 33*, 62–68.

Kuete, V., Omosa, L. K., Tala, V. R., Midiwo, J. O., Mbaveng, A. T., Swaleh, S., et al. (2016). Cytotoxicity of plumbagin, rapanone and 12 other naturally occurring quinones from Kenyan flora towards human carcinoma cells. *BMC Pharmacology and Toxicology, 17*(1), 60.

Kuete, V., Tangmouo, J. G., Marion Meyer, J. J., & Lall, N. (2009). Diospyrone, crassiflorone and plumbagin: three antimycobacterial and antigonorrhoeal naphthoquinones from two *Diospyros* spp. *Int J Antimicrob Ag, 34*(4), 322–325.

Kuete, V., Wabo, H. K., Eyong, K. O., Feussi, M. T., Wiench, B., Krusche, B., et al. (2011). Anticancer activities of six selected natural compounds of some Cameroonian medicinal plants. *PLoS One, 6*(8), e21762.

Kuo, Y. H., Chang, C. I., Li, S. Y., Chou, C. J., Chen, C. F., Kuo, Y. H., et al. (1997). Cytotoxic constituents from the stems of *Diospyros maritima*. *Planta Medica, 63*(4), 363–365.

Lenaz, G. & Genova, M. L., 2013. Quinones. In *The Encyclopedia of Biological Chemistry*, (2nd edition), vol. 3 (pp. 722–729). Academic Press Elsevier.

Liu, H. (2011). Extraction and isolation of compounds from herbal medicines. In J. H. Lui (Ed.). *Traditional herbal medicine research methods*. John Wiley and Sons.

Majiene, D., Kuseliauskyte, J., Stimbirys, A., & Jekabsone, A. (2019). Comparison of the effect of Native 1,4-naphthoquinones plumbagin, menadione, and lawsone on viability, redox status, and mitochondrial functions of C6 glioblastoma cells. *Nutrients, 11*(6).

Mbaveng, A. T., Kuete, V., & Efferth, T. (2017). Potential of Central, Eastern and Western Africa medicinal plants for cancer therapy: spotlight on resistant cells and molecular targets. *Frontiers in Pharmacology, 8*, 343.

Mbaveng, A. T., Noulala, C. G. T., Samba, A. R. M., Tankeo, S. B., Fotso, G. W., Happi, E. N., et al. (2020). Cytotoxicity of botanicals and isolated phytochemicals from *Araliopsis soyauxii* Engl. (Rutaceae) towards a panel of human cancer cells. *Journal of Ethnopharmacology, 267*, 113535.

McGaw, L. J., Lall, N., Meyer, J. J., & Eloff, J. N. (2008). The potential of South African plants against *Mycobacterium* infections. *Journal of Ethnopharmacology, 119*(3), 482–500.

Ndonsta, B. L., Tatsimo, J. S. N., Csupor, D., Forgo, P., Berkecz, R., Berényi, Á., et al. (2011). Alkylbenzoquinones with antiproliferative effect against human cancer cell lines from stem of *Ardisia kivuensis*. *Phytochemistry Letters, 4*(3), 227–230.

O'Brien, P. J. (1991). Molecular mechanisms of quinone cytotoxicity. *Chemico-Biological Interactions, 80*(1), 1–41.

Omosa, L. K., Midiwo, J. O., Mbaveng, A. T., Tankeo, S. B., Seukep, J. A., Voukeng, I. K., et al. (2016). Antibacterial activity and structure-activity relationships of a panel of 48 compounds from kenyan plants against multidrug resistant phenotypes. *SpringerPlus, 5*, 901.

Padmaa, M. P., Dileep, C. S., Shruthi, S. D., Sujan Ganapathy, P. S., & Vedamurthy, A. B. (2016). *In vitro* antiproliferative and in silico activity of rubiadin isolated from roots of *Rubia cordifolia*. *Mintage Journal of Pharmaceutical and Medical Sciences, 5*(1), 20–23.

Panrit, L., Plengsuriyakarn, T., Martviset, P., & Na-Bangchang, K. (2018). Inhibitory activities of plumbagin on cell migration and invasion and inducing activity on cholangiocarcinoma cell apoptosis. *Asian Pacific. Journal of Tropical Medicine, 11*(7), 430.

Paul, D. J., Laure, N. B., Guru, S. K., Khan, I. A., Ajit, S. K., Vishwakarma, R. A., et al. (2014). Antiproliferative and antimicrobial activities of alkylbenzoquinone derivatives from *Ardisia kivuensis*. *Pharmaceutical Biology, 52*(3), 392–397.

Sandjo, L. P., de Moraes, M. H., Kuete, V., Kamdoum, B. C., Ngadjui, B. T., & Steindel, M. (2016). Individual and combined antiparasitic effect of six plant metabolites against *Leishmania amazonensis* and *Trypanosoma cruzi*. *Bioorganic & Medicinal Chemistry Letters, 26*(7), 1772–1775.

Savenkov, P. M., & Sabirova, R. K. (1967). Dimecarbine treatment of patients with hypertensive disease. *Sovereign Medical, 30*(4), 121–123.

Semwal, R. B., Semwal, D. K., Combrinck, S., & Viljoen, A. (2021). Emodin - A natural anthraquinone derivative with diverse pharmacological activities. *Phytochemistry, 190*, 112854.

Thani, W., Vallisuta, O., Siripong, P., & Ruangwises, N. (2010). Anti-proliferative and antioxidative activities of Thai noni/Yor (*Morinda citrifolia* Linn.) leaf extract. *Southeast. Southeast Asian Journal of Tropical Medicine and Public Health, 41*(2), 482–489.

Tsaffack, M., Nguemeving, J. R., Kuete, V., Ndejouong Tchize Ble, S., Mkounga, P., Penlap Beng, V., et al. (2009). Two new antimicrobial dimeric compounds: febrifuquinone, a vismione-anthraquinone coupled pigment and adamabianthrone, from two *Psorospermum* species. *Chemical and Pharmaceutical Bulletin, 57*(10), 1113–1118.

Van der Kooy, F., Meyer, J. J. M., & Lall, N. (2006). Antimycobacterial activity and possible mode of action of newly isolated neodiospyrin and other naphthoquinones from *Euclea natalensis*. *South African Journal of Botany, 72*(3), 349–352.

Wabo, H. K., Kouam, S. F., Krohn, K., Hussain, H., Tala, M. F., Tane, P., et al. (2007). Prenylated anthraquinones and other constituents from the seeds of *Vismia laurentii*. *Chemical and Pharmaceutical Bulletin, 55*(11), 1640–1642.

Zhang, L., Dong, R., Wang, Y., Wang, L., Zhou, T., Jia, D., et al. (2021). The anti-breast cancer property of physcion via oxidative stress-mediated mitochondrial apoptosis and immune response. *Pharmaceutical Biology, 59*(1), 303–310.

CHAPTER FOUR

Benzophenones from African plants to fight cancers and cancer drug resistance

Vaderament-A. Nchiozem-Ngnitedem[a,*], Daniel Buyinza[b], and Victor Kuete[c]
[a]Institute of Chemistry, University of Potsdam, Potsdam-Golm, Germany
[b]Department of Chemistry, Kabale University, Kabale, Uganda
[c]Department of Biochemistry, Faculty of Science, University of Dschang, Dschang, Cameroon
*Corresponding author. e-mail address: n.vaderamentalexe@gmail.com

Contents

Abstract

In response to the cancer threat, several interventions have been sought including the search for small molecules from natural sources that may arrest cancer at any of its different cycle stages. Such molecules include the benzophenones from fungi or higher plants (Clusiaceae and Guttiferae families). This book chapter will systematically cover the emerging areas in the field of benzophenones and their polyprenylated benzophenone analogs isolated from herbal medicines found in Africa with application in cancer treatment. Initially, benzophenones have been introduced as chemical entities, then an overview of their biosynthesis has been introduced, followed by phytochemical characterization from various tissues in the plant kingdom, and lastly a comprehensive updated study on benzophenones and their prenylated derivatives against various cancer cell lines has been provided.

Abbreviations

BPS	benzophenone synthase
DMAPP	dimethylallyl pyrophosphate
HPLC	high-performance liquid chromatography
IC_{50}	half inhibitory concentration

Advances in Botanical Research, Volume 115
ISSN 0065-2296, https://doi.org/10.1016/bs.abr.2024.02.007

IPP isopentenyl pyrophosphate
MPLC medium pressure liquid chromatography
PIBs polyisoprenylated benzophenones
PKS polyketide synthases
PPBS polyprenylated benzophenones
VLC vacuum liquid chromatography

1. Introduction

A benzophenone is chemically a diphenylketone having structural diversity based on substitution and cyclization that characterize its bioactive properties. Over three hundred natural benzophenones of varying structures have been isolated (Wu, Long, & Kennelly, 2014). Natural benzophenones are nonpolar phenolic compounds having many prenyl functional groups that increase their hydrophobicity (Muñoz, Jancovski, & Kennelly, 2009; Santa-Cecília et al., 2012). The basic structure of benzophenones consists of a Ph_2CO or $(C_6H_5)_2CO$ (Fig. 1-I). The natural benzophenones have great structural diversity though the phenyl-keto-phenyl skeleton is a sharable feature. Ring-A is derived from the shikimic acid pathway and may have 1 or 2 substituents. While ring-B is derived from the acetate-malonate pathway. Prenylation and cyclization are common to this ring producing a variety of compounds with unique structures of polycyclic ring systems having either a bi-, tri-, and/or tetra-cycle (Wu et al., 2014). The plant-derived polyprenylated benzophenones (PPBS) tend to have a bicyclo [3.3.1]-nonane-2,4,10-trione

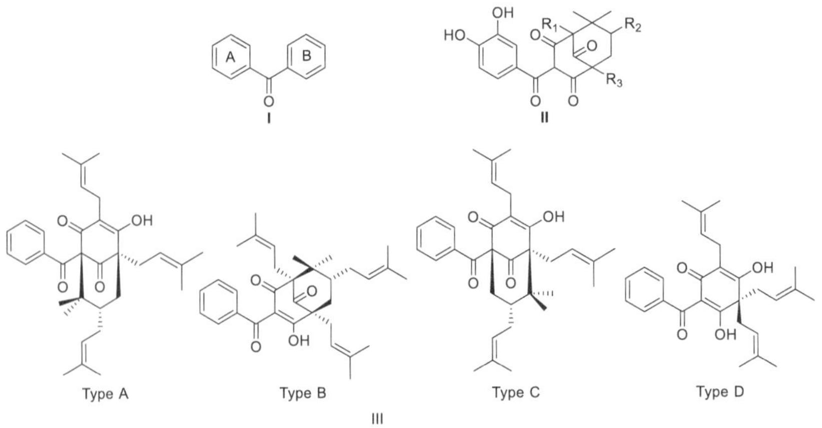

A B O I

HO OH O R1 R2 O R3 O O II

O OH O O Type A

O O OH Type B

O OH O O Type C

O OH O OH Type D

III

Fig. 1 (I) Basic structure of benzophenone; (II) General structure of polyisoprenylated benzophenones; and (III) Types of natural complex polyprenylated benzophenones.

core structure linked to a 13,14-dihydroxy substituted phenyl ring (Fig. 1-II). This planar structure having polar, carbonyl, and hydroxyl groups can either be simple or substituted with isoprene units (Muñoz et al., 2009). Such poly-isoprenylated benzophenones (PIBS) are a major intermediate in the biosynthetic pathway of xanthones common in the plant families like Clusiaceae. Wu et al., 2014 classify natural complex PPBS under types-A, -B, -C, and -D structures (Fig. 1-III).

Many benzophenones have been reported from higher plants or fungi. PIBS have been isolated from the Clusiaceae family, including the widely used medicinal herb, *Hypericum perforatum* L., commonly known as St. John's wort, traditionally used for its antidepressant activity (Muñoz et al., 2009). *Garcinia brasiliensis* (Mart.) Planch. and Triana (Syn. *Rheedia brasiliensis*) (Guttifeare), commonly known as bacupari, is also a rich source of PIBS. These plants have been used in traditional medicine for the treatment of urinary diseases and several types of tumors (Santa-Cecília et al., 2012).

The natural benzophenones exhibit a range of biological activities including antimicrobial (Li, Jiang, Guo, Zhang, & Che, 2008; Lokvam, Braddock, Reichardt, & Clausen, 2000; Zendah et al., 2012), anti-inflammatory (Chen, Ting, Hwang, & Chen, 2009; Xin et al., 2012), antioxidant (Acuña et al., 2010; Baggett et al., 2005), antiviral (Abdel-Mageed et al., 2014), anti-infective (Ochora et al., 2022; Osman et al., 2020), anti-diabetic (Liu et al., 2021) activities. Benzophenone and its derivatives such as 2,4-dihydroxybenzophenone and 2-hydroxy- 4-methoxybenzophenone are UV filters and have been used as ingredients for pharmaceuticals, insecticides, agricultural chemicals, and fragrances in medicine and industry and as UV light absorber in plastic and polymers. Benzophenones are also widely used in cosmetic products such as photostabilizers to protect the skin and hair from UV irradiation (Chen, Guo, Wang, & Zhang, 2015). However, benzophenones are also known as pervasive pollutants in the urban aquatic environment, found in both surface water and sediment in the order of parts per billion to parts per thousand and it is also a toxin in aquatic systems and a contact allergen. Consequently, they have been listed among endocrine-disrupter chemicals by the World Wildlife Fund and the Japanese Ministry of Environment (Chen et al., 2015).

2. Biosynthesis of benzophenones

From biosynthesis consideration, benzophenones are derived from benzoyl-CoA derivative which in turn originate from L-phenylalanine

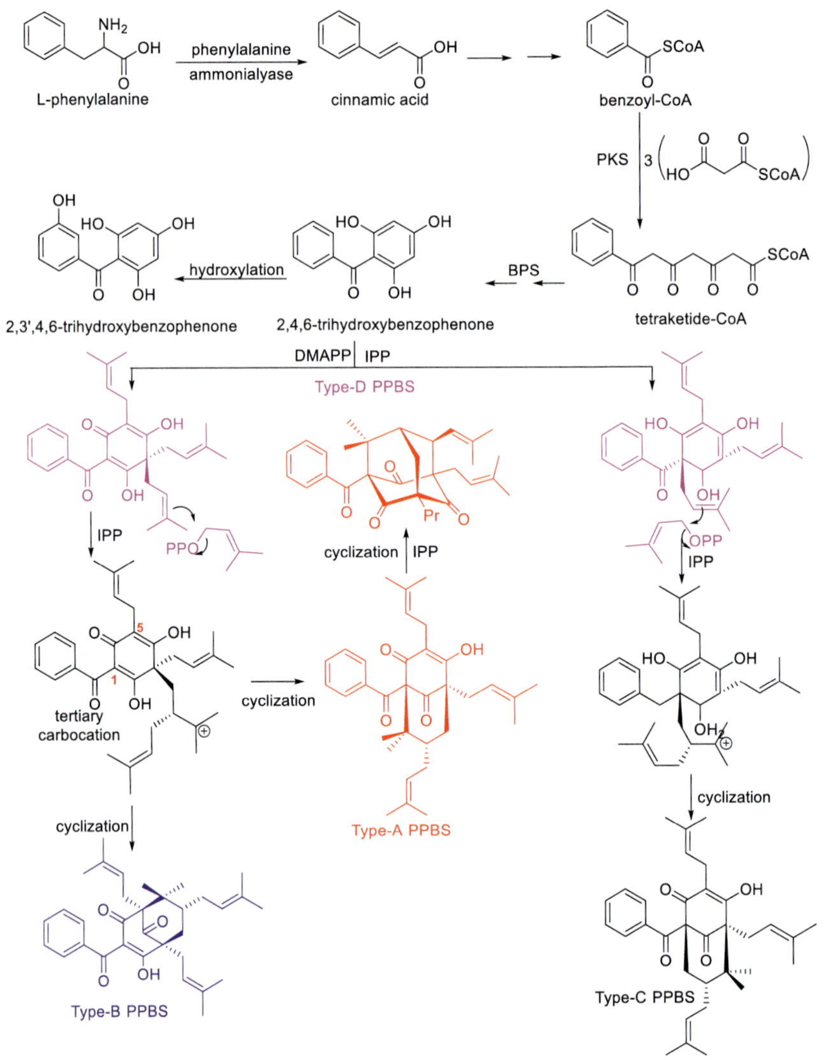

Scheme 1 Biosynthetic pathways of simple and of polyisoprenylated benzophenones. *DMAPP*, dimethylallyl pyrophosphate; *IPP*, isopentenyl pyrophosphate; *PPBS*, polyprenylated benzophenones; *BPS*, benzophenone synthase; *PKS*, polyketide synthases.

following a series of reactions (Scheme 1). Polyketide synthases enzymes then catalyze the sequential condensation of benzoyl-CoA with three molecules of malonyl-CoA to form a linear tetraketide-CoA. The latter undergoes a series of reactions intramolecular Claisen condensation to form 2,4,6-trihydroxybenzophenone (maclurin). Stereochemical selectivity over the course of this coupling is governed by the enzyme (benzophenone synthase (BPS)). The product of the BPS reaction, 2,4,6-trihydroxybenzophenone, and its hydroxylated derivatives can further undergo a series of prenylation, cyclization reactions using C_5 (isoprenyl) and C_{10} (geranyl) isoprenoid units as prenyl donors giving rise to various PIBS. In some cases, phlorobenzophenone can be coupled with dimethylallyl pyrophosphate in the presence of isopentenyl pyrophosphate yielding to the biosynthesis of various polyprenylated benzophenones (Types A–D). Then, the type-D PPBS, which just has two or three prenyls or geranyls in ring B is obtained. Cuesta-Rubio et al. reported that both type-A and -B PPBS are synthesized from a common precursor. An Attack of one of the geminal prenyl groups of this precursor on prenyl pyrophosphate gives the tertiary carbocation (TC). In this intermediate, attacking C-1 of TC on the pendant carbocation (or the corresponding pyrophosphate) would provide a type-A PPBS, whereas attacking C-5 of TC would provide a type-B PPBS. The 7-prenyl group of type-A PPBS may further repeat this route and attack its C-5 and obtain another adamantly type-A PPBS configuration. The TC intermediate is a diastereoisomer, and as a result, most type-A PPBS have an exoprenyl group and most type-B PPBS have endo-7-prenyl groups. In contrast, during the biosynthesis of type-C PPBS, the precursor requires another phloroglucinol moiety which should have one prenyl or geranyl group at C-1. The pending carbocation can then attack the C-3 location to obtain type-C PPBS, Scheme 1. Severe conditions are necessary to obtain this C-1 quaternary center precursor, and thus the occurrence of type-C PPBS is rarer than types-A, -B, and -D in nature (Wu et al., 2014).

2.1 Phytochemical characterization from a crude plant extract

PIBs are mainly found in various tissues of plants belonging to Clusiaceae and Hypericaceae families. The usual isolation techniques for natural products still apply in obtaining these components from their natural sources always involving several steps. These techniques include column chromatography, medium-pressure liquid chromatography, vacuum column chromatography, and preparative HPLC as a single technique or in combination. However, high-speed counter-current chromatography, free of irreversible adsorption, can separate multiple natural products from plant extracts, with minimal losses and lower solvent usage. In addition, bioassay-guided fractionation through methylation reactions directly on the floral resins has been used to identify PIBs. In the extraction, partition, separation, isolation, and purification process, different polar and nonpolar solvents are used to get the different secondary metabolites. This is also dependent on the nature of the purification technique employed either the traditional or the recent analytical techniques (MPLC, semi-preparative HPLC, VLC) (Kumar, Sharma, & Chattopadhyay, 2013).

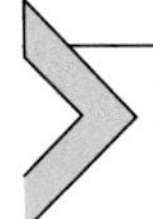

3. Cell lines used in the screening of the cytotoxicity of benzophenones from African medicinal plants

Several cancer cell lines were used to assess the cytotoxicity of benzophenones isolated from African medicinal plants. They include lung adenocarcinoma cells (A549), human breast carcinoma cells (MCF-7; SKBr-3; MDAMB-231-*pcDNA3* and its resistant subline MDA-MB-231-*BCRP* clone 23), human ovarian carcinoma cells (SKOV-3), leukemia cells (THP-1, JURKAT, CCRF-CEM and its multidrug-resistant (MDR) P-glycoprotein-over-expressing subline CEM/ADR5000), colorectal cancer cells (Caco-2, HT-29, HCT116, HCT116 $p53^{+/+}$ and its knockout clone HCT116 $p53^{-/-}$), murine alveolar macrophage cells (MH-S), prostate carcinoma cells (DU145 and PC-3), cervical carcinoma cells (HeLa), glioblastoma cells (U87MG and its resistant subline U87MG.Δ*EGFR*), stomach cancer cells (BGC-823), and human epidermoid carcinoma cells (A431) (Table 1).

4. Cytotoxic benzophenones from the African medicinal plants towards drug sensitive and MDR cancer cells

Our attempt to summarize the main finding of different research groups engaged in the search for new biologically active benzophenones from African medicinal plants against cancer revealed that nearly thirteen natural benzophenones have been disclosed so far, and most of them originate from *Garcinia* plants with few exceptions from *Pentadesma* and *Hypericum* genus (Fig. 2, and Table 1). Out of thirteen benzophenones recorded, twelve were natural PIBs that exhibited significant cytotoxic activity against some cancer cell lines such as drug sensitive leukemia (CCRF-CEM) and its MDR subline (CEM/ADR5000), breast cancer cells (MDAMB-231-*pcDNA3*) and its resistant subline (MDA-MB-231-*BCRP* clone 23), colon cancer cells (HCT116 $p53^{+/+}$), and its knockout clone (HCT116 $p53^{-/-}$), glioblastoma cells (U87MG) and its resistant subline (U87MG.Δ*EGFR*), and others. They include 2,2′,5,6′-tetrahydroxybenzophenone (**1**), isogarcinol (**2**), isoxanthochymol (**3**),

Table 1 Cytotoxic benzophenones from African medicinal plants and their effects on sensitive and drug-resistant cancer cell lines.

Compounds names	Source (Plant part)	Country of plants origin	Cancer Cell Lines and IC_{50} values (μM) and degree of resistance (D.R.) in bracket	References
2,2′,5,6′-Tetrahydroxybenzophenone (1)	*Hypericum lanceolatum* (Leaves)	Cameroon	40.28 (CCRF-CEM) *vs* 66.30 (CEM/ADR5000) (D.R.:1.65); 4.59 (HL60) *vs* 67.32 (HL60AR) (D.R.:14.67); 7.97 (MDA-MB231-*pcDNA*) *vs* 44.11 (MDA-MB231/*BCRP*) (D.R.: 5.53); 8.33 (HCT116 $p53^{+/+}$) *vs* 94.47 (HCT116 $p53^{-/-}$) (D.R.:11.34); 50.61 (U87MG) *vs* 11.99 (U87MG.Δ*EGFR*) (D.R.: 0.24)	Kuete et al. (2013)
30-*Epi*-cambogin (12)	*Pentadesma butyracea* (Leaves)	Cameroon	4.45 (BGC-823); 10.64 (HeLa); 13.60 (A549); >16.6 (THP-1); 9.4 (HCT116)	Tala et al. (2013); Wabo, Kikuchi, Katou, Tane, and Oshima (2010)
7-*Epi*-clusianone (10)	*Garcinia preussii* (Fruits)	Cameroon	33.4 (DU145); 49.5 (HeLa); 32.6 (HT-29); 38.2 (A431)	Biloa et al. (2014)
Epunctanone (5)	*Garcinia epunctata* (Stem bark)	Cameroon	11.81 (CCRF-CEM) *vs* 8.66 (CEM/ADR5000) (D.R.:0.73); 26.66 (MDA-MB231-*pcDNA*) *vs* 5.88 (MDA-MB231/*BCRP*) (D.R.: 0.22); 9.74 (HCT116 $p53^{+/+}$) *vs* 4.84 (HCT116 $p53^{-/-}$) (D.R.:0.50); 8.63 (U87MG) *vs* 7.46 (U87MG.Δ*EGFR*) (D.R.: 0.86)	Mbaveng et al. (2018)

(continued)

Table 1 Cytotoxic benzophenones from African medicinal plants and their effects on sensitive and drug-resistant cancer cell lines. (*cont'd*)

Compounds names	Source (Plant part)	Country of plants origin	Cancer Cell Lines and IC_{50} values (μM) and degree of resistance (D.R.) in bracket	References
Garcimangosone (11)	*Garcinia preussii* (Leaves)	Cameroon	>150 (DU145); >150 (HeLa); >150 (HT-29); >150 (A431)	Biloa et al. (2014)
Garcimultiflorone E (7)	*Garcinia preussii* (Fruits)	Cameroon	14.5 (DU145); 21.0 (HeLa); 15.9 (HT-29); 17.1 (A431)	Biloa et al. (2014)
Garciniagifolone (6)	*Garcinia preussii* (Fruits)	Cameroon	7.7 (DU145); 14.6 (HeLa); 7.4 (HT-29); 9.9 (A431)	Biloa et al. (2014)
Garcinialiptone B (8)	*Garcinia preussii* (Fruits)	Cameroon	11.5 (DU145); 13.0 (HeLa); 14.0 (HT-29); 11.0 (A431)	Biloa et al. (2014)
Garcinol (9)	*Garcinia preussii* (Fruits): *Garcinia nobilis* (Leaves)	Cameroon	14.9 (DU145); 16.4 (HeLa); 16.1 (HT-29); 9.9 (A431); 13.77 (MCF-7)	Biloa et al. (2014); Ngwoke et al. (2017)
Guttiferone E (4)	*Garcinia punctate* (Stem bark)	Cameroon	6.86 (CCRF-CEM) *vs* 13.57 (CEM/ADR5000) (D.R.:1.98); 11.69 (HL60) *vs* 11.69 (HL60AR) (D.R.:1.0); 11.69 (MDA-MB231-*pcDNA*) *vs* 13.92 (MDA-MB231/*BCRP*) (D.R.: 1.19); 12.74 (HCT116 $p53^{+/+}$) *vs* 7.87 (HCT116 $p53^{-/-}$) (D.R.:0.62); 7.87 (U87MG) *vs* 3.39 (U87MG.Δ*EGFR*) (D.R.: 0.43)	Kuete et al. (2013)

Isogarcinol (2)	*Hypericum lanceolatum* (Leaves); *Garcinia nobilis* (Leaves)	Cameroon	1.38 (CCRF-CEM) *vs* 6.28 (CEM/ADR5000) (D.R.:4.55); 4.70 (HL60) *vs* 4.78 (HL60AR) (D.R.:1.02); 14.70 (MDA-MB231-*pcDNA*) *vs* 4.52 (MDA-MB231/*BCRP*) (D.R.: 0.31); 0.86 (HCT116 $p53^{+/+}$) *vs* 7.31 (HCT116 $p53^{-/-}$) (D.R.:8.46); 7.97 (U87MG) *vs* 6.58 (U87MG.Δ*EGFR*) (D.R.: 0.83); 10.76 (MCF-7)	Kuete et al. (2013); Ngwoke et al. (2017)
Isoxanthochymol (3)	*Garcinia punctate* (Stem bark)	Cameroon	9.55 (CCRF-CEM) *vs* 10.33 (CEM/ADR5000) (D.R.:1.08); 8.92 (HL60) *vs* 8.92 (HL60AR) (D.R.:1.0); 6.30 (MDA-MB231-*pcDNA*) *vs* 3.42 (MDA-MB231/*BCRP*) (D.R.: 0.54); 3.24 HCT116 $p53^{+/+}$ *vs* 3.62 (HCT116 $p53^{-/-}$) (D.R.:1.12); 6.40 (U87MG) *vs* 3.12 (U87MG.Δ*EGFR*) (D.R.: 0.49)	Kuete et al. (2013)

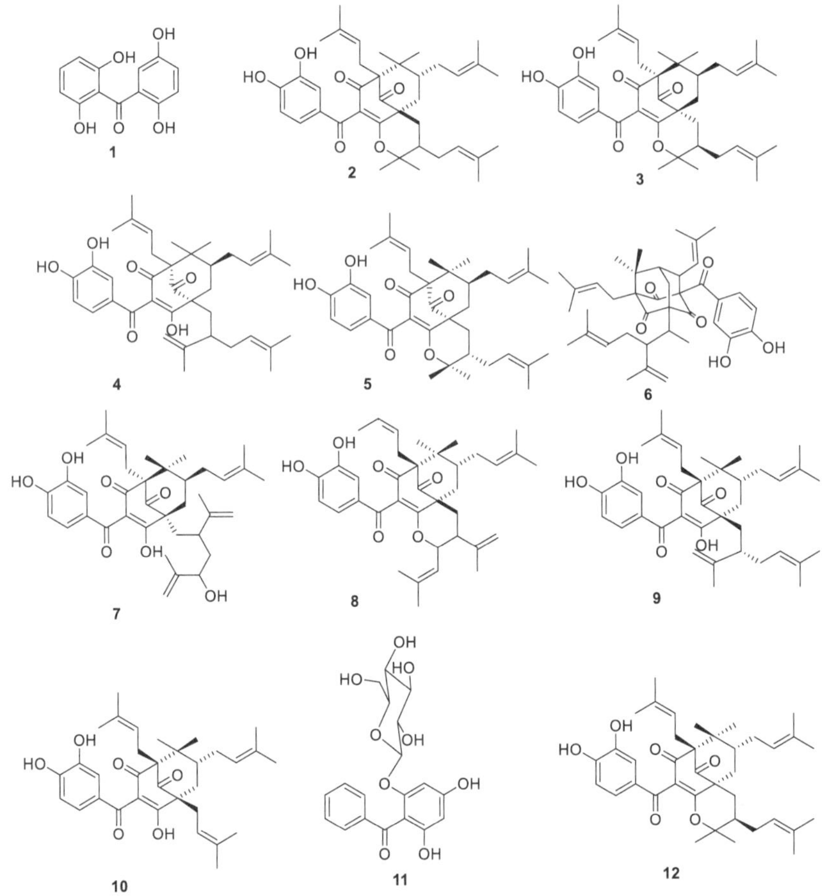

Fig. 2 Chemical structures of cytotoxic benzophenones isolated from African medicinal plants: **1:** 2,2′,5,6′-Tetrahydroxybenzophenone; **2:** isogarcinol; **3:** isoxanthochymol; **4:** guttiferone E; **5:** epunctanone; **6:** garciniagifolone; **7:** garcimultiflorone E; **8:** garcinialiptone B; **9:** garcinol; **10:** 7-*epi*-clusianone; **11:** garcimangosone; **12:** 30-*epi*-cambogin.

guttiferone E (**4**), epunctanone (**5**), garciniagifolone (**6**), garcimultiflorone E (**7**), garcinialiptone B (**8**), garcinol (**9**), 7-*epi*-clusianone (**10**), garcimangosone (**11**), and 30-*epi*-cambogin (**12**) (Table 1 and Fig. 2). As part of their investigation campaign in the search of cytotoxic agents from Cameroonian medicinal plants, Kuete et al., reported the cytotoxicity potency of 2,2′,5,6′-tetrahydroxybenzophenone (**1**), isogarcinol (**2**), isoxanthochymol (**3**) and guttiferone E (**4**) from *Garcinia punctata* (Stem bark) against the aforementioned cancer cell lines. Collateral sensitivity (the ability of compounds to kill MDR cells selectively over the parental cells) was observed for **1–4**. Out of the tested samples (**2**), displayed the best activity with IC_{50} ranging from 1 μM (against HCT116 *p53*$^{+/+}$) to 14.70 μM (towards MDAMB-231-*pcDNA3*). Further, the mechanistic study demonstrated that compounds **2–4** strongly induced apoptosis in CCRF–CEM cells *via* caspases 3/7, caspase 8, and caspase 9 activation and disruption of mitochondrial membrane potential (MMP) (Kuete et al., 2013). Similarly, isoxanthochymol (**3**) obtained from the leaves of *Garcinia polyantha* displayed cytotoxic activities against HeLa (cervix), A549 (lung), and PC-3 (prostate) with IC_{50} values of 16.7, 34.0, >82.9 μM, respectively (Fig. 2 and Table 1). The cytotoxic activity was found to be related to apoptosis induction. Compound **3** further exhibited a strong inhibitory effect on THP-1 cell proliferation with IC_{50} inhibition values of 2.5 μM, inducing apoptotic cell death characterized by DNA fragmentation (Lannang et al., 2014). Using MTT assays, it was demonstrated that, guttiferone E (**4**) was cytotoxic against five human cancer cell lines with IC_{50} values of 20.4 μM (human breast carcinoma cell SKBr-3), 66.7 μM (human ovarian carcinoma SKOV-3), 16.2 μM (lung A549 adenocarcinoma), 9.1 μM (human T cell leukemia), 22.3 μM (human epithelial colorectal adenocarcinoma cell CaCo-2), 36.2 μM (murine alveolar macrophage cell MH-S) (Dzoyem et al., 2022). It was also reported that epunctanone (**5**) from the stem bark *Garcinia epunctata* exhibited antiproliferative activity against a panel of cancer cell lines, including drug sensitive and their MRD sublime (Fig. 2 and Table 1) (Mbaveng et al., 2018). The IC_{50} values ranged from 5.88 μM (MDA-MB-231-*BCRP*) to 26.66 μM (MDAMB-231-*pcDNA3*). Hypersensitivity (collateral sensitivity or D.R.: < 1) of P-gp-overexpressing CEM/ADR5000 cells and *BCRP*-expressing MDA-MB-231 cells was observed for **5** (Table 1). Epunctanone further induced apoptosis in CCRF-CEM leukemia cells through alteration of MMP, increase ROS production, and induced ferroptotic cell death (Mbaveng et al., 2018). Other medicinal spices *Garcinia preussii* produced PIBs such as **6**, **7**, **8**, **9**, and **10**, which displayed cytotoxic activity against DU145, HeLa, HT-29, and A431 with IC_{50} values as low as 7.7 to 49.5 μM, Fig. 2, and Table 1 (Biloa et al., 2014). In a comparable study, Sales and co-workers further demonstrates that 7-*epi*-clusianone (**10**) had antineoplastic effect in a glioblastoma cell line (U251MG and U138MG). At low concentrations, 7-epiclusianone altered cell cycle progression, and decreased the S and G2/M populations while at higher concentrations increased the number of cells at sub-G1, in harmony with the increase of apoptotic cells (Sales et al., 2015). The cytotoxicity of phytochemicals will be appreciated according to the classification criteria established as follows: outstanding activity ($IC_{50} \leq 0.5$ μM), excellent activity

(0.5 < IC_{50} ≤ 2 μM), very good activity (2 < IC_{50} ≤ 5 μM), good activity (5 < IC_{50} ≤ 10 μM), average activity (10 < IC_{50} ≤ 20 μM), weak activity (20 < IC_{50} ≤ 60 μM), very weak activity (60 < IC_{50} ≤ 150 μM), and not active (IC_{50} > 150 μM) (Kuete, 2025). This basis of classification will be used to discuss the cytotoxicity of benzophenones isolated from African medicinal plants. From the data summarized in Table 1, it appears that excellent cytotoxic effects were obtained with compound **2** against CCRF-CEM cells and HCT116 *p53*$^{+/+}$ cells; very good cytotoxic effects were obtained with compound **1** against HL60 cells and HL60AR cells, **2** and **3** against MDA-MB231/*BCRP* cells, **3** against HCT116 *p53*$^{+/+}$ cells, **3** and **5** against HCT116 *p53*$^{-/-}$ cells, **3** and **4** against U87MG.Δ*EGFR* cells, and **12** against BGC-823 cells; good cytotoxic effects were obtained with compounds **1** and **3** against MDA-MB231-*pcDNA* cells, **1** and **5** against HCT116 *p53*$^{+/+}$ cells, **2**, **3**, and **5** against CEM/ADR5000 cells, **2** and **3** against HCT116 *p53*$^{-/-}$ cells, **2** and **3** against U87MG cells, **2** and **5** against U87MG.Δ*EGFR* cells, **3** and **4** against CCRF-CEM cells, **3** against HL60 cells, **3** and **5** against U87MG cells, **5** against MDA-MB231/*BCRP* cells, **6** against DU145 cells and HT-29 cells, **6** and **9** against A431 cells, and **12** against HCT116 cells. Other samples displayed moderate to non-antiproliferative effects on the studied cancer cell lines. Interestingly, collateral sensitivity of MDA-MB231/*BCRP* cells *vs* MDA-MB231-*pcDNA* cells to compounds **2** (D.R: 0.31), and **5** (D.R: 0.22), U87MG.Δ*EGFR* cells *vs* U87MG cells to compounds **2** (D.R: 0.83), **3** (D.R: 0.49), **4** (D.R: 0.43), and **5** (D.R: 0.86), CEM/ADR5000 cells *vs* CCRF-CEM cells to compounds **5** (D.R: 0.73), and HCT116 *p53*$^{-/-}$ cells *vs* HCT116 *p53*$^{+/+}$ cells to **5** (D.R: 0.50) was achieved; meanwhile, normal sensitivity HL60AR cells *vs* HL60 cells to compound **2** (D.R: 1.02), and **3** (D.R: 1.00), CEM/ADR5000 cells *vs* CCRF-CEM cells to **3** (D.R: 1.08), and HCT116 *p53*$^{-/-}$ cells *vs* HCT116 *p53*$^{+/+}$ cells to **3** (D.R: 1.12) was also recorded (Table 1). This is an indication that compounds **1**, **2**, **3**, **4**, **5**, **6**, **9**, and **12** are potent cytotoxic benzophenones that could be used to fight cancer meanwhile **2**, **3**, **4**, and **5** are more suitable to combat cancer drug resistance.

5. Conclusion

This study aimed to investigate the arsenal of natural benzophenones from African folklore medicine as potential cytotoxic agents against cancer. Overall, our finding unveiled that, Clusiaceae family is an important source of bioactive prenylated benzophenones with cytotoxic/antineoplastic potencies. The results further indicated that the most potent cytotoxic benzophenones were 2,2′,5,6′-tetrahydroxybenzophenone (**1**), isogarcinol (**2**), isoxanthochymol (**3**), guttiferone E (**4**), epunctanone (**5**), garciniagifolone (**6**), garcinol (**9**), and 30-*epi*-cambogin (**12**). Those with the ability to combat cancer drug resistance included **2**, **3**, **4**, and **5**. These natural benzophenones could therefore represent valuable cytotoxic molecules, that can be further investigated as potent anticancer drugs to fight sensitive and resistant phenotypes. Finally, despite being natural products, the toxicity of these PIBs against normal cell lines cannot be ignored. This can be overcome through structural modification of biologically active PIBs aimed at increasing activity, in addition to decreasing toxicity, as well as improving other pharmacological potencies.

References

Abdel-Mageed, W. M., Bayoumi, S. A. H., Chen, C., Vavricka, C. J., Li, L., Malik, A., ... Zhang, L. (2014). Benzophenone C-glucosides and gallotannins from mango tree stem bark with broad-spectrum anti-viral activity. *Bioorganic and Medicinal Chemistry, 22*(7), 2236–2243.

Acuña, U. M., Figueroa, M., Kavalier, A., Jancovski, N., Basile, M. J., & Kennelly, E. J. (2010). Benzophenones and biflavonoids from *Rheedia edulis*. *Journal of Natural Products, 73*(11), 1775–1779.

Baggett, S., Protiva, P., Mazzola, E. P., Yang, H., Ressler, E. T., Basile, M. J., ... Kennelly, E. J. (2005). Bioactive benzophenones from *Garcinia xanthochymus* fruits. *Journal of Natural Products, 68*(3), 354–360.

Biloa, M. B., Ho, R., Meli Lannang, A., Cressend, D., Perron, K., Nkengfack, A. E., ... Cuendet, M. (2014). Isolation and biological activity of compounds from *Garcinia preussii*. *Pharmaceutical Biology, 52*(6), 706–711.

Chen, D. Y., Guo, X. F., Wang, H., & Zhang, H. S. (2015). The natural degradation of benzophenone at low concentration in aquatic environments. *Water Science and Technology, 72*(4), 503–509.

Chen, J. J., Ting, C. W., Hwang, T. L., & Chen, I. S. (2009). Benzophenone derivatives from the fruits of *Garcinia multiflora* and their anti-inflammatory activity. *Journal of Natural Products, 72*(2), 253–258.

Dzoyem, J. P., Pinnapireddy, S. R., Fouotsa, H., Brüßler, J., Runkel, F., & Bakowsky, U. (2022). Liposome-encapsulated bioactive guttiferone E exhibits anti-inflammatory effect in lipopolysaccharide-stimulated MHS macrophages and cytotoxicity against human cancer cells. *Mediators of Inflammation, 2022*, 8886087.

Kuete, V. (2025). Chapter Four-African medicinal plants and their derivative as the source of potent anti-leukemic products: Rationale classification of naturally occurring anticancer agents. *Advances in Botanical Research, 113*, https://doi.org/10.1016/bs.abr.2023.12.010.

Kuete, V., Tchakam, P. D., Wiench, B., Ngameni, B., Wabo, H. K., Tala, M. F., ... Efferth, T. (2013). Cytotoxicity and modes of action of four naturally occuring benzophenones: 2,2′,5,6′-Tetrahydroxybenzophenone, guttiferone E, isogarcinol and isoxanthochymol. *Phytomedicine, 20*(6), 528–536.

Kumar, S., Sharma, S., & Chattopadhyay, S. K. (2013). The potential health benefit of polyisoprenylated benzophenones from *Garcinia* and related genera: Ethnobotanical and therapeutic importance. *Fitoterapia, 89*(1), 86–125.

Lannang, A. M., Tatsimo, S. J. N., Fouotsa, H., Dzoyem, J. P., Saxena, A. K., & Sewald, N. (2014). Cytotoxic compounds from the leaves of *Garcinia polyantha. Chemistry & Biodiversity, 11*(6), 975–981.

Li, E., Jiang, L., Guo, L., Zhang, H., & Che, Y. (2008). Pestalachlorides A-C, antifungal metabolites from the plant endophytic fungus *Pestalotiopsis adusta. Bioorganic and Medicinal Chemistry, 16*(17), 7894–7899.

Liu, B., Chen, N., Chen, Y., Xiang Shen, J., Jing Xu, Y., & Ji, Y. B. (2021). A new benzophenone with biological activities purified from *Aspergillus fumigatus* SWZ01. *Natural Product Research, 35*(24), 5710–5719.

Lokvam, J., Braddock, J. F., Reichardt, P. B., & Clausen, T. P. (2000). Two poly-isoprenylated benzophenones from the trunk latex of *Clusia grandiflora* (Clusiaceae). *Phytochemistry, 55*(1), 29–34.

Mbaveng, A. T., Fotso, G. W., Ngnintedo, D., Kuete, V., Ngadjui, B. T., Keumedjio, F., ... Efferth, T. (2018). Cytotoxicity of epunctanone and four other phytochemicals isolated from the medicinal plants *Garcinia epunctata* and *Ptycholobium contortum* towards multi-factorial drug resistant cancer cells. *Phytomedicine, 48*, 112–119.

Muñoz, A. U., Jancovski, N., & Kennelly, E. J. (2009). Polyisoprenylated benzophenones from clusiaceae: Potential drugs and lead compounds. *Current Topics in Medicinal Chemistry, 9*, 1560–1580.

Ngwoke, K. G., Orame, N., Liu, S., Okoye, F. B. C., Daletos, G., & Proksch, P. (2017). A new benzophenone glycoside from the leaves of *Mitracarpus villosus. Natural Product Research, 31*(20), 2354–2360.

Ochora, D. O., Kakudidi, E., Namukobe, J., Heydenreich, M., Coghi, P., Yang, L. J., ... Yenesew, A. (2022). A new benzophenone, and the antiplasmodial activities of the constituents of *Securidaca longipedunculata* fresen (Polygalaceae). *Natural Product Research, 36*(11), 2758–2766.

Osman, A. G., Ali, Z., Fantoukh, O., Raman, V., Kamdem, R. S. T., & Khan, I. (2020). Glycosides of ursane-type triterpenoid, benzophenone, and iridoid from *Vangueria agrestis (Fadogia agrestis)* and their anti-infective activities. *Natural Product Research, 34*(5), 683–691.

Sales, L., Pezuk, J. A., Borges, K. S., Sol Brassesco, M., Scrideli, C. A., Tone, L. G., ... De Oliveira, J. C. (2015). Anticancer activity of 7-epiclusianone, a benzophenone from *Garcinia brasiliensis*, in glioblastoma. *BMC Complementary and Alternative Medicine, 15*(1), 1–8.

Santa-Cecília, F. V., Santos, G. B., Fuzissaki, C. N., Derogis, P. B. M. C., Freitas, L. A. S., Gontijo, V. S., ... Santos, M. H. D. (2012). 7-Epiclusianone, the natural prenylated benzophenone, inhibits superoxide anions in the neutrophil respiratory burst. *Journal of Medicinal Food, 15*(2), 200–205.

Tala, M. F., Wabo, H. K., Zeng, G. Z., Ji, C. J., Tane, P., & Tan, N. H. (2013). A prenylated xanthone and antiproliferative compounds from leaves of *Pentadesma butyracea. Phytochemistry Letters, 6*(3), 326–330.

Wabo, H. K., Kikuchi, H., Katou, Y., Tane, P., & Oshima, Y. (2010). Xanthones and a benzophenone from the roots of *Pentadesma butyracea* and their antiproliferative activity. *Phytochemistry Letters, 3*(2), 104–107.

Wu, S. B., Long, C., & Kennelly, E. J. (2014). Structural diversity and bioactivities of natural benzophenones. *Natural Product Reports, 31*(9), 1158–1174.

Xin, W. B., Man, X. H., Zheng, C. J., Jia, M., Jiang, Y. P., Zhao, X. X., ... Qin, L. P. (2012). Prenylated phloroglucinol derivatives from *Hypericum sampsonii. Fitoterapia, 83*(8), 1540–1547.

Zendah, I., Riaz, N., Nasr, H., Frauendorf, H., Schüffler, A., Raies, A., & Laatsch, H. (2012). Chromophenazines from the terrestrial Streptomyces sp. Ank 315. *Journal of Natural Products, 75*(1), 2–8.

CHAPTER FIVE

Flavonoids from African medicinal plants as potential pharmaceuticals to tackle cancers and their refractory phenotypes

Jenifer R.N. Kuete[a] and Victor Kuete[b,*]
[a]Department of Chemistry, Faculty of Science, University of Dschang, Dschang, Cameroon
[b]Department of Biochemistry, Faculty of Science, University of Dschang, Dschang, Cameroon
*Corresponding author. e-mail address: kuetevictor@yahoo.fr

Contents

Advances in Botanical Research, Volume 115
ISSN 0065-2296, https://doi.org/10.1016/bs.abr.2024.02.009

Abstract

Flavonoids are polyphenolic secondary metabolites of plants, with many of them having pharmacological activities against a variety of health disorders including chronic inflammation, cancer, cardiovascular complications, and hypoglycemia. In the present report, we have compiled the cytotoxicity of 52 flavonoids isolated from African medicinal plants on various human cancer cell lines. They include chalcones, flavones, flavonols, flavanones, biflavonoids and flavonolignans. The most active chalcones were 4-hydroxylonchocarpin (**7**), cardamomin (**8**), isobavachalcone (**9**), licoagrochalcone A (**10**), and poinsettifolin B (**11**), while the most active flavones were 3,4′,5-trihydroxy-6″,6″-dimethylpyrano[2,3-*g*]flavone (**12**), canniflavone (**19**), ciyrsiliol (**20**), and 6,8-diprenyleriodictyol (**27**). The best flavanones were identified as 5,4′-dihydroxy-7-methoxyflavanone (**43**), dorsmanin F (**46**), and naringenin (**47**). 7,7″-Di-*O*-methylchamaejasmin (**48**) and hydnocarpin (**52**) were identified as the most active biflavonoid and flavonolignan, respectively. These active flavonoids generally displayed apoptotic cell death, mediated by caspase activation, alteration of mitochondrial membrane potential (MMP), and enhanced reactive oxygen species (ROS) production in cancer cells. The most potent flavonoids herein should be further explored to develop novel cytotoxic products to fight cancer.

Abbreviations

4CL	4-coumaroyl-coenzyme A ligase.
ANR	anthocyanidin reductase.
ANS	anthocyanidin synthase.
AS	anthocianin synthase.
AT	acyltransferase.
C4H	cinnamate 4-hydroxylase.
CGT	C-glycosyltransferase.
CH4′GT	chalcone 4′-*O*-glucosyltransferase.
CHI	chalcone isomerase.
CHR	chalcone reductase.
CHS	chalcone synthase.
CoA	co-enzyme.
D.R.	degree of resistance.
DFR	dihydroflavonol 4-reductase.
DHK	dihydrokaempferol.
DHM	dihydromyricetin.
DHQ	dihydroquercetin.
F2H	flavanone-2-hydroxylase.
F3′5′H	flavanone 3′,5′-hydroxylase.
F3′H	flavanone 3′-hydroxylase.
F3H	flavanone 3β-hydroxylase.
F3H	flavanone 3β-hydroxylase.
FLS	flavonol synthase.
FNS	flavone synthase.
GTs	glucosyltransferases.
HID	hydroxyisoflavanone dehydratase.

IC_{50}	inhibitory concentration 50.
IFS	isoflavone synthase.
IFS	isoflavonoid synthase.
LAR	Leucoanthocyanidin reductase.
PAL	phenylalanine amonialyase.
PPO	polyphenol oxidase.
STS	stilbene synthase.
THC	4,2′,4′,6′-tétrahydroxychalcone.
UF3GT	UDP glucose flavonoid 3-O-glucosyltransferase.

1. Introduction

Flavonoids are a class of polyphenolic secondary metabolites found in plants, bearing the general structure of a 15-carbon skeleton, which consists of two phenyl rings (A and B) and a heterocyclic ring (C), the ring containing the embedded oxygen. They are ubiquitous in photosynthesizing cells and are commonly found in fruit, vegetables, nuts, seeds, stems, flowers, tea, wine, propolis, and honey (Cushnie & Lamb, 2005). They include flavonoids or bioflavonoids, isoflavonoids, and neoflavanoids (Ngameni et al., 2013). However, in this review, we will focus on flavonoids or biflavonoids. The antiproliferative potential of isoflavonoids from African medicinal plants is discussed in Chapter 6. The pharmacological activities of flavonoids as well as associated mechanisms of action against a variety of health disorders including chronic inflammation, cancer, cardiovascular complications, and hypoglycemia have been reported (Kuete & Sandjo, 2012; Kuete, 2013; Ngameni et al., 2013; Wen et al., 2021). Many naturally occurring flavonoids have been documented for their anti-inflammatory (Al-Khayri et al., 2022; Dzoyem, McGaw, Kuete, & Bakowsky, 2017; Koorbanally, Randrianarivelojosia, Mulholland, van Ufford, & van den Berg, 2003; Njamen et al., 2003), anti-diabetic (Ndip, Tanih, & Kuete, 2013), antiviral (Kuete, Ngameni, Mbaveng, et al., 2010), and antimicrobial (Cushnie & Lamb, 2005; Jepkoech et al., 2021; Kuete, Ngameni, Tangmouo, et al., 2010; Mbaveng et al., 2008), as well as cytotoxic activities. The role of flavonoids as cytotoxic agents against human cancer cells has been largely described. Some well-known cytotoxic flavonoids derived from plants include luteolin (Lin, Shi, Wang, & Shen, 2008), licochalcones, xanthohumol, panduretin, and loncocarpine (Constantinescu & Lungu, 2021), quercetin (Ahmed, Moawad, Owis, AbouZid, & Ahmed, 2016), naringin (Yadav, Vats, Bano, Vashishtha, & Bhardwaj, 2022), and kaempferol (Nejabati & Roshangar, 2022). In the present chapter, the overview of the cytotoxic potential of flavonoids isolated from African medicinal plants against human cancer cell lines including the multidrug-resistant (MDR) phenotypes is provided.

2. Biosynthesis of flavonoids

The biogenetic mechanism which leads to the formation of the C15 carbon skeleton of these compounds has been described by many authors. The biosynthesis of flavonoids proceeds through two different pathways: the acetate pathway (ring A) and the shikimate pathway (ring B along with the linking chain (ring C) making the C6-C3 component). Ring A is synthesized from three malonyl-CoA molecules generated *via* the transformations of glucose while ring B is synthesized from 4-coumaroyl-CoA produced from phenylalanine *via* the shikimate pathway. Condensation of rings A and B generates chalcone, which subsequently undergoes isomerase-catalyzed cyclization to form flavanone. The latter compound is utilized as the starting compound for the synthesis of other flavonoids. All known flavonoid compounds, approximately 7000, have this common biosynthetic pathway and therefore share the same basic structural skeleton. Further structural diversity stems from the degree of unsaturation and oxidation of ring C: 2-phenylchromenyliums (anthocyanidins/anthocyanins); 2-phenylochromones (flavones, flavonols, flavanones, di-OH-flavonols); 2-phenylchromanes (flavans), flavan-3-ols, and flavan-3,4-diols (proanthocyanidins); chalcones/dihydrochalcones; 2-benzylidene coumaranones (aurones) and some others (Khoo, Azlan, Tang, & Lim, 2017). The later compounds undergo further modifications such as hydroxylation, glycosylation, or methylation resulting in an enormous range of flavonoid colors.

The first three steps of the phenylpropanoid pathway are called the general phenylpropanoid pathway (Scheme 1). In this pathway, phenylalanine, an aromatic amino acid, is converted to p-coumaroyl-CoA by the activity of phenylalanine ammonia lyase (PAL), cinnamic acid 4-hydroxylase (C4H), and 4-coumarate: CoA ligase (4CL) (Shah & Smith, 2020). PAL catalyzes the first step in the general phenylpropanoid pathway, namely the deamination of phenylalanine to trans-cinnamic acid. The second step in the general phenylpropanoid pathway involves the activity of C4H, a cytochrome P450 monooxygenase in plants, which catalyzes the hydroxylation of trans-cinnamic acid to generate p-coumaric acid. It is also

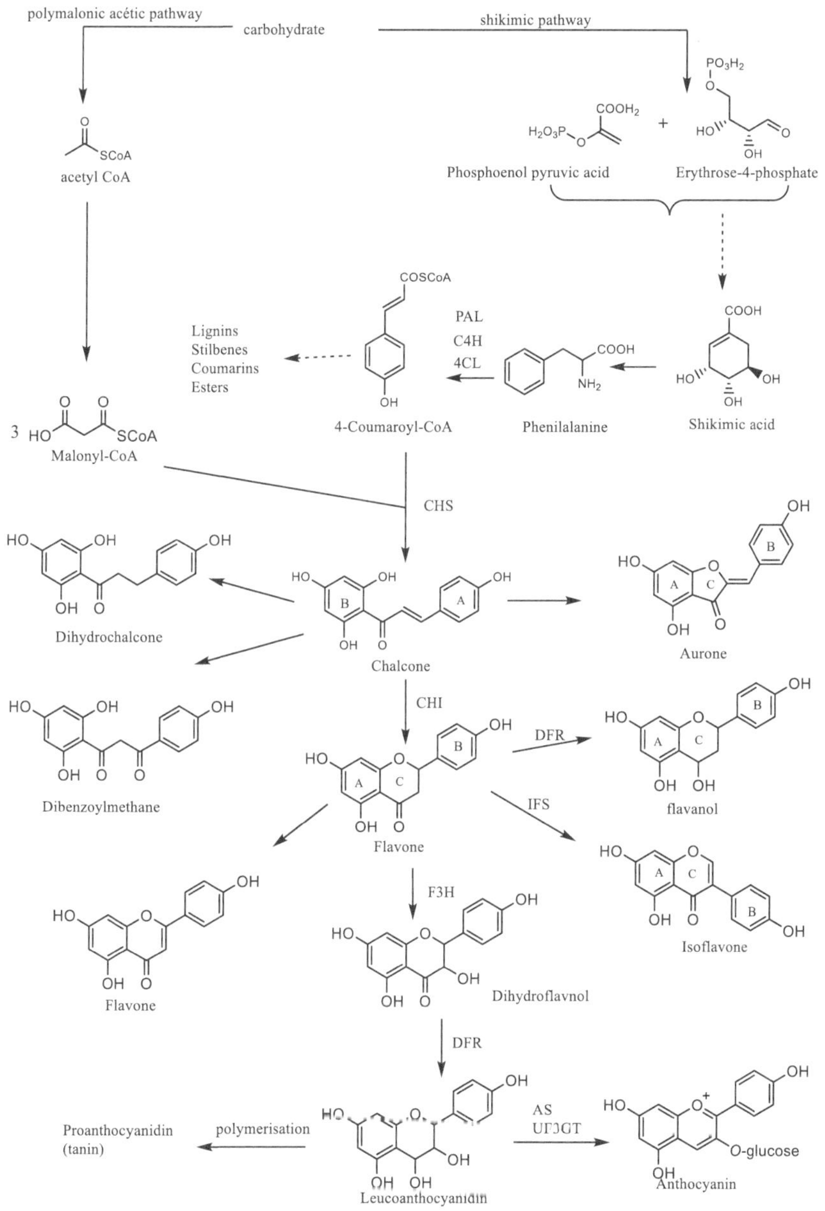

Scheme 1 A diagram of the flavonoid biosynthetic pathway (Belkacem, 2009; Herbert, 1981).

the first oxidation reaction in the flavonoid synthesis pathway. In the third step of the general phenylpropanoid pathway, 4CL catalyzes the formation of p-coumaroyl-CoA by the addition of a co-enzyme A (CoA) unit to p-coumaric acid. In plants, the 4CL gene usually exists as a family whose members exhibit mostly substrate specificity. In plants, 4CL activity is positively correlated with anthocyanin and flavonol content in response to stress, while PAL, C4H, and 4CL are often expressed in a coordinated fashion. The general phenylpropanoid pathway is common to all down-stream metabolites, such as flavonoids and lignin, and has previously been reported by Ngameni and collaborators (Ngameni et al., 2013). Due to their structural differences, flavonoids are mainly classified into different groups, such as flavanones, flavanols, flavones, and flavonols. Other classes of flavonoid compounds include isoflavones, biflavonoids, flavonolignans, prenylflavonoids, flavonoids, glycosidoesters, aurones, or chalcones. It is worth noting that individual flavonoid compound differs in ring substituent (s) formed in the process of hydroxylation (a large number of hydroxyl groups: –OH), methylation (methoxy groups being more common as substituents of ring B, compared with ring A), acylation, and glycosidation with mono- or oligosaccharides (*e.g.* glucose, galactose rhamnose, xylose, arabinose) at various ring position (Chen et al., 2022).

PAL, phenylalanine amonialyase; *C4H*, cinnamate 4-hydroxylase; *4CL*, 4-coumaroyl-coenzyme A ligase; *CHS*, chalcone synthase; *CHI*, chalcone isomerase; *F3H*, flavanone 3β-hydroxylase; *DFR*, dihydroflavonol 4-reductase; *IFS*, isoflavonoid synthase; *AS*, anthocianin synthase and *UF3GT*, UDP glucose flavonoid 3-O-glucosyltransferase.

2.1 Chalcone: The first key intermediate metabolite in flavonoid biosynthesis

The entry of 4-coumaroyl-CoA into the flavonoid biosynthetic pathway represents the onset of specific flavonoid synthesis, which begins with chalcone formation (Nabavi et al., 2020). One molecule of 4-coumaroyl-CoA and three molecules of malonyl-COA, derived from acetyl-CoA *via* acetyl-CoA carboxylase (ACCase) activity, generate naringenin chalcone (4,2′,4′,6′-tetrahydroxychalcone [THC] [chalcone]) by the action of chalcone synthase (CHS) (Wang et al., 2021). CHS, a polyketide synthase, is the key and first limiting enzyme in the flavonoid biosynthetic pathway. Chalcone reductase (CHR), a member of the aldo-keto reductase super-family, acts on an intermediate of the CHS reaction, catalyzing its dehydroxylation at C-6′, yielding isoliquiritigenin (4,2′,4′-trihydroxychalcone [deoxychalcone]) (Bomati, Austin, Bowman, Dixon, & Noel, 2005).

2.2 Stilbene biosynthesis: The first branch of the flavonoid biosynthetic pathway

Stilbene synthase (STS) also uses 4-coumaroyl-CoA and malonyl-CoA as substrates and catalyzes the formation of the stilbene backbone, such as resveratrol. The stilbene pathway is the first branch of the flavonoid biosynthetic pathway and exists only in a few plants, such as grapevine, pine, sorghum, and groundnut (Chong, Poutaraud, & Hugueney, 2009). However, STS generates a compound with a different C14 backbone (C_6-C_2-C_6) with the release of 4 carbon dioxide (CO_2), whereas CHS catalyzes the formation of C15 backbones (C_6-C_3-C_6), with only 3 molecules of CO_2 released (Sparvoli, Martin, Scienza, Gavazzi, & Tonelli, 1994). The biosynthesis of stilbene will be discussed in deep in Chapter 7.

2.3 Aurone biosynthesis: The bright yellow pigment pathway

Aurones, important yellow pigments in plants, comprise a class of flavonoids derived from chalcone. Aurone pigments produce a brighter yellow color than chalcones and are responsible for the golden color of some popular ornamental plants (Zhou, Wang, & Peng, 2009). First, chalcone 4′-*O*-glucosyltransferase (CH4′GT) catalyzes the formation of THC (4,2′,4′,6′-tetrahydroxychalcone [THC][chalcone]) 4′-*O*-glucoside from THC in the plant cytoplasm. The former is then transferred to the vacuole and converted to aureusidin 6-*O*-glucoside (aurone) by the action of aureusidine synthase (AS) (Ono & Nakayama, 2007). AS, a homolog of plant polyphenol oxidase (PPO), catalyzes 4-monohydroxylation or 3,4-dihydroxylation of ring B to produce aurone, followed by oxidative cyclization by oxygenation. Additionally, various classical substitution patterns, such as hydroxylation, methoxylation, and glycosylation, lead to the formation of a series of aurone compounds, with over 100 structures reported to date (Boucherle, Peuchmaur, Boumendjel, & Haudecoeur, 2017).

2.4 Flavanones: The central point of the flavonoid biosynthetic pathway

CHI catalyzes the intramolecular cyclization of chalcones to form flavanones in the cytoplasm, resulting in the formation of the heterocyclic ring

C in the flavonoid pathway (Lin, Singh, Moehninsi, & Navarre, 2021). In general, CHIs can be classified into two types in plants based on the substrate used. Type I CHIs, ubiquitous in vascular plants, are responsible for the conversion of THC (4,2′,4′,6′-tétrahydroxychalcone) to naringenin (Zhu et al., 2021). Type II CHIs are found primarily in leguminous plants and can use either THC or isoliquiritigenin to generate naringenin and liquiritigenin. In addition, naringenin can be converted to eriodictyol and pentahydroxyflavanone (two flavanones) by the action of flavanone 3′-hydroxylase (F3′H) and flavanone 3′,5′-hydroxylase (F3′5′H) in position C-3 and/or C-5 of ring B (Grotewold, 2006). The flavanones (naringenin, liquiritigenin, pentahydroxyflavanone, and eriodictyol) represent the central branch point in the flavonoid biosynthetic pathway, acting as common substrates for the flavone, isoflavone, and phlobaphene branches, as well as the downstream flavonoid pathway (Casas et al., 2016; Lin, Singh, & Moehninsi Navarre, 2021).

2.5 Biosynthesis of flavone

Flavone biosynthesis is an important branch of the flavonoid pathway in all higher plants. Flavones are produced from flavanones by flavone synthase (FNS); for example, naringenin, liquiritigenin, eriodictyol, and pentahydroxyflavanone can be converted to apigenin, dihydroxyflavone, luteolin, and tricetin, respectively (Dinkins et al., 2021; Qian et al., 2021; Zuk, Szperlik, Hnitecka, & Szopa, 2019). FNS catalyzes the formation of a double bond between the C-2 and C-3 positions of the C ring in flavanones and can be divided into two classes: FNSI and FNSII. FNSIs are soluble 2-oxoglutarate and Fe^{2+}-dependent dioxygenases mainly found in Apiaceae. Meanwhile, FNSII members belong to membrane-bound monooxygenases of NADPH and oxygen-dependent cytochrome P450 and are widely distributed in higher plants (Wei, Zhang, Fu, & Zhang, 2021; Yun, Yamamoto, Nozawa, & Tozawa, 2008). FNS is the key enzyme in the formation of flavones, FNSI can use both naringenin and eriodictyol as substrates to generate the corresponding flavones. Flavanones can also be converted into C-glycosyl flavones (Dong & Lin, 2021). Naringenin and eriodictyol are converted to apigenin C-glycosides and luteolin C-glycosides by the action of flavanone-2-hydroxylase (F2H), C-glycosyltransferase (CGT) and dehydratase (Lam et al., 2019).

2.6 Biosynthesis of isoflavones

The isoflavone biosynthesis pathway is mainly distributed in leguminous plants. Isoflavone synthase (IFS) leads flavanone to the isoflavone pathway and appears to be able to use both naringenin and liquiritigenin as substrates to generate 2-hydroxy-2,3-dihydrogenistein and 2,7,4′-trihydroxyiso-flavanone, respectively (Pandey et al., 2014). The detailed biosynthesis of isoflavones is given in Chapter 6.

2.7 Dihydroflavonol: A key branch point in the flavonoid biosynthesis pathway

Dihydroflavonol (or flavanonol) is an important intermediate metabolite and a key branch point in the flavonoid biosynthesis pathway. Dihydroflavonol is generated from flavanone under the catalysis of flavanone 3β-hydroxylase (F3H) and is the common precursor for flavonol, anthocyanin, and proanthocyanin (Song et al., 2016). F3H acts onnaringenin, eriodictyol, and pentahydroxyflavanone to form the corresponding dihydroflavonols, namely, dihydrokaempferol (DHK), dihydroquercetin (DHQ), and dihydromyricetin (DHM) (Lin et al., 2021). F3H, a FeII/2-oxoglutarate-dependent dioxygenase, catalyzes the dydroxylation of flavanones at position C-3 and is the key enzyme in dihydroflavonol synthesis. Because flavanones are also the substrates in the flavone, isoflavone, and phlobaphene biosynthetic pathways, F3H competes with FNS, IFS, and FNR for these common substrates (Cao, Xing, Xu, & Li, 2018). F3′H and F3′5′H, both cytochrome P450 enzymes, catalyze the hydroxylation of flavonoids at position C-3′ or C-3′ and C-5′ of ring B, respectively, to the formation of substrates of different pathways (Wen, Tsao, Wang, & Chu, 2021).

2.8 Biosynthesis of flavonol

Flavonols are flavonoid metabolites that are hydroxylated at position C-3 of ring C. Their C-3 position is highly prone to glycosidation; accordingly, they often exist in plant cells in glycosylated forms (Cao et al., 2018). The dihydroflavonols DHK, DHQ, and DHM are respectively converted to the flavonols kaempferol, quercetin, and myricetin by flavonol synthase (FLS). F3′H can also catalyze the conversion of kaempferol to quercetin, while F3′5′H activity generates myricetin from kaempferol or quercetin. Kaempferol, quercetin, and myricetin are further modified to various flavonol derivatives through the activities of enzymes such as methyl transferases, glucosyltransferases (GTs), and acyltransferase (AT), among others (Chapman & Muday, 2021). FLS, a

FeII/2-oxoglutarate-dependent dioxygenase, is the key and rate-limiting enzyme in the flavonol biosynthesis pathway and catalyzes the desaturation of dihydroflavonol to form a C-2 and C-3 double bond in ring C (Irmisch et al., 2019).

2.9 Leucoanthocyanidin and anthocyanin biosynthesis

Leucoanthocyanidin is an important intermediate product in the flavonoid pathway and the direct synthetic precursor of anthocyanidin and proanthocyanidin. The colorless leucopelargonidin, leucocyanidin, and leucodelphinidin are transformed into the corresponding anthocyanidins (the colored pelargonidin, cyanidin, and delphinidin) under the catalysis of anthocyanidin synthase (ANS), also known as leucoanthocyanidin dioxygenase (LDOX) (Naing & Kim, 2018; Solfanelli, Poggi, Loreti, Alpi, & Perata, 2006). Like FLS, F3H, and FNSI, ANS/LDOX is also a FeII/2-oxoglutarate-dependent dioxygenase and catalyzes the dehydroxylation of C-4 and the formation of a double bond in ring C (Xu et al., 2008).

2.10 Biosynthesis of proanthocyanidins

Proanthocyanidins, also known as condensed tannins, are an important type of flavonoid synthesized from leucoanthocyanidins and anthocyanidins. Leucoanthocyanidin reductase (LAR) converts leucoanthocyanidins, leucopelargonidin, leucocyanidin, and leucodelphinidin to *trans*-flavan-3-ols, afzelechin, catechin, and gallocatechin, respectively (Lepiniec et al., 2006). LAR, an NADPH-dependent reductase, drives the C-4 dehydroxylation of the C ring. Anthocyanidin reductase (ANR) can convert anthocyanidins, pelargonidin, cyanidin, and delphinidin, into the corresponding *cis*-flavan-3-ols, epiafzelechin, epicatechin, and epigallocatechin. ANR is also an NADPH-dependent reductase and catalyzes the removal of a double bond at ring C. Flavan-3-ols, *trans*-flavan-3-ols, and *cis*-flavan-3-ols are the basic proanthocyanidin units. Proanthocyanidins are synthesized *via* the polymerization (or condensation) of flavan-3-ols (Henry-Kirk, McGhie, Andre, Hellens, & Allan, 2012). Colorless proanthocyanidins are transferred to plant vacuoles and can be oxidized to generate colored tannins (yellow to brown) by PPO (Boucherle et al., 2017).

Table 1 Cytotoxic flavonoids from African medicinal plants and their effects on sensitive and drug-resistant cancer cell lines.

Compound names	Source	Country	Cancer Cell Lines and IC_{50} values (μM) and degree of resistance in bracket	References
Chalcones				
2′,4′,6′-Trihydroxy-5′-methoxychalcone (1)	*Polygonum senegalense*	Kenya	9.90 (MCF-7); 12.38 (Caco-2); 31.40 (DLD-1); 28.53 (HepG2); 58.67 (A549); 28.29 (SPC212)	Kuete et al. (2019)
2′,4′-Dihydroxy-6′-methoxychalcone (3)	*Polygonum senegalense*	Kenya	6.30 (MCF-7); 12.77 (Caco-2); 19.00 (DLD-1); 14.87 (HepG2); 46.23 (A549); 20.97 (SPC212)	Kuete et al. (2019)
2′,6′-Diacetate-4′-methoxychalcone (4)	*Polygonum senegalense*	Kenya	23.51 (MCF-7); 15.21 (Caco-2); 59.51 (DLD-1); 59.51 (HepG2); 84.27 (A549); 104.33 (SPC212)	Kuete et al. (2019)
2′,4′-Dihydroxy-3′,6′-dimethoxychalcone (2)	*Polygonum limbatum*	Cameroon	10.36 (CCRF-CEM) *vs* 18.60 (CEM/ADR5000) [1.74]; 49.50 (MDA-MB-231-*pcDNA*); 54.93 (HCT116 *p53*$^{+/+}$); 112.30 (U87MG.Δ*EGFR*); 82.23 (HepG2)	Kuete Nkuete et al. (2014)

(*continued*)

Table 1 Cytotoxic flavonoids from African medicinal plants and their effects on sensitive and drug-resistant cancer cell lines. (*cont'd*)

Compound names	Source	Country	Cancer Cell Lines and IC_{50} values (μM) and degree of resistance in bracket	References
4′,6′-Dihydroxy-2′,5′-dimethoxychalcone (5)	*Polygonum senegalense*	Kenya	6.11 (MCF-7); 15.59 (Caco-2); 25.37 (DLD-1); 21.78 (HepG2); 37.81 (A549); 44.59 (SPC212)	Kuete et al. (2019)
4′-Hydroxy-2′,6′-dimethoxychalcone (6)	*Polygonum limbatum*	Cameroon	9.37 (CCRF-CEM) *vs* 2.54 (CEM/ADR5000) [0.27]; 19.58 (MDA-MB-231-*pcDNA*) *vs* 6.48 (MDA-MB-231-*BCRP*) [0.33]; 6.80 (HCT116 $p53^{+/+}$) *vs* 6.27 (HCT116 $p53^{-/-}$) [0.64]; 35.25 (U87MG) *vs* 41.09 (U87MG.Δ*EGFR*) [1.17]; 58.63 (HepG2)	Kuete Nkuete et al. (2014)
4-Hydroxylonchocarpin (7)	*Dorstenia dinklagei. Lonchocarpus bussei*	Kenya	88.91 (CCRF-CEM) *vs* 3.47 (CEM/ADR5000) [0.03]; 67.89 (MDA-MB-231-*pcDNA*); 11.2 (PF-382); 103.30 (HCT116 $p53^{+/+}$); 34.3 (U87MG); 80.19 (U87MG.Δ*EGFR*) [2.34]; 59.1 (colo-38)	Adem Kuete et al. (2019); Kuete et al. (2011)

Cardamomin (8)	*Polygonum limbatum*	Cameroon	8.59 (CCRF-CEM) *vs* 17.56 (CEM/ADR5000) [2.04]; 62.74 (HCT116 $p53^{+/+}$) *vs* 124.48 (HCT116 $p53^{-/-}$) [1.98]; 138.48 (U87MG.Δ*EGFR*); 90.04 (HepG2); 6.34 (THP-1); 172.54 (Panc-1)	Dzoyem et al. (2012); Kuete Nkuete et al. (2014)
Isobavachalcone (9)	*Dorstenia barteri*	Cameroon	2.90 (CCRF-CEM) *vs* 3.73 (CEM/ADR5000) [1.29]; 22.56 (MDA-MB-231-*pcDNA*) *vs* 2.93 (MDA-MB-231-*BCRP*) [0.13]; 23.80 (HCT116 $p53^{+/+}$) *vs* 19.94 (HCT116 $p53^{-/-}$) [0.84]; 23.78 (U87MG) *vs* 17.43 (U87MG.Δ*EGFR*) [0.73]; 24.04 (HepG2); 15.06 (Panc-1)	Jing et al. (2010); Kuete Mbaveng Zeino Fozing et al. (2015); Mbaveng et al. (2008)
Licoagrochalcone A (10)	*Dorstenia kameruniana*	Kenya	5.16 (CCRF-CEM) *vs* 17.38 (CEM/ADR5000) [3.36]; 34.84 (MDA-MB-231-*pcDNA*) *vs* 49.57 (MDA-MB-231-*BCRP*) [1.42]; 21.84 (HCT116 $p53^{+/+}$) *vs* 31.25 (HCT116 $p53^{-/-}$) [1.43]; 37.63 (U87MG) *vs* 22.75 (U87MG.Δ*EGFR*) [0.60]; 42.95 (HepG2)	Adem et al. (2018)

(*continued*)

Table 1 Cytotoxic flavonoids from African medicinal plants and their effects on sensitive and drug-resistant cancer cell lines. (*cont'd*)

Compound names	Source	Country	Cancer Cell Lines and IC_{50} values (μM) and degree of resistance in bracket	References
Poinsettifolin B (11)	*Dorstenia poinsettifolia*	Cameroon	1.94 (CCRF-CEM) *vs* 14.87 (CEM/ADR5000) [7.66]; 24.35 (MDA-MB-231-*pcDNA*) *vs* 28.92 (MDA-MB-231-*BCRP*) [1.19]; 15.46 (HCT116 $p53^{+/+}$) *vs* 9.89 (HCT116 $p53^{-/-}$) [0.64]; 25.36 (U87MG) *vs* 22.13 (U87MG.Δ*EGFR*) [0.87]; 22.83 (HepG2)	Kuete Mbaveng Zeino Ngameni et al. (2015)
Flavones and flavone glycosides				
3,4′,5-Trihydroxy-6″,6″-dimethylpyrano[2,3-*g*]flavone (12)	*Xylopia aethiopica*	Cameroon	2.61 (CCRF-CEM) *vs* 9.54 (CEM/ADR5000) [3.66]; 14.71 (MDA-MB-231-*pcDNA*) *vs* 17.84 (MDA-MB-231-*BCRP*) [1.21]; 18.01 (HCT116 $p53^{+/+}$) *vs* 17.27 (HCT116 $p53^{-/-}$) [0.96]; 18.01 (U87MG) *vs* 18.60 (U87MG.Δ*EGFR*) [1.03]; 16.76 (HepG2)	Kuete et al. (2015)

5-Hydroxy-7,4′-dimethoxyflavone (13)	*Piper capense*	Cameroon	9.6 (MDA-MB-231-*pcDNA*) *vs* 29.4 (MDA-MB-231-*BCRP*) [3.1]; 33.0 (HCT116 $p53^{+/+}$); 4.6 (U87MG) *vs* 11.6 (U87MG.Δ*EGFR*) [2.5]	Mbaveng et al. (2021)
6,8-Diprenyleriodictyol (27)	*Dorstenia dinklagei*	Cameroon	4.8 (CCRF-CEM) *vs* 10.8 (CEM/ADR5000) [2.25]; 41.6 (HL60); 21.9 (U87MG); 19.3 (colo-38); 11.3 (Capan-1)	Kuete et al. (2011)
6-Hydroxy-2-(4-hydroxyphenyl)-4*H*-chromen-4-one (14)	*Dichrostachys cinerea*	Cameroon	18.90 (CCRF-CEM) *vs* 38.82 (CEM/ADR5000) [2.05]; 45.75 (MDA-MB-231-*pcDNA*) *vs* 40.46 (MDA-MB-231-*BCRP*) [0.88]; 48.86 (HCT116 $p53^{+/+}$) *vs* 48.62 (HCT116 $p53^{-/-}$) [1.00]; 43.86 (U87MG) *vs* 44.57 (U87MG.Δ*EGFR*) [1.02]; 34.02 (HepG2)	Mbaveng et al. (2019)
6-Prenylapigenin (15)	*Dorstenia mannii*	Cameroon	11.6 (CCRF-CEM) *vs* 14.9 (CEM/ADR5000) [1.28]; 15.8 (U87MG); 35.2 (HL60); 18.3 (PF-382); 32.1 (colo-38); 10.3 (Capan-1)	Kuete et al. (2011)

(continued)

Table 1 Cytotoxic flavonoids from African medicinal plants and their effects on sensitive and drug-resistant cancer cell lines. (*cont'd*)

Compound names	Source	Country	Cancer Cell Lines and IC_{50} values (μM) and degree of resistance in bracket	References
7-Hydroxy-2-(4-hydroxyphenyl)-4*H*-chromen-4-one (16)	*Dichrostachys cinerea*	Cameroon	31.18 (CCRF-CEM) *vs* 124.21 (CEM/ADR5000) [3.98]; 75.55 (MDA-MB-231-*pcDNA*) *vs* 80.00 (MDA-MB-231-*BCRP*) [1.06]; 66.30 (HCT116 $p53^{-/-}$); 61.30 (U87MG) *vs* 53.78 (U87MG.Δ*EGFR*) [0.88]	Mbaveng et al. (2019)
Artocarpesin (17)	*Morus mesozygia*	Cameroon	86.55 (CCRF-CEM) *vs* 67.85 (CEM/ADR5000) [0.78]; 59.75 (MDA-MB-231-*pcDNA*) *vs* 94.04 (MDA-MB-231-*BCRP*) [1.57]; 79.94 (HCT116 $p53^{+/+}$) *vs* 105.51 (HCT116 $p53^{-/-}$) [1.32]; 76.84 (U87MG) *vs* 102.18 (U87MG.Δ*EGFR*) [1.33]; 23.95 (HepG2)	Kuete et al. (2009); Kuete Mbaveng Zeino Fozing et al. (2015)
Atalantoflavone (18)	*Erythrina sigmoidea*	Cameroon	44.69 (CCRF-CEM) *vs* 33.35 (CEM/ADR5000) [0.75]; 24.55 (MDA-MB-231-*pcDNA*); 64.83 (HCT116 $p53^{+/+}$); 113.12 (U87MG) *vs* 106.49 (U87MG.Δ*EGFR*) [0.94]; 88.99 (HepG2)	Kuete, Sandjo, et al. (2014)

Canniflavone (19)	*Dorstenia psilurus*	Cameroon	2.34 (CCRF-CEM) *vs* 4.01 (CEM/ADR5000) [3.21]; 4.95 (HL60); 21.11 (HL60AR); 5.49 (MDA-MB-231-*pcDNA*) *vs* 10.23 (MDA-MB-231-*BCRP*) [1.86]; 4.23 (HCT116 $p53^{+/+}$) *vs* 2.43 (HCT116 $p53^{-/-}$) [0.57]; 6.82 (U87MG) *vs* 5.23 (U87MG.Δ*EGFR*) [0.77]; 3.57 (HepG2)	Efferth and Kuete (2014)
Chrysoeriol-7-*O*-β-D-glucopyranoside (28)	*Acacia sieberiana*	Cameroon	20.80 (CCRF-CEM) *vs* 18.68 (CEM/ADR5000) [0.90]; 30.69 (HCT116 $p53^{+/+}$) *vs* 28.43 (HCT116 $p53^{-/-}$) [0.93]; 42.48 (U87MG) *vs* 38.12 (U87MG.Δ*EGFR*) [0.90]; 34.25 (HepG2)	Ngaffo et al. (2020)
Chrysosplenol D (31)	*Achillea fragrantissima*	Egypt	23.14 (MCF-7); 57.78 (HepG2); 10.64 (PC-3)	Awad et al. (2017)
Cirsiliol (20)	*Achillea fragrantissima*	Egypt	9.79 (MCF-7): 72.12 (HepG2); 12.06 (PC-3)	Awad et al. (2017)

(*continued*)

Table 1 Cytotoxic flavonoids from African medicinal plants and their effects on sensitive and drug-resistant cancer cell lines. (*cont'd*)

Compound names	Source	Country	Cancer Cell Lines and IC_{50} values (μM) and degree of resistance in bracket	References
Cirsimaritin (21)	*Achillea fragrantissima*	Egypt	12.20 (MCF-7); 75.80 (HepG2)	Awad et al. (2017)
Cycloartocarpesin (22)	*Morus mesozygia*	Cameroon	15.51 (CCRF-CEM) *vs* 16.19 (CEM/ADR5000) [1.04]; 23.15 (MDA-MB-231-*pcDNA*) *vs* 23.27 (MDA-MB-231-*BCRP*) [1.05]; 22.30 (HCT116 $p53^{+/+}$) *vs* 36.56 (HCT116 $p53^{-/-}$) [1.64]; 20.94 (U87MG) *vs* 49.83 (U87MG.Δ*EGFR*) [2.37]; 21.37 (HepG2)	Kuete et al. (2009); Kuete Mbaveng Zeino Fozing et al. (2015)
Eupatilin 7-methyl ether (23)	*Achillea fragrantissima*	Egypt	26.54 (MCF-7); 79.05 (HepG2)	Awad et al. (2017)
Flavonol and flavonol glycosides				
Gancaonin Q (24)	*Dorstenia angusticornis*	Cameroon	39.8 (CCRF-CEM) *vs* 26.3 (CEM/ADR5000) [0.66]; 15.3 (HL60); 21.5 (U87MG); 15.2 (PF-382); 46.5 (colo-38); 46.6 (SW-680); 3.4 (Capan-1); 21.6 (MiaPaCa-2); 51.5 (A549)	Kuete et al. (2011)

Kaempferol 3,4′-di-*O*-α-L-rhamnopyranoside (35)	*Albizia zygia*	Cameroon	16.8 (MiaPaCa-2)	Koagne et al. (2020)
Kaempferol-3,7,4′-trimethylether (32)	*Aframomum arundinaceum*	Cameroon	18.38 (CCRF-CEM) *vs* 18.22 (CEM/ADR5000) [0.99]; 33.14 (MDA-MB-231-*BCRP*) [< 0.83]; 36.74 (HCT116 *p53*$^{-/-}$) [< 0.82]	Kuete Ango et al. (2014)
Kaempferol-3-*O*-rhamnoside (36)	*Euphorbia sanctae-catharinae*	Egypt	46.4 (MiaPaCa-2); 50.2 (Caco-2)	Hegazy et al. (2018); Koagne et al. (2020)
Luteolin (25)	*Acacia sieberiana; Tetrapleura tetraptera*	Cameroon	10.97 (CCRF-CEM) *vs* 27.04 (CEM/ADR5000) [2.46]; 34.45 (MDA-MB-231-*pcDNA*) *vs* 31.28 (MDA-MB-231-*BCRP*) [0.91]; 28.32 (HCT116 *p53*$^{+/+}$) *vs* 30.35 (HCT116 *p53*$^{-/-}$) [1.07]; 22.71 (U87MG) *vs* 18.95 (U87MG.Δ*EGFR*) [0.83]; 28.98 (HepG2)	Mbaveng Chi et al. (2020); Ngaffo et al. (2020)

(continued)

Table 1 Cytotoxic flavonoids from African medicinal plants and their effects on sensitive and drug-resistant cancer cell lines. *(cont'd)*

Compound names	Source	Country	Cancer Cell Lines and IC_{50} values (μM) and degree of resistance in bracket	References
Luteolin-3′,4′-dimethoxylether-7-*O*-β-D-glucoside (29)	*Acacia sieberiana*	Cameroon	58.52 (CCRF-CEM) *vs* 18.68 (CEM/ADR5000) [0.32]; 27.98 (HCT116 *p53*$^{-/-}$); 33.54 (U87MG) *vs* 27.34 (U87MG.Δ*EGFR*) [0.82]	Ngaffo et al. (2020)
Luteolin-7-*O*-rutinoside (30)	*Acacia sieberiana*	Cameroon	64.65 (CCRF-CEM); 46.71 (U87MG) *vs* 48.22 (U87MG.Δ*EGFR*) [1.03]	Ngaffo et al. (2020)
Myricetin-3-O-rhamnoside (37)	*Euphorbia sanctae-catharinae*	Egypt	44.7 (Caco-2)	Hegazy et al. (2018)
Neocyclomorusin (26)	*Erythrina sigmoidea*	Cameroon	59.02 (CCRF-CEM) *vs* 69.98 (CEM/ADR5000) [1.19]; 78.51 (MDA-MB-231-*pcDNA*); 75.44 (HCT116 *p53*$^{+/+}$); 70.53 (U87MG.Δ*EGFR*)	Kuete Sandjo et al. (2014)
Quercetin 3,4′-di-*O*-α-L-rhamnopyranoside (38)	*Albizia zygia*	Cameroon	17.3 (MiaPaCa-2)	Koagne et al. (2020)
Quercetin 3-*O*-α-L-rhamnopyranoside (39)	*Albizia zygia*	Cameroon	87.5 (MiaPaCa-2)	Koagne et al. (2020)

Quercetin-3-*O*-galactopyranoside (40)	*Euphorbia sanctae-catharinae*	Egypt	79.4 (Caco-2)	Hegazy et al. (2018)
Thonningiol (33)	*Ptycholobium contortum*	Cameroon	0.36 (MCF7); 5.28 (Caco-2); 5.65 (DLD-1); 3.88 (HepG2)	Kuete et al. (2018)
Trans-tiliroside (34)	*Xylopia aethiopica*	Cameroon	31.95 (CCRF-CEM) *vs* 24.47 (CEM/ADR5000) [0.76]; 41.63 (MDA-MB-231-*pcDNA*) *vs* 47.48 (MDA-MB-231-*BCRP*) [1.14]	Kuete et al. (2015)
Flavanones				
(S)-(−)-Onysilin (42)	*Polygonum limbatum*	Cameroon	156.67 (Panc-1)	Dzoyem et al. (2012); Kuete Nkuete et al. (2014)
(S)-(−)-Pinostrobin (41)	*Cajanus cajan; Polygonum limbatum*	Cameroon; Nigeria	116.41 (CCRF-CEM) *vs* 116.59 (CEM/ADR5000) [1.00]	Ashidi et al. (2010); Kuete Nkuete et al. (2014)
5,4′-Dihydroxy-7-methoxyflavanone (43)	*Dodonaea angustifolia*	Kenya	5.98 (MCF-7); 59.27 (HepG2); 43.43 (A549); 111.89 (SPC212)	Kuete et al. (2019)

(continued)

Table 1 Cytotoxic flavonoids from African medicinal plants and their effects on sensitive and drug-resistant cancer cell lines. (*cont'd*)

Compound names	Source	Country	Cancer Cell Lines and IC_{50} values (μM) and degree of resistance in bracket	References
Abyssinone IV (44)	*Erythrina sigmoidea*	Cameroon	15.07 (CCRF-CEM) *vs* 18.23 (CEM/ADR5000) [1.21]; 14.43 (MDA-MB-231-*pcDNA*) *vs* 16.45 (MDA-MB-231-*BCRP*) [1.14]; 20.65 (HCT116 $p53^{+/+}$) *vs* 17.26 (HCT116 $p53^{-/-}$) [0.84]; 17.82 (U87MG) *vs* 16.37 (U87MG.Δ*EGFR*) [0.92]; 20.42 (HepG2)	Djeussi et al. (2015); Kuete Sandjo et al. (2014)
Alpinetin (45)	*Polygonum limbatum*	Cameroon	88.22 (CCRF-CEM) *vs* 116.07 (CEM/ADR5000) [1.32]	Kuete Nkuete et al. (2014)
Dorsmanin F (46)	*Dorstenia mannii*	Cameroon	5.34 (CCRF-CEM) *vs* 11.73 (CEM/ADR5000) [2.20]; 5.34 (MDA-MB-231-*pcDNA*) *vs* 11.73 (MDA-MB-231-*BCRP*) [2.20]; 25.02 (HCT116 $p53^{+/+}$) *vs* 32.66 (HCT116 $p53^{-/-}$) [1.31]; 28.82 (U87MG) *vs* 17.64 (U87MG.Δ*EGFR*) [0.61]; 30.23 (HepG2)	Kuete Mbaveng Zeino Ngameni et al. (2015)

Naringenin (47)	*Aframomum arundinaceum*	Cameroon	12.20 (CCRF-CEM) *vs* 7.86 (CEM/ADR5000) [0.64]; 9.51 (MDA-MB-231-*pcDNA*) *vs* 18.12 (MDA-MB-231-*BCRP*) [1.91]; 13.65 (HCT116 $p53^{+/+}$) *vs* 13.86 (HCT116 $p53^{-/-}$) [1.02]; 29.81 (U87MG) *vs* 18.02 (U87MG.Δ*EGFR*) [0.60]; 23.46 (HepG2)	Kuete Ango et al. (2014)
Biflavonoid				
7,7″-Di-*O*-methylchamaejasmin (48)	*Ormocarpum kirkii*	Kenya	3.58 (CCRF-CEM) *vs* 5.69 (CEM/ADR5000) [1.58]; 5.97 (MDA-MB-231-*pcDNA*) *vs* 7.76 (MDA-MB-231-*BCRP*) [1.29]; 6.02 (HCT116 $p53^{+/+}$) *vs* 5.48 (HCT116 $p53^{-/-}$) [0.91]; 6.16 (U87MG) *vs* 4.96 (U87MG.Δ*EGFR*) [0.80]	Adem Mbaveng et al. (2019)
Amentoflavone (49)	*Dorstenia barteri*	Cameroon	63.18 (CCRF-CEM) *vs* 48.63 (CEM/ADR5000) [0.77]; 56.91 (MDA-MB-231-*pcDNA*) *vs* 27.92 (MDA-MB-231-*BCRP*) [0.49]; 68.86 (HCT116 $p53^{+/+}$)	Kuete et al. (2016); Kuete Ngameni Mbaveng et al. (2010); Kuete et al. (2007); Mbaveng et al. (2008)

(continued)

Table 1 Cytotoxic flavonoids from African medicinal plants and their effects on sensitive and drug-resistant cancer cell lines. (*cont'd*)

Compound names	Source	Country	Cancer Cell Lines and IC_{50} values (μM) and degree of resistance in bracket	References
Campylospermone A (50)	*Ormocarpum kirkii*	Kenya	34.25 (CCRF-CEM)	Adem Mbaveng et al. (2019)
Chamaejasmin (51)	*Ormocarpum kirkii*	Kenya	65.28 (CEM/ADR5000)	Adem Mbaveng et al. (2019)
Flavonolignan				
Hydnocarpin (52)	*Brucea antidysenterica*	Cameroon	4.28 (CCRF-CEM) *vs* 3.07 (CEM/ADR5000) [0.72]; 4.07 (MDA-MB-231-*pcDNA*) *vs*; 6.65 (MDA-MB-231-*BCRP*) [1.63]; 11.44 (HCT116 $p53^{+/+}$) *vs* 6.74 (HCT116 $p53^{-/-}$) [0.59]; 9.29 (U87MG) *vs* 5.42 (U87MG.Δ*EGFR*) [0.58]; 2.73 (HepG2); 7.87 (L1210); 70.99 (Tmolt3); 20.3 (SW480)	Lee, Kim, Park, Kang, and Lee (2013); Sharma and Hall (1991); Youmbi et al. (2023)

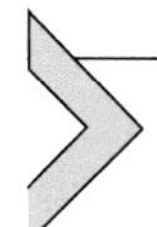

3. Cancer cell lines used in the screening of the cytotoxicity of flavonoids isolated from African medicinal plants

Several cancer cell lines were used to assess the cytotoxicity of flavonoids isolated from African medicinal plants. They include cell lines from breast cancer (MCF-7, MDA-MB-231-*pcDNA*, and MDA-MB-231-*BCRP*), colon cancer (Caco-2, colo-38, DLD-1, HCT116 $p53^{+/+}$, HCT116 $p53^{-/-}$, and SW-680), glioblastoma (U87MG and U87MG.Δ*EGFR*), hepatocarcinoma (HepG2), leukemia (CCRF-CEM, CEM/ADR5000, HL60, HL60AR, PF-382, THP-1, Tmolt3 and L1210 murine leukemia), lung cancer (A549, SPC212), pancreatic cancer (Capan-1, Panc-1, MiaPaCa-2), and prostate cancer (PC-3) (Table 1).

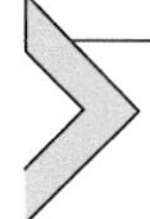

4. Cytotoxic flavonoids from African medicinal plants towards drug sensitive and multidrug-resistant cancer cells

A total of 52 cytotoxic flavonoids belonging to chalcones, flavones, and their glycosides have been isolated from African medicinal plants. Their effects on various human cancer cell lines are summarized in Table 1. The degree of resistance (D.R.) determined as the ratio of the IC_{50} value in resistant cells *versus* the IC_{50} values in the corresponding sensitive cell line: CEM/ADR5000 cells *vs* CCRF-CEM cells, MDA-MB-231-*BCRP* *vs* MDA-MB-231-pcDNA3, HCT116 $p53^{-/-}$ cells *vs* HCT116 $p53^{+/+}$ cells, and U87MG.Δ*EGFR* *vs* U87MG, is also given in Table 1 (Mbaveng, Noulala, et al., 2020). In the laboratory scale, collateral sensitivity is achieved when the D.R., defined as the ratio of the IC_{50} values of the cytotoxic agent towards the resistant cell line *versus* that in its sensitive counterpart, is below 1 whilst the D.R. above 1 defines normal sensitivity. It has been established that the D.R. < 0.9 defines the collateral sensitivity, whilst D.R. between 0.9 and 1.2 defines normal sensitivity. The cross-resistance is noted if the cytotoxic agent is more active in the sensitive cell line than its resistant subline, with D.R. above 1.2 (Efferth et al., 2020; Efferth et al., 2021; Mbaveng, Kuete, & Efferth, 2017). Collateral or normal sensitivities should be achieved for samples with the ability to combat drug resistant cancer cells (Efferth et al., 2020; Kuete & Efferth, 2015). In this section, the activity of these flavonoids will be discussed according to the cut-off points of cytotoxic phytochemicals defined as follows: outstanding activity ($IC_{50} \le 0.5\ \mu M$), excellent activity ($0.5 < IC_{50} \le 2\ \mu M$), very good activity ($2 < IC_{50} \le 5\ \mu M$), good activity ($5 < IC_{50} \le 10\ \mu M$), average activity ($10 < IC_{50} \le 20\ \mu M$), weak activity ($20 < IC_{50} \le 60\ \mu M$), very weak activity ($60 < IC_{50} \le 150\ \mu M$), and not active ($IC_{50} > 150\ \mu M$) (Kuete, 2025).

4.1 Cytotoxic chalcones

Eleven cytotoxic chalcones have been isolated from African medicinal plants. They include 2′,4′,6′-trihydroxy-5′-methoxychalcone (**1**), 2′, 4′-dihydroxy-3′,6′-dimethoxychalcone (**2**), 2′,4′-dihydroxy-6′-methoxychalcone (**3**), 2′,6′-diacetate-4′-methoxychalcone (**4**), 4′,6′-dihydroxy-2′,5′-dimethoxychalcone (**5**), 4′-hydroxy-2′,6′-dimethoxychalcone (**6**), 4-hydroxylonchocarpin (**7**), cardamomin (**8**), isobavachalcone (**9**), licoagrochalcone A (**10**), and poinsettifolin B (**11**). Their chemical structures are shown in Fig. 1. The cytotoxicity in human cancer cell lines is recorded in Table 1. Very good cytotoxic activity ($2 < IC_{50} \le 5\ \mu M$) was obtained with chalcones **6** and **7** against CEM/ADR5000 leukemia cells, **9** and **10** against CCRF-CEM cells, as well as **9** against MDA-MB-231-*BCRP* breast cancer cells. Good cytotoxic activity ($5 < IC_{50} \le 10\ \mu M$) was obtained with chalcones **1, 3,** and **5** against MCF-7 breast cancer cells, **6** and **8** against CCRF-CEM leukemia cells, **8** against THP-1 leukemia cells, and **11** against HCT116 *p53*$^{-/-}$. Collateral sensitivity of CEM/ADR5000 cells *vs* CCRF-CEM cells (D.R. 0.27) and HCT116 *p53*$^{-/-}$ cells *vs* HCT116 *p53*$^{+/+}$ colon cancer cells (D.R. 0.64) to chalcone **6**, CEM/ADR5000 cells (D.R. 0.03) to **7**, MDA-MB-231-*BCRP* *vs* MDA-MB-231-*pcDNA* (D.R. 0.13), HCT116 *p53*$^{-/-}$ *vs* HCT116 *p53*$^{+/+}$ (D.R. 0.84), and U87MG.Δ*EGFR* *vs* U87MG to **9** (D.R. 0.73), as well as U87MG.Δ*EGFR* *vs* U87MG to **10** (D.R. 0.60) and **11** (D.R. 0.87) was observed (Table 1).

4.2 Cytotoxic flavones and flavone glycosides

A total of 19 flavones and their glycosides (Fig. 2) were isolated from the medicinal plants harvested in various African countries. They are 3,4′, 5-trihydroxy-6″,6″-dimethylpyrano[2,3-*g*]flavone (**12**), 5-hydroxy-7, 4′-dimethoxyflavone (**13**), 6-hydroxy-2-(4-hydroxyphenyl)-4*H*-chromen-4-one (**14**), 6-prenylapigenin (**15**), 7-hydroxy-2-(4-hydroxyphenyl)-4*H*-chromen-4-one (**16**), artocarpesin (**17**), atalantoflavone (**18**), canniflavone (**19**), ciyrsiliol (**20**), cirsimaritin (**21**), cycloartocarpesin (**22**), eupatilin 7-methyl ether (**23**), gancaonin Q (**24**), luteolin (**25**), neocyclomorusin (**26**),

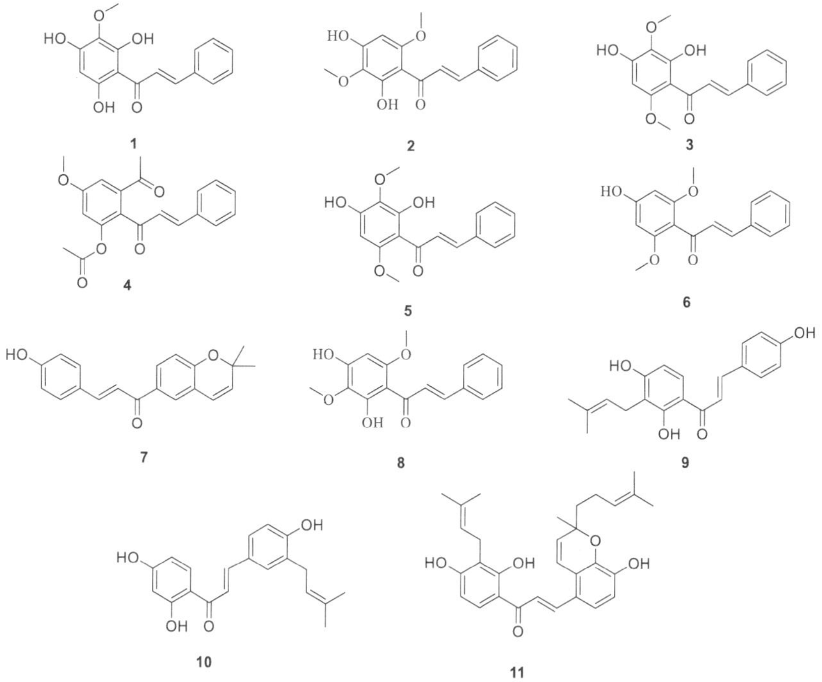

Fig. 1 Cytotoxic chalcones isolated from African medicinal plants. **1**: 2′,4′,6′-trihydroxy-5′-methoxychalcone; **2**: 2′,4′-dihydroxy-3′,6′-dimethoxychalcone: **3**: 2′,4′-dihydroxy-6′-methoxychalcone; **4**: 2′,6′-diacetate-4′-methoxychalcone; **5**: 4′,6′-dihydroxy-2′,5′-dimethoxychalcone; **6**: 4′-hydroxy-2′,6′-dimethoxychalcone; **7**: 4-hydroxylonchocarpin; **8**: cardamomin; **9**: isobavachalcone; **10**: licoagrochalcone A; **11**: poinsettifolin B.

6,8-diprenyleriodictyol (**27**), chrysoeriol-7-*O*-*β*-D-glucopyranoside (**28**), luteolin-3′,4′-dimethoxylether-7-*O*-*β*-D-glucoside (**29**), and luteolin-7-*O*-rutinoside (**30**). Their antiproliferative effects, recorded as IC_{50} values are shown in Table 1. Excellent cytotoxic activity ($0.5 < IC_{50} \leq 2\,\mu M$) was obtained with chalcone **19** against CCRF-CEM leukemia cells. Very good cytotoxic activity was obtained with flavones **12, 19,** and **27** against CCRF-CEM cells, **19** against HL60 leukemia cells, HCT116 *p53*$^{+/+}$ cells, and HCT116 *p53*$^{-/-}$ cells, and **24** against Capan-1 pancreatic cancer cells. Good cytotoxic activity was reported with chalcones **12** against CEM/ADR5000 cells, **13** and **19** against MDA-MB-231-*pcDNA* cells, **19** against U87MG cells, U87MG.Δ*EGFR* cells, and HepG2 cells, and **20** against MCH-7 cells. Collateral sensitivity of HCT116 *p53*$^{-/-}$ cells *vs* HCT116 *p53*$^{+/+}$ cancer cells (D.R. 0.64) to flavone **12** and HCT116 *p53*$^{-/-}$ *vs* HCT116 *p53*$^{+/+}$

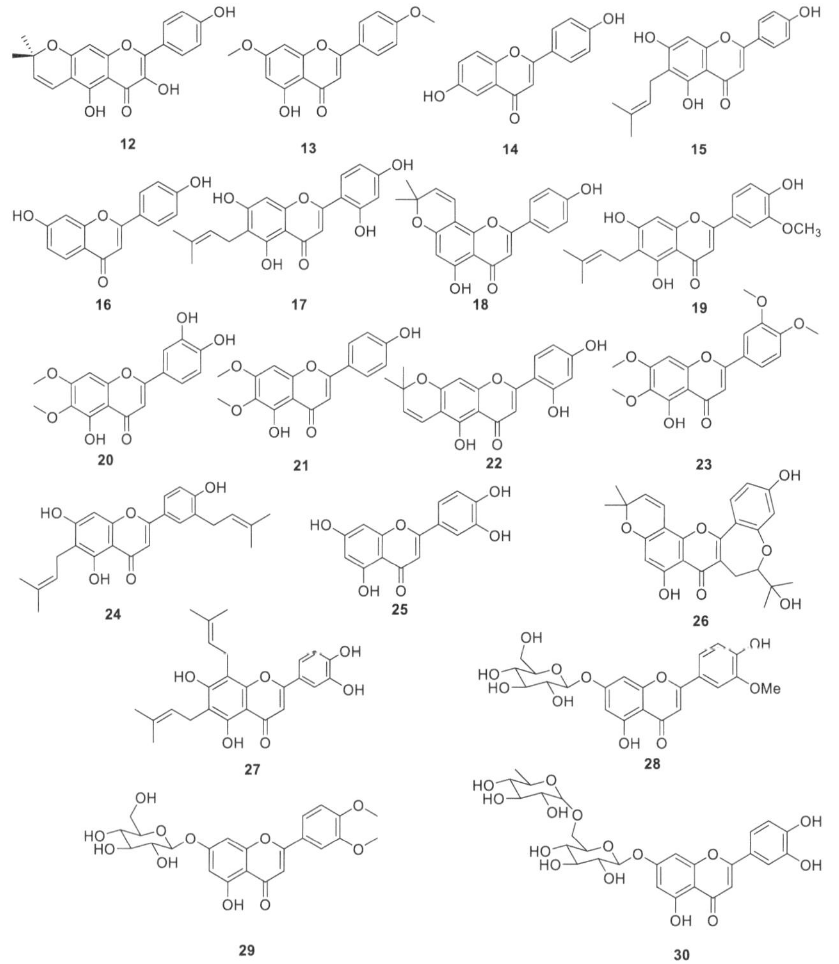

Fig. 2 Cytotoxic flavone and flavone glycosides isolated from African medicinal plants. **12:** 3,4′,5-trihydroxy-6″,6″-dimethylpyrano[2,3-*g*]flavone; **13:** 5-hydroxy-7,4′-dimethoxyflavone; **14:** 6-hydroxy-2-(4-hydroxyphenyl)-4*H*-chromen-4-one; **15:** 6-prenylapigenin; **16:** 7-hydroxy-2-(4-hydroxyphenyl)-4*H*-chromen-4-one; **17:** artocarpesin; **18:** atalantoflavone; **19:** canniflavone; **20:** cirsiliol; **21:** cirsimaritin; **22:** cycloartocarpesin; **23:** eupatilin 7-methyl ether; **24:** gancaonin Q; **25:** luteolin; **26:** neocyclomorusin; **27**: 6,8-diprenyleriodictyol; **28:** chrysoeriol-7-*O*-*β*-D-glucopyranoside; **29:** luteolin-3′,4′-dimethoxylether-7-*O*-*β*-D-glucoside; **30:** luteolin-7-*O*-rutinoside.

(D.R. 0.57) to **19** was achieved. Normal sensitivity of HCT116 $p53^{-/-}$ *vs* HCT116 $p53^{+/+}$ (D.R. 0.84) to **9** and U87MG.Δ*EGFR* *vs* U87MG to **9** (D.R. 1.03) was also obtained.

4.3 Cytotoxic flavonols and flavonol glycosides

Ten cytotoxic flavonols and their glycosides (Fig. 3) were isolated from several African medicinal plants. They include chrysosplenol D (**31**), kaempferol-3,7,4′-trimethylether (**32**), thonningiol (**33**), *trans*-tiliroside (**34**), kaempferol 3,4′-di-*O*-α-L-rhamnopyranoside (**35**), kaempferol-3-*O*-rhamnoside (**36**), myricetin-3-*O*-rhamnoside (**37**), quercetin 3,4′-di-*O*-α-L-rhamnopyranoside (**38**), quercetin 3-*O*-α-L-rhamnopyranoside (**39**),

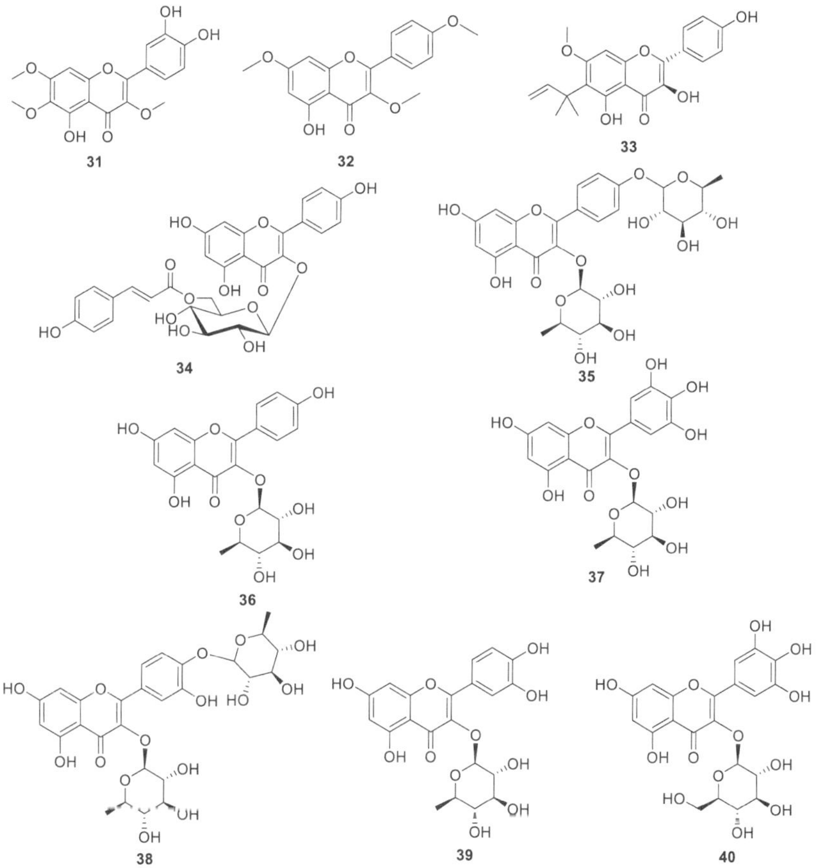

Fig. 3 Cytotoxic flavonol and flavonol glycosides isolated from African medicinal plants. **31:** chrysosplenol D; **32:** kaempferol-3,7,4′-trimethylether; **33:** thonningiol; **34:** *trans*-tiliroside; **35:** kaempferol 3,4′-di-*O*-α-L-rhamnopyranoside; **36:** kaempferol-3-*O*-rhamnoside; **37:** myricetin-3-*O*-rhamnoside; **38:** quercetin 3,4′-di-*O*-α-L-rhamnopyranoside; **39:** quercetin 3-*O*-α-L-rhamnopyranoside; **40:** quercetin-3-*O*-galactopyranoside.

and quercetin-3-*O*-galactopyranoside (**40**). The IC_{50} values of these flavonol and flavonol glycosides are summarized in Table 1. Most of them rather displayed average ($10 < IC_{50} \leq 20\,\mu M$) to very weak activity ($60 < IC_{50} \leq 150\,\mu M$). However, outstanding cytotoxic activity ($IC_{50} \leq 0.5\,\mu M$) was obtained with flavonol **33** against MCF7 cells, meanwhile, very good cytotoxic activities were also reported with **33** against Caco-2 cells and **33** against DLD-1 cells and HepG2 cells.

4.4 Cytotoxic flavanones

Seven cytotoxic flavonols and their glycosides (Fig. 4) were isolated from African medicinal plants. They include (S)-(−)-pinostrobin (**41**), (S)-(−)-onysilin (**42**), 5,4′-dihydroxy-7-methoxyflavanone (**43**), abyssinone IV (**44**), alpinetin (**45**), dorsmanin F (**46**), and naringenin (**47**). The IC_{50} values of compounds of these flavanones are shown in Table 1. Most of them rather displayed average ($10 < IC_{50} \leq 20\,\mu M$) to weak activity ($20 < IC_{50} \leq 60\,\mu M$). Nonetheless, good cytotoxic activities were obtained with flavanone **43** against MCF-7 cells, **46** against CCRF-CEM cells, **47** against CEM/ADR5000 cells, and **46** and **47** against MDA-MB-231-*pcDNA* cells.

4.5 Cytotoxic biflavonoids and flavonolignan

Four cytotoxic biflavonoid, 7,7″-di-*O*-methylchamaejasmin (**48**), amentoflavone (**49**), campylospermone A (**50**), and chamaejasmin (**51**), and one flavonolignan hydnocarpin (**52**) (Fig. 5) were isolated from several African medicinal plants. Their cytotoxicity in the human cancer cell line is given

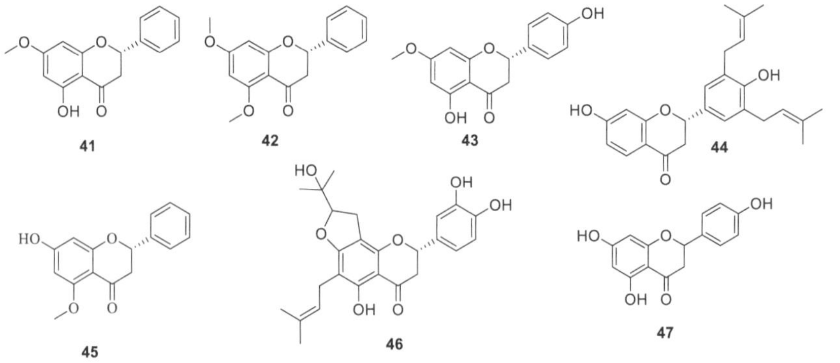

Fig. 4 Cytotoxic flavanone isolated from African medicinal plants. **41:** (S)-(−)-pinostrobin; **42:** (S)-(−)-onysilin; **43:**5,4′-dihydroxy-7-methoxyflavanone; **44:** abyssinone IV; **45:** alpinetin; **46:** dorsmanin F; **47:** naringenin.

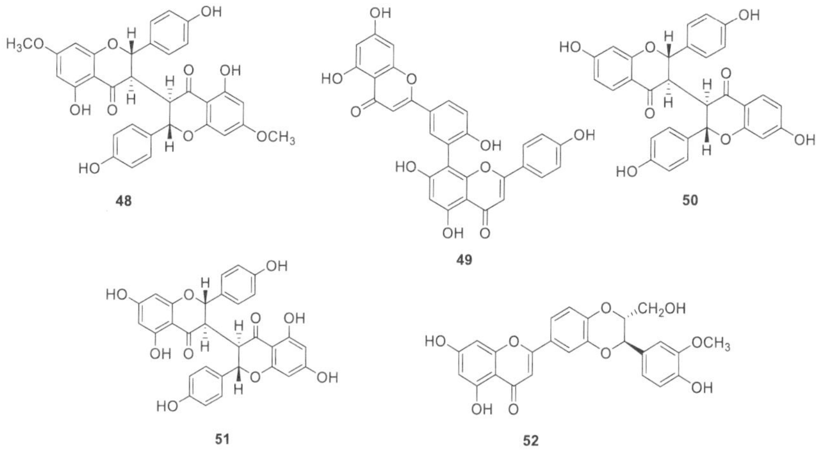

Fig. 5 Cytotoxic biflavonoid (**48–51**) and flavonolignan (**52**) isolated from African medicinal plants. **48:** 7,7″-di-*O*-methylchamaejasmin; **49:** amentoflavone; **50:** campylospermone A; **51:** chamaejasmin; **52:** hydnocarpin.

in Table 1. Amongst biflavonoids, **49, 50,** and **51** displayed activities in the range of weak to very weak. In contrast, good cytotoxic effects were achieved with **48** against CCRF-CEM cells, CEM/ADR5000 cells, MDA-MB-231-*pcDNA* cells, MDA-MB-231-*BCRP* cells, HCT116 *p53*$^{+/+}$ cells, HCT116 *p53*$^{-/-}$ cells, U87MG cells, and U87MG.Δ*EGFR* cells. Collateral sensitivity of U87MG.Δ*EGFR* *vs* U87MG (D.R. 0.80), as well as normal sensitivity of HCT116 *p53*$^{-/-}$ cells *vs* HCT116 *p53*$^{+/+}$ cells (D.R. 0.91) to **48**, were achieved. The flavonolignan **52** displayed very good cytotoxic effects against CCRF-CEM cells, CEM/ADR5000 cells, MDA-MB-231-*pcDNA* cells, and HepG2 cells, and good activity against MDA-MB-231-*BCRP* cells, HCT116 *p53*$^{-/-}$ cells, U87MG cells, and U87MG.Δ*EGFR*. Collateral sensitivity of CEM/ADR5000 cells *vs* CCRF-CEM cells (D.R. 0.72), HCT116 *p53*$^{-/-}$ cells *vs* HCT116 *p53*$^{+/+}$ cells (D.R. 0.59), and U87MG.Δ*EGFR* *vs* U87MG (D.R. 0.58) to **52** was also achieved.

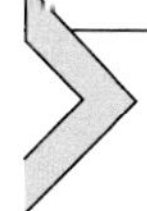

5. Modes of action of the cytotoxic flavonoids from the African medicinal plants

Several potent cytotoxic flavonoids from African medicinal induced apoptotic cell death in cancer cell lines. These include chalcones

4′-hydroxy-2′,6′-dimethoxychalcone (**6**), 4-hydroxylonchocarpin (**7**), isobavachalcone (**9**), and poinsettifolin B (**11**), the flavones gancaonin Q (**24**), cycloartocarpesin (**22**), 6,8-diprenyl eriodictyol (**27**), and 3,4′, 5-trihydroxy-6″,6″-dimethylpyrano[2,3-*g*]flavone (**47**), the flavanones abyssinone IV (**44**) and dorsmanin F (**46**), the biflavonoid 7,7″-di-*O*-methylchamaejasmin (**48**), and the flavonolignan hydnocarpin (**52**). Compound **6**-induced apoptosis in MCF-7 cells is mediated by increased reactive oxygen species (ROS) production and mitochondrial membrane potential (MMP) loss (Kuete, Omosa, Midiwo, Karaosmanoğlu, & Sivas, 2019). Chalcone **11** and flavanone **46** induced apoptosis in CCRF-CEM leukemia cells, mediated by MMP disruption and increased ROS production (Kuete, Mbaveng, Zeino, B. Ngameni, et al., 2015). Compound **44** induced apoptotic cell death in CCRF-CEM cells, mediated by the loss of MMP as well as an increase in ROS production (Kuete, Sandjo, et al., 2014). Chalcone **9** and flavone **22** induced apoptosis in CCRF-CEM leukemia cells, mediated by caspase activation and the disruption of MMP (Kuete, Mbaveng, Zeino, Fozing, et al., 2015). Compound **9** also induced apoptotic cell death in two human neuroblastoma cell lines (IMR-32 and NB-39) *via* the mitochondrial pathway (Nishimura et al., 2007). The studies of Nishimura et al. demonstrated that compound **9** induced apoptotic cell death with caspase-3 and -9 activation and Bax upregulation in neuroblastoma cell lines (Nishimura et al., 2007). Compound **9** also inhibited more than 70% MMP-2 secretion from U87 glioblastoma cells at the concentration of 250 μM (Ngameni et al., 2007). It was demonstrated that caspase 3/7 activation is one of the modes of induction of apoptosis of flavones **24** and **27**, and one chalcone **7** in CCRF-CEM cells (Kuete et al., 2011). Flavone **47** also induced apoptosis in leukemia CCRF-CEM cells mediated by the disruption of the MMP (Kuete, Sandjo, Mbaveng, Zeino, & Efferth, 2015). Biflavonoid **48** induced apoptosis in CCRF-CEM cells through MMP alteration and increased ROS production (Adem, Mbaveng, et al., 2019) while the **52** also induced apoptotic cell death in CCRF-CEM cells *via* caspase activation, the alteration of MMP, and increased ROS levels (Youmbi et al., 2023).

6. Conclusion

In the present report, we have identified 52 flavonoids isolated from African medicinal plants with cytotoxic effects on various human cancer cell lines. They include chalcones, flavones and flavone glycosides, flavonols and flavonol glycosides, flavanones, biflavonoids and flavonolignans. The most active chalcones were 4-hydroxylonchocarpin (**7**), cardamomin (**8**), isobavachalcone (**9**), licoagrochalcone A (**10**), and poinsettifolin B (**11**), while the most active flavones were 3,4′,5-trihydroxy-6″,6″-dimethylpyrano[2,3-*g*]flavone (**12**), canniflavone (**19**), ciyrsiliol (**20**), and 6, 8-diprenyleriodictyol (**27**). The best flavanones include 5,4′-dihydroxy-7-methoxyflavanone (**43**), dorsmanin F (**46**), and naringenin (**47**). 7,7″-Di-*O*-methylchamaejasmin (**48**) and hydnocarpin (**52**) were identified as the most active biflavonoid and flavonolignan, respectively. Some of the active flavonoids displayed apoptotic cell death, mainly mediated by caspase activation, alteration of MMP, and enhanced ROS production in cancer cells. The most potent flavonoids herein should be further explored to develop novel cytotoxic products to fight cancer.

References

Adem, F. A., Kuete, V., Mbaveng, A. T., Heydenreich, M., Ndakala, A., Irungu, B., et al. (2018). Cytotoxic benzylbenzofuran derivatives from *Dorstenia kameruniana*. *Fitoterapia, 128*, 26–30.

Adem, F. A., Mbaveng, A. T., Kuete, V., Heydenreich, M., Ndakala, A., Irungu, B., et al. (2019). Cytotoxicity of isoflavones and biflavonoids from *Ormocarpum kirkii* towards multi-factorial drug resistant cancer. *Phytomedicine, 58*, 152853.

Adem, F. A., Kuete, V., Mbaveng, A. T., Heydenreich, M., Koch, A., Ndakala, A., et al. (2019). Cytotoxic flavonoids from two *Lonchocarpus* species. *Natural Product Research, 33*(18), 2609–2617.

Ahmed, H., Moawad, A., Owis, A., AbouZid, S., & Ahmed, O. (2016). Flavonoids of *Calligonum polygonoides* and their cytotoxicity. *Pharmaceutical Biology, 54*(10), 2119–2126.

Al-Khayri, J. M., Sahana, G. R., Nagella, P., Joseph, B. V., Alessa, F. M., & Al-Mssallem, M. Q. (2022). Flavonoids as potential anti-inflammatory molecules: A review. *Molecules (Basel, Switzerland), 27*(9).

Ashidi, J. S., Houghton, P. J., Hylands, P. J., & Efferth, T. (2010). Ethnobotanical survey and cytotoxicity testing of plants of South-western Nigeria used to treat cancer, with isolation of cytotoxic constituents from Cajanus cajan Millsp. leaves. *Journal of Ethnopharmacology, 128*(2), 501–512.

Awad, B. M., Habib, E. S., Ibrahim, A. K., Wanas, A. S., Radwan, M. M., Helal, M. A., et al. (2017). Cytotoxic activity evaluation and molecular docking study of phenolic derivatives from *Achillea fragrantissima* (Forssk.) growing in Egypt. *Medicinal Chemistry Research, 26*(9), 2065–2073

Belkacem, S. (2009). *Investigation phytochimique de la phase n-butanol de l'extrait hydroalcoolique des parties aériennes de Centaurea parviflora (Compositae)*. Unpublished Doctorate thesis, *Université Mentouri-Constantine*, Constantine, Algeria.

Bomati, E. K., Austin, M. B., Bowman, M. E., Dixon, R. A., & Noel, J. P. (2005). Structural elucidation of chalcone reductase and implications for deoxychalcone biosynthesis. *Journal of Biological Chemistry, 280*(34), 30496–30503.

Boucherle, B., Peuchmaur, M., Boumendjel, A., & Haudecoeur, R. (2017). Occurrences, biosynthesis and properties of aurones as high-end evolutionary products. *Phytochemistry, 142*, 92–111.

Cao, Y. L., Xing, M. Y., Xu, C., & Li, X. (2018). Biosynthesis of flavonol and its regulation in plants. *Acta Horticulturae Sinica, 45*(1), 177–192.

Casas, M. I., Falcone-Ferreyra, M. L., Jiang, N., Mejía-Guerra, M. K., Rodríguez, E., Wilson, T., et al. (2016). Identification and characterization of maize salmon silks genes involved in insecticidal maysin biosynthesis. *The Plant Cell, 28*(6), 1297–1309.

Chapman, J. M., & Muday, G. K. (2021). Flavonols modulate lateral root emergence by scavenging reactive oxygen species in *Arabidopsis thaliana. The Journal of Biological Chemistry, 296*, 100222.

Chen, Y. Y., Lu, H. Q., Jiang, K. X., Wang, Y. R., Wang, Y. P., & Jiang, J. J. (2022). The flavonoid biosynthesis and regulation in Brassica napus: A review. *International Journal of Molecular Sciences, 24*(1), 357.

Chong, J., Poutaraud, A., & Hugueney, P. (2009). Metabolism and roles of stilbenes in plants. *Plant Science, 177*(3), 143–155.

Constantinescu, T., & Lungu, C. N. (2021). Anticancer activity of natural and synthetic chalcones. *International Journal of Molecular Sciences, 22*(21).

Cushnie, T. P., & Lamb, A. J. (2005). Antimicrobial activity of flavonoids. *International Journal of Antimicrobial Agents, 26*(5), 343–356.

Dinkins, R. D., Hancock, J., Coe, B. L., May, J. B., Goodman, J. P., Bass, W. T., et al. (2021). Isoflavone levels, nodulation and gene expression profiles of a CRISPR/Cas9 deletion mutant in the isoflavone synthase gene of red clover. *Plant Cell Reports, 40*(3), 517–528.

Djeussi, D. E., Sandjo, L. P., Noumedem, J. A., Omosa, L. K., B, T. N., & Kuete, V. (2015). Antibacterial activities of the methanol extracts and compounds from *Erythrina sigmoidea* against gram-negative multi-drug resistant phenotypes. *BMC Complementary and Alternative Medicine, 15*(1), 453.

Dong, N. Q., & Lin, H. X. (2021). Contribution of phenylpropanoid metabolism to plant development and plant-environment interactions. *Journal of Integrative Plant Biology, 63*(1), 180–209.

Dzoyem, J. P., McGaw, L. J., Kuete, V., & Bakowsky, U. (2017). Chapter 9—Anti-inflammatory and anti-nociceptive activities of African medicinal spices and vegetables. In Kuete (Ed.). *Medicinal spices and vegetables from Africa* (pp. 239–270). Oxford: Academic Press.

Dzoyem, J. P., Nkuete, A. H., Kuete, V., Tala, M. F., Wabo, H. K., Guru, S. K., et al. (2012). Cytotoxicity and antimicrobial activity of the methanol extract and compounds from *Polygonum limbatum. Planta Medica, 78*(8), 787–792.

Efferth, T., & Kuete, V. (2014). *Canniflavon: Zytotoxizität und Wirkmechanismen in Krebszellen.* Germany Patent No. DE201420006418.

Efferth, T., Saeed, M. E. M., Kadioglu, O., Seo, E. J., Shirooie, S., Mbaveng, A. T., et al. (2020). Collateral sensitivity of natural products in drug-resistant cancer cells. *Biotechnology Advances, 38*, 107342.

Efferth, T., Kadioglu, O., Saeed, M. E. M., Seo, E. J., Mbaveng, A. T., & Kuete, V. (2021). Medicinal plants and phytochemicals against multidrug-resistant tumor cells expressing ABCB1, ABCG2, or ABCB5: A synopsis of 2 decades. *Phytochemistry Reviews, 20*(1), 7–53.

Grotewold, E. (2006). The genetics and biochemistry of floral pigments. *Annual Review of Plant Biology, 57*, 761–780.

Hegazy, M. F., Hamed, A. R., Ibrahim, M. A. A., Talat, Z., Reda, E. H., Abdel-Azim, N. S., et al. (2018). Euphosantianane A–D: Antiproliferative premyrsinane diterpenoids from the endemic Egyptian plant *Euphorbia Sanctae-Catharinae. Molecules (Basel, Switzerland), 23*(9), 2221.

Henry-Kirk, R. A., McGhie, T. K., Andre, C. M., Hellens, R. P., & Allan, A. C. (2012). Transcriptional analysis of apple fruit proanthocyanidin biosynthesis. *Journal of Experimental Botany, 63*(15), 5437–5450.

Herbert, R. B. (1981). *The biosynthesis of secondary metabolites*. London: Chapman and Hall,.
Irmisch, S., Ruebsam, H., Jancsik, S., Man Saint Yuen, M., Madilao, L. L., & Bohlmann, J. (2019). Flavonol biosynthesis genes and their use in engineering the plant antidiabetic metabolite montbretin A. *Plant Physiology, 180*(3), 1277–1290.
Jepkoech, C., Omosa, L. K., Nchiozem-Ngnitedem, V. A., Kenanda, E. O., Guefack, M. F., Mbaveng, A. T., et al. (2021). Antibacterial secondary metabolites from *Vernonia auriculifera* Hiern (Asteraceae) against MDR phenotypes. *Natural Product Research, 36*(12), 3203–3206.
Jing, H., Zhou, X., Dong, X., Cao, J., Zhu, H., Lou, J., et al. (2010). Abrogation of Akt signaling by Isobavachalcone contributes to its anti-proliferative effects towards human cancer cells. *Cancer Letters, 294*(2), 167–177.
Khoo, H. E., Azlan, A., Tang, S. T., & Lim, S. M. (2017). Anthocyanidins and anthocyanins: Colored pigments as food, pharmaceutical ingredients, and the potential health benefits. *Food & Nutrition Research, 61*(1), 1361779.
Koagne, R. R., Annang, F., Cautain, B., Martín, J., Pérez-Moreno, G., Bitchagno, G. T. M., et al. (2020). Cytotoxycity and antiplasmodial activity of phenolic derivatives from *Albizia zygia* (DC.) J.F. Macbr. (Mimosaceae). *BMC Complementary Medicine and Therapies, 20*(1), 8.
Koorbanally, N. A., Randrianarivelojosia, M., Mulholland, D. A., van Ufford, L. Q., & van den Berg, A. J. (2003). Chalcones from the seed of *Cedrelopsis grevei* (Ptaeroxylaceae). *Phytochemistry, 62*(8), 1225–1229.
Kuete, V. (2013). Medicinal plant research in Africa. In V. Kuete (Ed.). *Pharmacology and chemistry*. Oxford: Elsevier.
Kuete, V. (2025). Chapter Four—African medicinal plants and their derivative as the source of potent anti-leukemic products: Rationale classification of naturally occurring anticancer agents. *Advances in Botanical Research, 113*. https://doi.org/10.1016/bs.abr.2023.12.010.
Kuete, V., & Sandjo, L. P. (2012). Isobavachalcone: An overview. *Chinese Journal of Integrative Medicine, 18*(7), 543–547.
Kuete, V., & Efferth, T. (2015). African flora has the potential to fight multidrug resistance of cancer. *BioMed Research International, 2015*, 914813.
Kuete, V., Sandjo, L. P., Mbaveng, A. T., Zeino, M., & Efferth, T. (2015). Cytotoxicity of compounds from *Xylopia aethiopica* towards multi-factorial drug-resistant cancer cells. *Phytomedicine, 22*, 1247–1254.
Kuete, V., Omosa, L. K., Midiwo, J. O., Karaosmanoğlu, O., & Sivas, H. (2019). Cytotoxicity of naturally occurring phenolics and terpenoids from Kenyan flora towards human carcinoma cells. *Journal of Ayurveda and Integrative Medicine, 10*(3), 178–184.
Kuete, V., Simo, I. K., Ngameni, B., Bigoga, J. D., Watchueng, J., Kapguep, R. N., et al. (2007). Antimicrobial activity of the methanolic extract, fractions and four flavonoids from the twigs of *Dorstenia angusticornis* Engl. (Moraceae). *Journal of Ethnopharmacology, 112*(2), 271–277.
Kuete, V., Fozing, D. C., Kapche, W. F., Mbaveng, A. T., Kuiate, J. R., Ngadjui, B. T., et al. (2009). Antimicrobial activity of the methanolic extract and compounds from *Morus mesozygia* stem bark. *Journal of Ethnopharmacology, 124*(3), 551–555.
Kuete, V., Ngameni, B., Tangmouo, J. G., Bolla, J. M., Alibert-Franco, S., Ngadjui, B. T., et al. (2010). Efflux pumps are involved in the defense of Gram-negative bacteria against the natural products isobavachalcone and diospyrone. *Antimicrobial Agents and Chemotherapy, 54*(5), 1749–1752.
Kuete, V., Ngameni, B., Mbaveng, A. T., Ngadjui, B., Meyer, J. J., & Lall, N. (2010). Evaluation of flavonoids from *Dorstenia barteri* for their antimycobacterial, antigonorrheal and anti-reverse transcriptase activities. *Acta Tropica, 116*(1), 100–104.

Kuete, V., Ngameni, B., Wiench, B., Krusche, B., Horwedel, C., Ngadjui, B. T., et al. (2011). Cytotoxicity and mode of action of four naturally occuring flavonoids from the genus *Dorstenia*: Gancaonin Q, 4-hydroxylonchocarpin, 6-prenylapigenin, and 6,8-diprenyleriodictyol. *Planta Medica, 77*(18), 1984–1989.

Kuete, V., Ango, P. Y., Yeboah, S. O., Mbaveng, A. T., Mapitse, R., Kapche, G. D., et al. (2014). Cytotoxicity of four *Aframomum* species (*A. arundinaceum, A. alboviolaceum, A. kayserianum and A. polyanthum*) towards multi-factorial drug resistant cancer cell lines. *BMC Complementary and Alternative Medicine, 14*, 340.

Kuete, V., Nkuete, A. H. L., Mbaveng, A. T., Wiench, B., Wabo, H. K., Tane, P., et al. (2014). Cytotoxicity and modes of action of 4′-hydroxy-2′,6′-dimethoxychalcone and other flavonoids toward drug-sensitive and multidrug-resistant cancer cell lines. *Phytomedicine, 21*(12), 1651–1657.

Kuete, V., Sandjo, L. P., Djeussi, D. E., Zeino, M., Kwamou, G. M., Ngadjui, B., et al. (2014). Cytotoxic flavonoids and isoflavonoids from *Erythrina sigmoidea* towards multi-factorial drug resistant cancer cells. *Investigational New Drugs, 32*, 1053–1062.

Kuete, V., Mbaveng, A. T., Zeino, M., Ngameni, B., Kapche, G. D. W. F., Kouam, S. F., et al. (2015). Cytotoxicity of two naturally occurring flavonoids (dorsmanin F and poinsettifolin B) towards multi-factorial drug-resistant cancer cells. *Phytomedicine, 22*(7–8), 737–743.

Kuete, V., Mbaveng, A. T., Zeino, M., Fozing, C. D., Ngameni, B., Kapche, G. D., et al. (2015). Cytotoxicity of three naturally occurring flavonoid derived compounds (artocarpesin, cycloartocarpesin and isobavachalcone) towards multi-factorial drug-resistant cancer cells. *Phytomedicine, 22*(12), 1096–1102.

Kuete, V., Mbaveng, A. T., Nono, E. C., Simo, C. C., Zeino, M., Nkengfack, A. E., et al. (2016). Cytotoxicity of seven naturally occurring phenolic compounds towards multi-factorial drug-resistant cancer cells. *Phytomedicine, 23*(8), 856–863.

Kuete, V., Ngnintedo, D., Fotso, G. W., Karaosmanoğlu, O., Ngadjui, B. T., Keumedjio, F., et al. (2018). Cytotoxicity of seputhecarpan D, thonningiol and 12 other phytochemicals from African flora towards human carcinoma cells. *BMC Complementary and Alternative Medicine, 18*(1), 36.

Lam, P. Y., Lui, A. C. W., Yamamura, M., Wang, L., Takeda, Y., Suzuki, S., et al. (2019). Recruitment of specific flavonoid B-ring hydroxylases for two independent biosynthesis pathways of flavone-derived metabolites in grasses. *The New Phytologist, 223*(1), 204–219.

Lee, M. A., Kim, W. K., Park, H. J., Kang, S. S., & Lee, S. K. (2013). Anti-proliferative activity of hydnocarpin, a natural lignan, is associated with the suppression of Wnt/β-catenin signaling pathway in colon cancer cells. *Bioorganic & Medicinal Chemistry Letters, 23*(20), 5511–5514.

Lepiniec, L., Debeaujon, I., Routaboul, J. M., Baudry, A., Pourcel, L., Nesi, N., et al. (2006). Genetics and biochemistry of seed flavonoids. *Annual Review of Plant Biology, 57*, 405–430.

Lin, S., Singh, R. K., & Moehninsi Navarre, D. A. (2021). R2R3-MYB transcription factors, StmiR858 and sucrose mediate potato flavonol biosynthesis. *Horticulture Research, 8*(1), 25.

Lin, Y., Shi, R., Wang, X., & Shen, H. M. (2008). Luteolin, a flavonoid with potential for cancer prevention and therapy. *Current Cancer Drug Targets, 8*(7), 634–646.

Mbaveng, A. T., Kuete, V., & Efferth, T. (2017). Potential of Central, Eastern and Western Africa medicinal plants for cancer therapy: Spotlight on resistant cells and molecular targets. *Frontiers in Pharmacology, 8*, 343.

Mbaveng, A. T., Kuete, V., Nguemeving, J. R., Beng, V. P., Nkengfack, A. E., Marion Meyer, J. J., et al. (2008). Antimicrobial activity of the extracts and compounds from *Vismia guineensis* (Guttiferae). *Asian Journal of Traditional Medicine, 3*, 211–223.

Mbaveng, A. T., Ngameni, B., Kuete, V., Simo, I. K., Ambassa, P., Roy, R., et al. (2008). Antimicrobial activity of the crude extracts and five flavonoids from the twigs of *Dorstenia barteri* (Moraceae). *Journal of Ethnopharmacology, 116*(3), 483–489.

Mbaveng, A. T., Damen, F., Simo Mpetga, J. D., Awouafack, M. D., Tane, P., Kuete, V., et al. (2019). Cytotoxicity of crude extract and isolated constituents of the *Dichrostachys cinerea* bark towards multifactorial drug-resistant cancer cells. *Evidence-Based Complementary and Alternative Medicine, 2019*, 8450158.

Mbaveng, A. T., Noulala, C. G. T., Samba, A. R. M., Tankeo, S. B., Fotso, G. W., Happi, E. N., et al. (2020). Cytotoxicity of botanicals and isolated phytochemicals from *Araliopsis soyauxii* Engl. (Rutaceae) towards a panel of human cancer cells. *Journal of Ethnopharmacology, 267*, 113535.

Mbaveng, A. T., Chi, G. F., Bonsou, I. N., Ombito, J. O., Yeboah, S. O., Kuete, V., et al. (2020). Cytotoxic phytochemicals from the crude extract of *Tetrapleura tetraptera* fruits towards multi-factorial drug resistant cancer cells. *Journal of Ethnopharmacology, 267*, 113632.

Mbaveng, A. T., Wamba, B. E. N., Bitchagno, G. T. M., Tankeo, S. B., Çelik, İ., Atontsa, B. C. K., et al. (2021). Bioactivity of fractions and constituents of *Piper capense* fruits towards a broad panel of cancer cells. *Journal of Ethnopharmacology, 271*, 113884.

Nabavi, S. M., Šamec, D., Tomczyk, M., Milella, L., Russo, D., Habtemariam, S., et al. (2020). Flavonoid biosynthetic pathways in plants: Versatile targets for metabolic engineering. *Biotechnology Advances, 38*, 107316.

Naing, A. H., & Kim, C. K. (2018). Roles of R2R3-MYB transcription factors in transcriptional regulation of anthocyanin biosynthesis in horticultural plants. *Plant Molecular Biology, 98*(1-2), 1–18.

Ndip, R. N., Tanih, N. F., & Kuete, V. (2013). 20—Antidiabetes activity of African medicinal plants. In Kuete (Ed.), *Medicinal plant research in Africa* (pp. 753–786). Oxford: Elsevier.

Nejabati, H. R., & Roshangar, L. (2022). Kaempferol: A potential agent in the prevention of colorectal cancer. *Physiological Reports, 10*(20), e15488.

Ngaffo, C. M. N., Tchangna, R. S. V., Mbaveng, A. T., Kamga, J., Harvey, F. M., Ngadjui, B. T., et al. (2020). Botanicals from the leaves of *Acacia sieberiana* had better cytotoxic effects than isolated phytochemicals towards MDR cancer cells lines. *Heliyon, 6*(11), e05412.

Ngameni, B., Touaibia, M., Belkaid, A., Ambassa, P., Watchueng, J., Patnama, R., et al. (2007). Inhibition of matrix metalloproteinase-2 secretion by chalcones from the twigs of *Dorstenia barteri* Bureau. *ARKIVOC: Free Online Journal of Organic Chemistry/Arkat-USA, Inc, 2007*(9), 91–103.

Ngameni, B., Fotso, G. W., Kamga, J., Ambassa, P., Abdou, T., Fankam, A. G., et al. (2013). 9—Flavonoids and related compounds from the medicinal plants of Africa. In Kuete (Ed.). *Medicinal plant research in Africa* (pp. 301–350). Oxford: Elsevier.

Nishimura, R., Tabata, K., Arakawa, M., Ito, Y., Kimura, Y., Akihisa, T., et al. (2007). Isobavachalcone, a chalcone constituent of *Angelica keiskei*, induces apoptosis in neuroblastoma. *Biological and Pharmaceutical Bulletin, 30*(10), 1878–1883.

Njamen, D., Talla, E., Mbafor, J. T., Fomum, Z. T., Kamanyi, A., Mbanya, J. C., et al. (2003). Anti-inflammatory activity of erycristagallin, a pterocarpene from *Erythrina mildbraedii*. *European Journal of Pharmacology, 468*(1), 67–74.

Ono, E., & Nakayama, T. (2007). Molecular breeding of novel yellow flowers by engineering the aurone biosynthetic pathway. *Transgenic Plant Journal, 1*, 66–80.

Pandey, B. P., Lee, N., Choi, K. Y., Kim, J. N., Kim, E. J., & Kim, B. G. (2014). Identification of the specific electron transfer proteins, ferredoxin, and ferredoxin reductase, for CYP105D7 in Streptomyces avermitilis MA4680. *Applied microbiology and biotechnology, 98*, 5009–5017.

Qian, W., Huijin, Z., Xiaohan, W., Wen, Z., Xian, Z., & Liangsheng, W. (2021). Research progress on flower color of waterlily (Nymphaea). *Acta Horticulturae Sinica, 48*(10), 2087.

Shah, A., & Smith, D. L. (2020). Flavonoids in agriculture: Chemistry and roles in, biotic and abiotic stress responses, and microbial associations. *Agronomy, 10*(8), 1209.

Sharma, D. K., & Hall, I. H. (1991). Hypolipidemic, anti-inflammatory, and antineoplastic activity and cytotoxicity of flavonolignans isolated from *Hydnocarpus wightiana* seeds. *Journal of Natural Products, 54*(5), 1298–1302.

Solfanelli, C., Poggi, A., Loreti, E., Alpi, A., & Perata, P. (2006). Sucrose-specific induction of the anthocyanin biosynthetic pathway in Arabidopsis. *Plant Physiology, 140*(2), 637–646.

Song, X., Diao, J., Ji, J., Wang, G., Guan, C., Jin, C., et al. (2016). Molecular cloning and identification of a flavanone 3-hydroxylase gene from *Lycium chinense*, and its overexpression enhances drought stress in tobacco. *Plant Physiology and Biochemistry, 98*, 89–100.

Sparvoli, F., Martin, C., Scienza, A., Gavazzi, G., & Tonelli, C. (1994). Cloning and molecular analysis of structural genes involved in flavonoid and stilbene biosynthesis in grape (*Vitis vinifera* L.). *Plant Molecular Biology, 24*(5), 743–755.

Wang, J., Li, G., Li, C., Zhang, C., Cui, L., Ai, G., et al. (2021). NF-Y plays essential roles in flavonoid biosynthesis by modulating histone modifications in tomato. *The New Phytologist, 229*(6), 3237–3252.

Wei, S., Zhang, W., Fu, R., & Zhang, Y. (2021). Genome-wide characterization of 2-oxoglutarate and Fe(II)-dependent dioxygenase family genes in tomato during growth cycle and their roles in metabolism. *BMC Genomics, 22*(1), 126.

Wen, C. H., Tsao, N. W., Wang, S. Y., & Chu, F. H. (2021). Color variation in young and senescent leaves of Formosan sweet gum (*Liquidambar formosana*) by the gene regulation of anthocyanidin biosynthesis. *Physiologia Plantarum, 172*(3), 1750–1763.

Wen, K., Fang, X., Yang, J., Yao, Y., Nandakumar, K. S., Salem, M. L., et al. (2021). Recent research on flavonoids and their biomedical applications. *Current Medicinal Chemistry, 28*(5), 1042–1066.

Xu, F., Cheng, H., Cai, R., Li, L. L., Chang, J., Zhu, J., et al. (2008). Molecular cloning and function analysis of an anthocyanidin synthase gene from *Ginkgo biloba*, and its expression in abiotic stress responses. *Molecules and Cells, 26*(6), 536–547.

Yadav, P., Vats, R., Bano, A., Vashishtha, A., & Bhardwaj, R. (2022). A Phytochemicals approach towards the treatment of cervical cancer using polyphenols and flavonoids. *Asian Pacific Journal of Cancer Prevention, 23*(1), 261–270.

Youmbi, L. M., Makong, Y. S. D., Mbaveng, A. T., Tankeo, S. B., Fotso, G. W., Ndjakou, B. L., et al. (2023). Cytotoxicity of the methanol extracts and compounds of *Brucea antidysenterica* (Simaroubaceae) towards multifactorial drug-resistant human cancer cell lines. *BMC Complementary Medicine and Therapies, 23*, 48.

Yun, C. S., Yamamoto, T., Nozawa, A., & Tozawa, Y. (2008). Expression of parsley flavone synthase I establishes the flavone biosynthetic pathway in *Arabidopsis thaliana*. *Bioscience, Biotechnology, and Biochemistry, 72*(4), 968–973.

Zhou, L., Wang, Y., & Peng, Z. (2009). Advances in study on formation mechanism and genetic engineering of yellow flowers. *Scientia Silvae Sinicae, 45*(2), 111–119.

Zhu, J., Zhao, W., Li, R., Guo, D., Li, H., Wang, Y., et al. (2021). Identification and characterization of chalcone isomerase genes involved in flavonoid production in *Dracaena cambodiana*. *Frontiers in Plant Science, 12*, 616396.

Zuk, M., Szperlik, J., Hnitecka, A., & Szopa, J. (2019). Temporal biosynthesis of flavone constituents in flax growth stages. *Plant Physiology and Biochemistry, 142*, 234–245.

CHAPTER SIX

Isoflavonoids from African medicinal plants can be useful in the fight against cancer and cancer drug resistance

Jenifer R.N. Kuete[a], Armelle T. Mbaveng[b], Leonidah K. Omosa[c], and Victor Kuete[b,*]

[a]Department of Chemistry, Faculty of Science, University of Dschang, Dschang, Cameroon
[b]Department of Biochemistry, Faculty of Science, University of Dschang, Dschang, Cameroon
[c]Department of Chemistry, Faculty of Science and Technology, University of Nairobi, Nairobi, Kenya
*Corresponding author. e-mail address: vk96701@gmail.com

Contents

Abstract

Isoflavonoids are a class of flavonoids also referred to as phytoestrogens, due to their biological effects *via* the estrogenic receptor. They possess interesting biological properties including antifungal, antibacterial, antiviral, anticancer, and antioxidant amongst others. In the present review, we have reported the cytotoxicity of 24 isoflavonoids isolated from African medicinal plants on various human cancer cell lines. They include nine isoflavones, five isoflavanones, and ten pterocarpans. The most potent cytotoxic isoflavones were 5,7-dihydroxy-4′-methoxy-6,8-diprenylisoflavone (**3**), 6,7,3′-Trimethoxy-4′,5′-methylenedioxyisoflavone (**4**), alpinumisoflavone (**5**), and osajin (**9**). The promising isoflavanones include seputheisoflavone (**12**) and sigmoidin I (**14**). The most potent pterocarpans were 6α-hydroxyphaseollidin (**15**), isoneorautenol (**19**), seputhecarpan A (**20**), seputhecarpan D (**23**), and sophorapterocarpan A (**24**). Isoflavonoids that can

Advances in Botanical Research, Volume 115
ISSN 0065-2296, https://doi.org/10.1016/bs.abr.2024.02.004

combat the multidrug resistance of cancer were identified as compounds **3, 5, 9, 15, 19,** and **24**. The most prominent mode of action for some of the active isoflavonoids was apoptotic cell death, mainly mediated by caspase activation, alteration of mitochondrial membrane potential, and enhanced reactive oxygen species production in cancer cells. Based on this review, the most potent cytotoxic isoflavonoids deserve in-depth studies to develop novel cytotoxic products to fight cancer.

Abbreviations

D.R. Degree of resistance.
HID Hydroxyisoflavanone dehydratase.
IC_{50} Inhibitory concentration 50.
IFS Isoflavone synthase.
MMP Mitochondrial membrane potential.
ROS Reactive oxygen species.

1. Introduction

Isoflavonoids are a class of flavonoids, also referred to with their derivatives as phytoestrogens, due to their biological effects *via* the estrogenic receptor. Genistein (7,4′-dihydroxy-6-methoxyisoflavone), daidzein (7,4′-dihydroxyisoflavone), glycitein (7,4′-dihydroxy-6-methoxyisoflavone), biochanin A (5,7-dihydroxy-4′-methoxyisoflavone), and formononetin (7-hydroxy-4′-methoxyisoflavone) belong to isoflavone phytoestrogens (Křížová, Dadáková, Kašparovská, & Kašparovský, 2019). Isoflavonoids constitute a large group of plant natural products and play important roles in plant defense (Wang, 2011). They also possess valuable health-promoting activities with significant health benefits for animals and humans (Wang, 2011). Medically, they have been used together with other compounds in many dietary supplements though the medical and scientific community is generally skeptical of their use. Some of them have been identified as toxins, including biliatresone which may cause biliary atresia in infants (Lorent et al., 2015). Isoflavonoids have been identified primarily in leguminous plants and are synthesized through the central phenylpropanoid pathway and the specific isoflavonoid branch pathways in legumes (Wang, 2011). There are many isoflavonoids skeletal structures including isoflavones, isoflavanones, isoflavans, pterocarpans, rotenoids amongst others (Fig. 1) belonging to a wide family of compounds known as flavonoids. Flavonoids in the narrow sense have the 2-phenylchromen-4-one backbone while isoflavonoids have the 3-phenylchromen-4-one backbone with no hydroxyl group substitution at position 2 (case of the isoflavones) or the 3-phenylchroman (isoflavan) backbone (case of

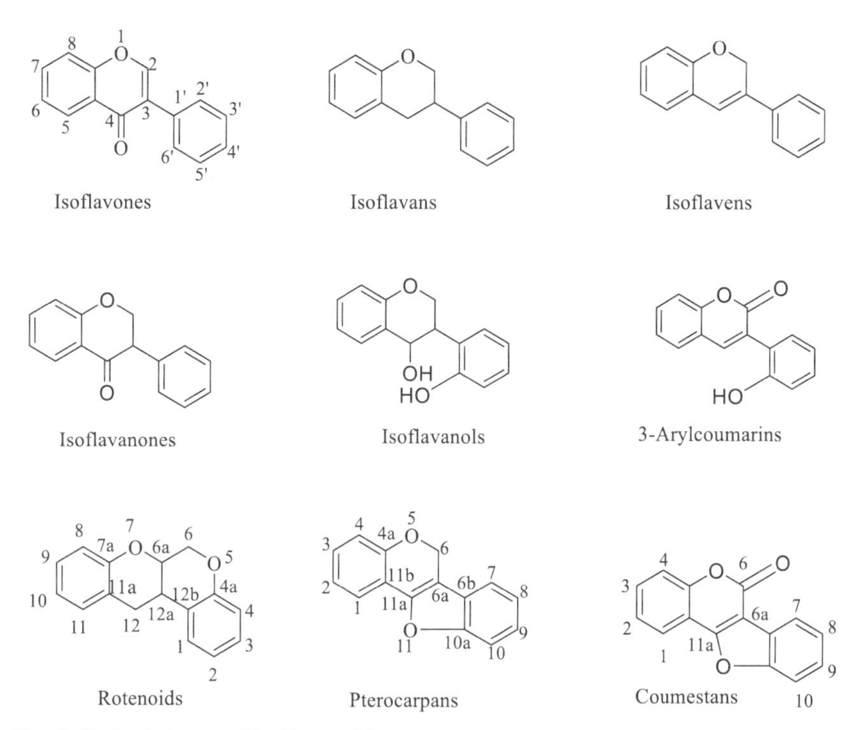

Fig. 1 Basis skeleton of isoflavonoids.

isoflavanes). They are derived from the flavonoid biosynthesis pathway *via* liquiritigenin or naringenin (Dixon & Ferreira, 2002; Ko, 2014; Křížová et al., 2019; Ralston, Subramanian, Matsuno, & Yu, 2005). The main dietary sources of isoflavones for humans are soybean and its products, which contain mainly daidzein and genistein (Křížová et al., 2019). Isoflavones have estrogenic and antiestrogenic effects as well as chemotherapeutic effects against several cancer types, cardiovascular diseases, osteoporosis, or menopausal symptoms (Křížová et al., 2019). They also have antifungal, antibacterial, antiviral, and antioxidant properties (Argenta et al., 2015; Kuete & Efferth, 2015; Kuete, 2010; Kuete, 2013; Li, Li, Yu, Jiao, & Cao, 2018; Mbaveng, Kuete, & Efferth, 2017; Ngameni et al., 2013). Nonetheless, they may also be considered endocrine disruptors with possible negative influences on the state of health in a certain part of the population or on the environment (Křížová et al., 2019). They also prevent angiogenesis, thereby being important in the fight against malignant tumors (Bellou et al., 2012). In the present review, the synopsis of the cytotoxic potential of isoflavonoids isolated from African medicinal plants against human cancer cell lines including the multidrug-resistant phenotypes is provided.

2. Biosynthesis of isoflavonoids: Biosynthesis of isoflavone

Flavonoids are synthesized by extension of *p*-hydroxycoumaroyl CoA with three molecules of malonyl CoA in a linear manner to form a tetraketide intermediate. This step is catalyzed by the enzyme chalcone synthase (CHS). The intermediate then folds and condenses further to give a chalcone. This reaction is the first committed step in flavonoid biosynthesis and is also catalyzed by CHS. Chalcone isomerase is the second enzyme involved in the biosynthesis of flavonoids and it catalyzes the stereospecific intramolecular cyclization of the chalcones isoliquiritigenin and naringeninchalcone into the flavanones (2S)-liquiritigenin and (2S)-narigenin, respectively. These flavanones are the precursors for the construction of 5-deoxyflavonoid and 5-hydroxyflavonoid skeletons. The isoflavone biosynthesis pathway is mainly distributed in leguminous plants. Isoflavone synthase (IFS) leads flavanone to the isoflavone pathway and appears to be able to use both naringenin and liquiritigenin as substrates to generate 2-hydroxy-2,3-dihydrogenistein and 2,7,4′-trihydroxyisoflavanone, respectively (Pandey et al., 2014). These are further converted to isoflavone genistein and daidzein under the action of hydroxyisoflavanone dehydratase (HID). IFS and HID catalyze two reactions to produce isoflavone, that is, the formation of a double bond between positions C-2 and C-3 of ring C and a shift of ring B from position C-2 to C-3 of ring C (He, Blount, Ge, Tang, & Dixon, 2011). The proposed mechanism for isoflavonoid formation involves two steps (Scheme 1). The first is the oxidation of a flavone and its rearrangement, then the second is the elimination of a water molecule, which leads to an isoflavone (ex: (2S)-liquiritigenine to daidzeine). The first step catalyzed by the IFS (cytochrome P450 dependent mono-oxidase) in the presence of NADPH and oxygen leads to 2-hydroxyisoflavonone. The mechanism would be radical, the hydroxylation accompanying the migration of the aryl. the isolation of the hydroxylated derivative at C-2 makes impossible the mechanism initially envisaged, namely the formation of an epoxide whose protonation and opening would allow the formation of a spirodienonic intermediate, then that of the isoflavone. The biosynthesis of pterocarpans and rotenones are shown respectively in Schemes 2 and 3.

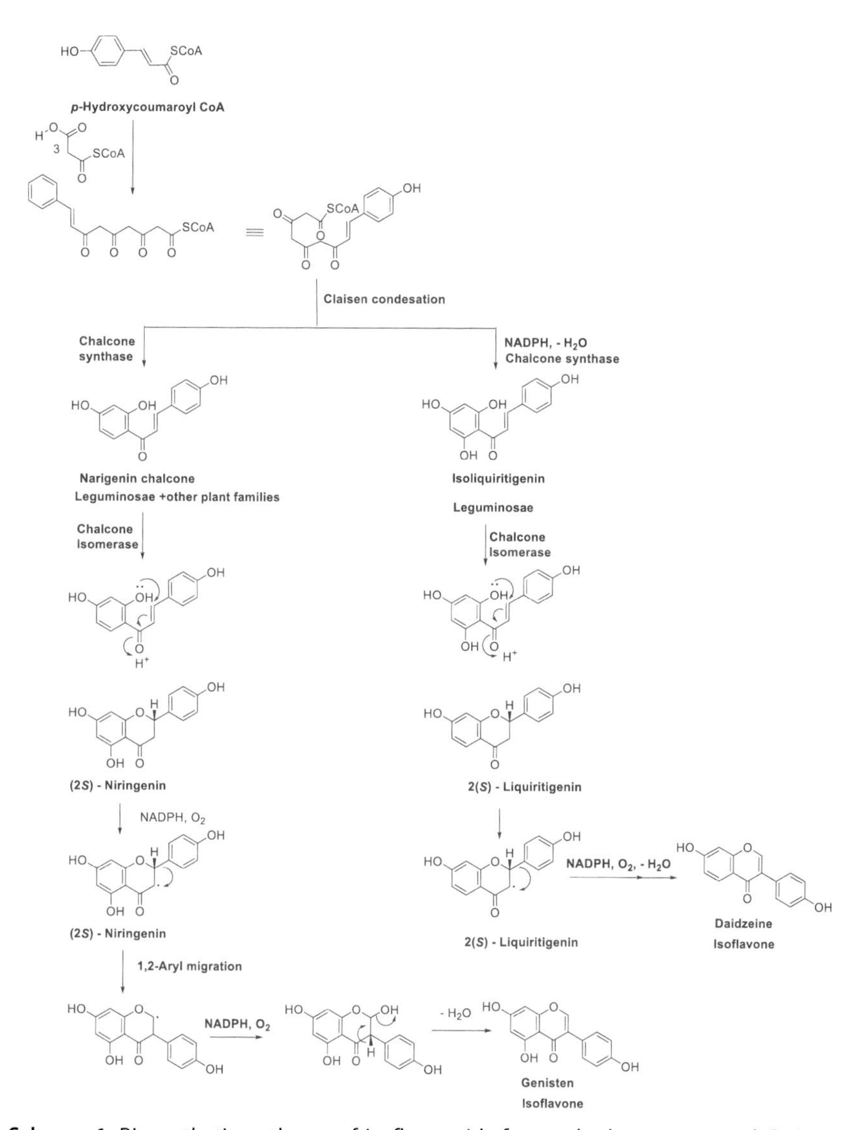

Scheme 1 Biosynthetic pathway of isoflavonoids from p-hydroxycoumaroyl CoA.

Scheme 2 Biosynthetic pathway of pterocarpans from isoflavones.

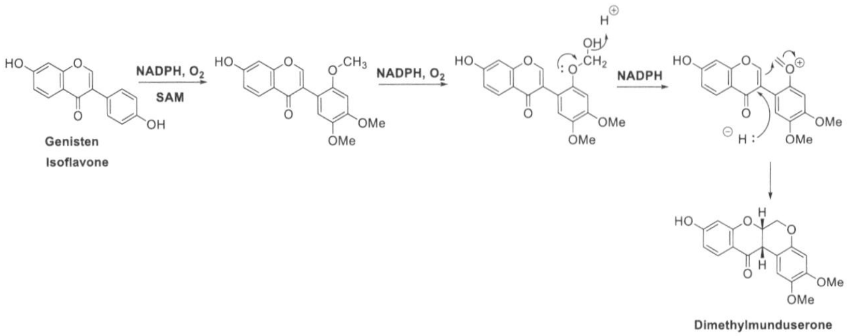

Scheme 3 Biosynthetic pathway of rotenones from isoflavones.

CoA = Coenzyme A; NADPH = Nicotine Adenine Dinucleotide; O_2 = Oxygen.
NADPH = Nicotine Adenine Dinucleotide; O_2 = Oxygen.
NADPH = Nicotine Adenine Dinucleotide; O_2 = Oxygen; SAM = S-Adenosylmethionie.

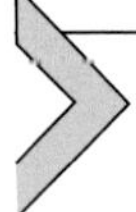

3. Cancer cell lines used in the screening of the cytotoxicity of isoflavonoids isolated from African medicinal plants

Several cancer cell lines were used to assess the cytotoxicity of isoflavonoids isolated from African medicinal plants. They include cell lines from breast cancer (MCF-7, MDA-MB-231-*pcDNA*, and MDA-MB-231-*BCRP*), colon cancer (Caco-2, DLD-1, HCT116 *p53*$^{+/+}$, and HCT116 *p53*$^{-/-}$), glioblastoma (U87MG and U87MG.Δ*EGFR*), hepatocarcinoma (HepG2), leukemia (CCRF-CEM and CEM/ADR5000), and lung cancer (A549 and SPC212) (Table 1).

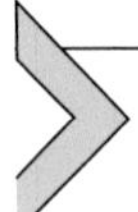

4. Cytotoxic isoflavonoids from the African medicinal plants towards drug sensitive and multidrug-resistant cancer cells

A total of 24 isoflavonoids belonging to isoflavones, isoflavanones, and pterocarpans displayed cytotoxic activities against various human cancer cell lines. Their effects on various tested human cancer cell lines are summarized in Table 1. The degree of resistance (D.R.) determined as the ratio of the IC_{50} value in resistant cells *versus* the IC_{50} values in the corresponding

Table 1 Cytotoxic isoflavonoids from African medicinal plants and their effects on sensitive and drug-resistant cancer cell lines.

Compound names	Source	Country	Cancer cell lines, IC_{50} values (μM) and degree of resistance in bracket	References
Isoflavones				
2′-Dimethoxy-3′,4′-methylenedioxyisoflavone (1)	*Lonchocarpus bussei*	Kenya	49.49 (CCRF-CEM) *vs* 32.47 (CEM/ADR5000) [0.65]	Adem, Kuete, et al. (2019)
4′-Prenyloxyvigvexin A (2)	*Lonchocarpus bussei*	Kenya	18.92 (CCRF-CEM) *vs* 25.53 (CEM/ADR5000) [1.34]; 83.08 (MDA-MB-231-*BCRP*); 57.49 (HCT116 *p53*$^{+/+}$) *vs* 67.42 (HCT116 *p53*$^{-/-}$) [1.17]; 85.71 (U87MG) *vs* 56.31 (U87MG.Δ*EGFR*) [0.65]	Adem, Kuete, et al. (2019)
5,7-Dihydroxy-4′-methoxy-6, 8-diprenylisoflavone (3)	*Ormocarpum kirkii*	Kenya	5.17 (CCRF-CEM) *vs* 3.87 (CEM/ADR5000) [0.74]; 9.95 (MDA-MB-231-*pcDNA*) *vs* 14.44 (MDA-MB-231-*BCRP*) [1.45]; 8.49 (HCT116 *p53*$^{+/+}$) *vs* 8.76 (HCT116 *p53*$^{-/-}$) [1.03]; 12.25 (U87MG) *vs* 7.85 (U87MG.Δ*EGFR*) [0.64]; 12.86 (HepG2)	Adem, Mbaveng, et al. (2019)
6,7,3′-Trimethoxy-4′,5′-methylenedioxyisoflavone (4)	*Lonchocarpus bussei*	Kenya	6.27 (CCRF-CEM) *vs* 29.51 (CEM/ADR5000) [6.70]	Adem, Kuete, et al. (2019)

(*continued*)

Table 1 Cytotoxic isoflavonoids from African medicinal plants and their effects on sensitive and drug-resistant cancer cell lines. (*cont'd*)

Compound names	Source	Country	Cancer cell lines, IC_{50} values (μM) and degree of resistance in bracket	References
Alpinumisoflavone (5)	*Ficus chlamydocarpa*	Cameroon	9.60 (CCRF-CEM) *vs* 5.91 (CEM/ADR5000) [0.62];.42.57 (MDA-MB-231-*pcDNA*) vs 65.65 (MDA-MB-231-*BCRP*) [1.54]; 42.37 (HCT116 $p53^{+/+}$) vs 36.40 (HCT116 $p53^{-/-}$) [0.86]; 46.77 (U87MG) vs 42.04 (U87MG.Δ*EGFR*) [0.90]; 37.99 (HepG2)	Kuete et al. (2016); Kuete et al. (2008)
Durmillone (6)	*Lonchocarpus bussei*	Kenya	0.54 (CCRF-CEM) vs 0.86 (CEM/ADR5000) [1.59]	Adem, Kuete, et al. (2019)
Laburnetin (7)	*Ficus chlamydocarpa*	Cameroon	27.63 (CCRF-CEM) vs 43.93 (CEM/ADR5000) [1.59]; 107.57 (MDA-MB-231-*pcDNA*) vs 105.31 (MDA-MB-231-*BCRP*) [0.98]; 70.25 (HCT116 $p53^{+/+}$) *vs* 51.81 (HCT116 $p53^{-/-}$) [0.74]; 61.29 (U87MG) *vs* 52.42 (U87MG.Δ*EGFR*) [0.86]; 66.17 (HepG2)	Kuete et al. (2016); Kuete et al. (2008)
Neobavaisoflavone (8)	*Erythrina senegalensis*	Cameroon	42.93 (CCRF-CEM) *vs* 51.36 (CEM/ADR5000) [1.20]; 114.64 (HCT116 $p53^{+/+}$) *vs* 99.57 (HCT116 $p53^{-/-}$) [0.87]; 72.67 (U87MG.Δ*EGFR*); 110.27 (HepG2)	Kuete, Sandjo, Kwamou, et al. (2014)

Osajin (9)	*Ormocarpum kirkii*	Kenya	15.05 (CCRF-CEM) *vs* 5.81 (CEM/ADR5000) [0.38]; 24.28 (MDA-MB-231-*pcDNA*) *vs* 17.82 (MDA-MB-231-*BCRP*) [0.73]; 19.67 (HCT116 $p53^{+/+}$) *vs* 23.15 (HCT116 $p53^{-/-}$) [1.17]; 42.23 (U87MG) *vs* 16.90 (U87MG.Δ*EGFR*) [0.40]; 48.67 (HepG2)	Adem, Mbaveng, et al. (2019)
Isoflavanones				
Bidwillon A (10)	*Erythrina sigmoidea*	Cameroon	13.70 (CCRF-CEM) *vs* 16.17 (CEM/ADR5000) [1.18]; 16.37 (MDA-MB-231-*BCRP*); 19.01 (HCT116 $p53^{+/+}$) *vs* 16.64 (HCT116 $p53^{-/-}$) [0.88]; 25.53 (U87MG) *vs* 9.90 (U87MG.Δ*EGFR*) [0.39]; 17.28 (HepG2)	Kuete, Sandjo, Djeussi, et al. (2014)
Glyasperin F (11)	*Ptycholobium contortum*	Cameroon	11.14 (MCF-7); 17.60 (Caco-2); 20.89 (DLD-1); 13.19 (A549); 16.38 (SPC212)	Kuete et al. (2018); Ngnintedo et al. (2016)
Seputheisoflavone (12)	*Ptycholobium contortum*	Cameroon	40.80 (CCRF-CEM) *vs* 28.08 (CEM/ADR5000) [0.69]; 45.05 (MDA-MB-231-*pcDNA*) *vs* 42.64 (MDA-MB-231-*BCRP*) [0.95]; 35.86 (HCT116 $p53^{+/+}$) *vs* 68.66 (HCT116 $p53^{-/-}$) [1.91]; 29.27 (U87MG) *vs* 23.80 (U87MG.Δ*EGFR*) [0.81]; 52.80 (HepG2); 38.68 (A549); 0.59 (SPC212)	Mbaveng et al. (2018); Ngnintedo et al. (2016)

(continued)

Table 1 Cytotoxic isoflavonoids from African medicinal plants and their effects on sensitive and drug-resistant cancer cell lines. (*cont'd*)

Compound names	Source	Country	Cancer cell lines, IC_{50} values (μM) and degree of resistance in bracket	References
Sigmoidin H (13)	*Erythrina senegalensis*	Cameroon	97.59 (CCRF-CEM) *vs* 99.83 (CEM/ADR5000) [1.02]; 110.51 (HCT116 *p53*$^{+/+}$); 25.59 (U87MG); 24.37 (HepG2)	Kuete, Sandjo, Kwamou, et al. (2014)
Sigmoidin I (14)	*Erythrina sigmoidea*	Cameroon	4.24 (CCRF-CEM) *vs* 24.65 (CEM/ADR5000) [5.81]; 29.48 (HCT116 *p53*$^{+/+}$) *vs* 25.67 (HCT116 *p53*$^{-/-}$) [0.87]; 24.37 (U87MG) *vs* 15.90 (U87MG.Δ*EGFR*) [0.65]	Djeussi et al. (2015); Kuete, Sandjo, Djeussi, et al. (2014)
Pterocarpans				
6α-Hydroxyphaseollidin (15)	*Erythrina sigmoidea*	Cameroon	3.36 (CCRF-CEM) *vs* 5.51 (CEM/ADR5000) [1.64]; 5.70 (MDA-MB-231-*pcDNA*) *vs* 5.87 (MDA-MB-231-*BCRP*) [1.03]; 5.68 (HCT116 *p53*$^{+/+}$) *vs* 4.60 (HCT116 *p53*$^{-/-}$) [0.81]; 4.91 (U87MG) *vs* 4.91 (U87MG.Δ*EGFR*) [1.00]; 6.44 (HepG2)	Kuete, Sandjo, Djeussi, et al. (2014)
(6*aR*,11*aR*)-3,8-Dimethoxybitucarpin B (16)	*Lonchocarpus eriocalyx*	Kenya	31.82 (CCRF-CEM) *vs* 16.87 (CEM/ADR5000) [0.53]	Adem, Kuete, et al. (2019)

(6*aR*,11*aR*)-Edunol (17)	*Lonchocarpus bussei. Lonchocarpus eriocalyx*	Kenya	34.11 (CCRF-CEM) *vs* 33.22 (CEM/ADR5000) [0.97]	Adem, Kuete, et al. (2019)
(6*aR*,11*aR*)-Maackiain (18)	*Lonchocarpus bussei*	Kenya	81.34 (CEM/ADR5000)	Adem, Kuete, et al. (2019)
Isoneorautenol (19)	*Erythrina excelsa*	Cameroon	7.51 (CCRF-CEM) *vs* 14.86 (CEM/ADR5000) [1.98]; 15.85 (MDA-MB-231-*pcDNA*) *vs* 2.67 (MDA-MB-231-*BCRP*) [0.17]; 12.62 (HCT116 $p53^{+/+}$) *vs* 9.89 (HCT116 $p53^{-/-}$) [0.78]; 21.84 (U87MG) *vs* 20.22 (U87MG.Δ*EGFR*) [0.93]; 19.46 (HepG2)	Djeussi et al. (2015); Kuete, Sandjo, Kwamou, et al. (2014); Kwamou et al. (2014)
Seputhecarpan A (20)	*Ptycholobium contortum*	Cameroon	47.39 (CCRF-CEM) *vs* 12.17 (CEM/ADR5000) [0.26]; 65.08 (MDA-MB-231-*pcDNA*) *vs* 40.23 (MDA-MB-231-*BCRP*) [0.62]; 29.67 (HCT116 $p53^{+/+}$) *vs* 34.16 (HCT116 $p53^{-/-}$) [1.15]; 26.65 (U87MG) *vs* 30.65 (U87MG.Δ*EGFR*) [1.15]; 46.51 (HepG2); 46.70 (A549); 9.35 (SPC212)	Mbaveng et al. (2018)
Seputhecarpan B (21)	*Ptycholobium contortum*	Cameroon	11.92 (MCF-7); 84.41 (Caco-2); 57.93 (DLD-1); 61.92 (HepG2)	Kuete et al. (2018)

(continued)

Table 1 Cytotoxic isoflavonoids from African medicinal plants and their effects on sensitive and drug-resistant cancer cell lines. (*cont'd*)

Compound names	Source	Country	Cancer cell lines, IC_{50} values (μM) and degree of resistance in bracket	References
Seputhecarpan C (22)	*Ptycholobium contortum*	Cameroon	13.30 (MCF-7); 67.29 (Caco-2); 37.84 (DLD-1); 76.72 (HepG2); 73.49 (A549), 63.47 (SPC212)	Kuete et al. (2018); Ngnintedo et al. (2016)
Seputhecarpan D (23)	*Ptycholobium contortum*	Cameroon	9.78 (MCF-7); 10.46 (Caco-2); 10.78 (DLD-1); 67.68 (HepG2); 26.39 (A549); 12.99 (SPC212)	Kuete et al. (2018); Ngnintedo et al. (2016)
Sophorapterocarpan A (24)	*Erythrina sigmoidea*	Cameroon	3.73 (CCRF-CEM) *vs* 6.63 (CEM/ADR5000) [1.78]; 12.59 (MDA-MB-231-*pcDNA*) *vs* 11.35 (MDA-MB-231-*BCRP*) [0.90]; 9.35 (HCT116 $p53^{+/+}$) *vs* 8.89 (HCT116 $p53^{-/-}$) [0.95]; 14.81 (U87MG) *vs* 10.15 (U87MG.Δ*EGFR*) [0.69]; 11.63 (HepG2)	Kuete, Sandjo, Djeussi, et al. (2014)

sensitive cell lines: CEM/ADR5000 cells *vs* CCRF-CEM cells, MDA-MB-231-*BCRP* *vs* MDA-MB-231-pcDNA3, HCT116 *p53*$^{-/-}$ cells *vs* HCT116 *p53*$^{+/+}$ cells, and U87MG.Δ*EGFR* *vs* U87MG, is also depicted in Table 1 (Mbaveng et al., 2020). In the laboratory scale, collateral sensitivity is achieved when the D.R., defined as the ratio of the IC_{50} values of the cytotoxic agent towards the resistant cell line *versus* its sensitive counterpart, is below 1 while the D.R. above 1 defines normal sensitivity. It has been established that the D.R.< 0.9 defines collateral sensitivity, while D.R. between 0.9 and 1.2 defines normal sensitivity. The cross-resistance is noted if the cytotoxic agent is more active in the sensitive cell line than its resistant subline, with D.R above 1.2 (Efferth et al., 2020; Efferth et al., 2021; Mbaveng et al., 2017). Collateral or normal sensitivities should be achieved for samples that can combat drug resistant cancer cells (Efferth et al., 2020; Kuete & Efferth, 2015;). In this section, the activity of 24 isoflavonoids will be discussed according to the cut-off points of cytotoxic phytochemicals defined as follows: outstanding activity ($IC_{50} \leq 0.5\,\mu M$), excellent activity ($0.5 < IC_{50} \leq 2\,\mu M$), very good activity ($2 < IC_{50} \leq 5\,\mu M$), good activity ($5 < IC_{50} \leq 10\,\mu M$), average activity ($10 < IC_{50} \leq 20\,\mu M$), weak activity ($20 < IC_{50} \leq 60\,\mu M$), very weak activity ($60 < IC_{50} \leq 150\,\mu M$), and not active ($IC_{50} > 150\,\mu M$) (Kuete, 2025).

4.1 Cytotoxic isoflavones

Nine potent cytotoxic isoflavones were identified in African medicinal plants. They include 2′-dimethoxy-3′,4′-methylenedioxyisoflavone (**1**), 4′-prenyloxyvigvexin A (**2**), 5,7-dihydroxy-4′-methoxy-6,8-diprenylisoflavone (**3**), 6,7,3′-Trimethoxy-4′,5′-methylenedioxyisoflavone (**4**), alpinumisoflavone (**5**), durmillone (**6**), laburnetin (**7**), neobavaisoflavone (**8**), and osajin (**9**). Their chemical structures are depicted in Fig. 2 while their cytotoxic activities in human cancer cell lines are shown in Table 1. Very good antiproliferative activity ($2 < IC_{50} \leq 5\,\mu M$) was obtained by isoflavone **3** against CEM/ADR5000 cells, while good activities ($5 < IC_{50} \leq 10\,\mu M$) were shown by isoflavones **3, 4,** and **5** against CCRF-CEM cells, **5** and **9** against CEM/ADR5000 cells, as well as **3** against MDA-MB-231-*pcDNA* cells, HCT116 *p53*$^{+/+}$ cells, HCT116 *p53*$^{-/-}$ cells, and U87MG.Δ*EGFR* cells. Collateral sensitivities of CEM/ADR5000 cells *vs* CCRF-CEM cancer cells were observed for **3, 5,** and **9** with the D.R. of 0.74, 0.62, and 0.38, while that of U87MG.Δ*EGFR* cells *vs* U87MG cells was observed for **3** with the D.R. of 0.47. Normal sensitivity of HCT116 *p53*$^{-/-}$ *vs* HCT116 *p53*$^{+/+}$ was illustrated for **3** with the D.R. of 1.03.

4.2 Cytotoxic isoflavanones

Five potent cytotoxic isoflavanones including bidwillon A (**10**), glyasperin F (**11**), seputheisoflavone (**12**), sigmoidin H (**13**), and sigmoidin I (**14**) were isolated from African plants. Their chemical structures are shown in Fig. 2 meanwhile the cytotoxic activities in human cancer cell lines are elaborated in Table 1. They mostly displayed cytotoxic effects in the range of average ($10 < IC_{50} \leq 20\,\mu M$) to very weak activity ($60 < IC_{50} \leq 150\,\mu M$).

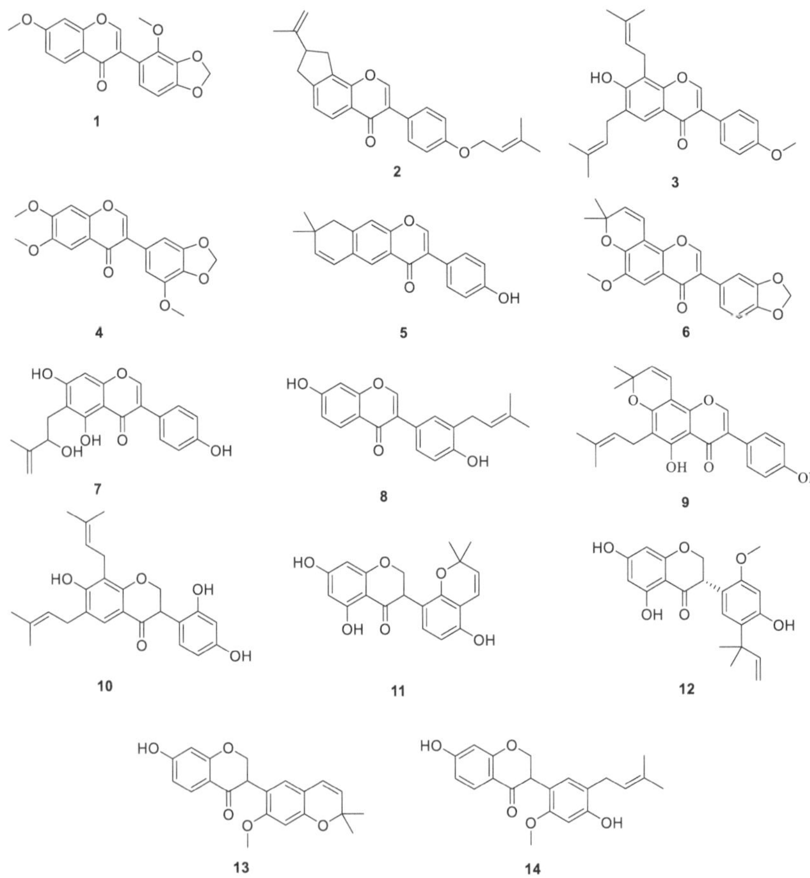

Fig. 2 Chemical structures of potent isoflavones (1–9) and isoflavanone (10–14) isolated from African medicinal plants. 1: 2′-dimethoxy-3′,4′-methylenedioxyisoflavone; 2: 4′-prenyloxyvigvexin A; 3: 5,7-dihydroxy-4'-methoxy-6,8-diprenylisoflavone; 4: 6,7,3′-Trimethoxy-4′,5′-methylenedioxyisoflavone; 5: alpinumisoflavone; 6: durmillone; 7: laburnetin; 8: neobavaisoflavone; 9: osajin; 10: bidwillon A; 11: glyasperin F; 12 seputheisoflavone; 13: sigmoidin H; 14: sigmoidin I.

However, excellent cytotoxic activity was obtained with isoflavanone **12** against SCP212 cells, while very good antiproliferative activity was obtained for isoflavanone **14** against CCRF-CEM cells.

4.3 Cytotoxic pterocarpans

Ten potent cytotoxic pterocarpans were identified in African medicinal plants. They were 6α-hydroxyphaseollidin (**15**), (6aR,11aR)-3,8-dimethoxybitucarpin B (**16**), (6aR,11aR)-edunol (**17**), (6aR,11aR)-maackiain (**18**), isoneorautenol (**19**), seputhecarpan A (**20**), seputhecarpan B (**21**), seputhecarpan C (**22**), seputhecarpan D (**23**), and sophorapterocarpan A (**24**). Their chemical structures are depicted in Fig. 3 meanwhile their cytotoxic activities in human cancer cell lines are shown in Table 1. Amongst them, very good antiproliferative activity was obtained with pterocarpans **15** against HCT116 *p53*$^{-/-}$ cells, U87MG cells, and

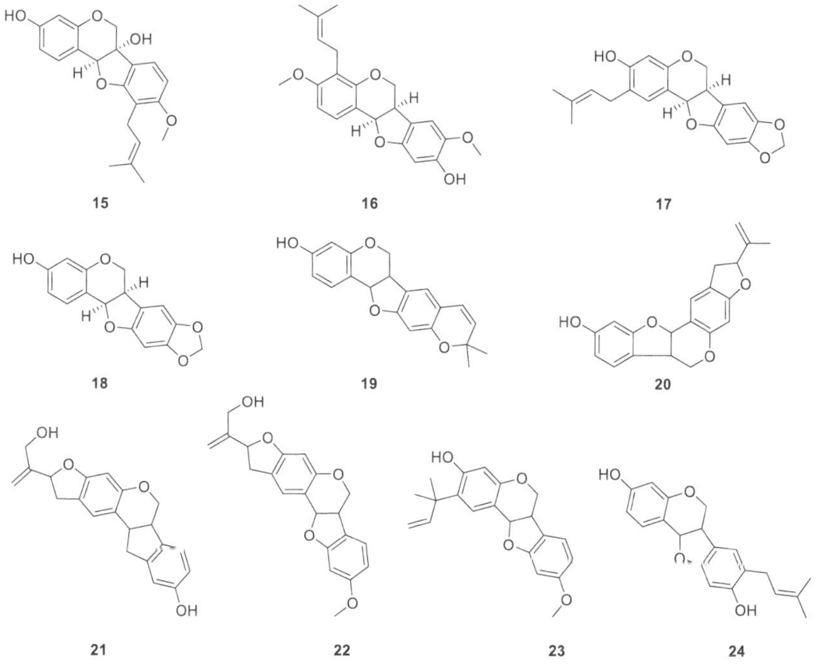

Fig. 3 Chemical structures of potent pterocarpans (15–24) isolated from African medicinal plants. 15: 6α-hydroxyphaseollidin; 16: (6aR,11aR)-3,8-dimethoxybitucarpin B; 17: (6aR,11aR)-edunol; 18: (6aR,11aR)-maackiain; 19: isoneorautenol; 20: seputhecarpan A; 21: seputhecarpan B; 22: seputhecarpan C; 23: seputhecarpan D; 24: sophorapterocarpan A.

U87MG.Δ*EGFR* cells, **19** against MDA-MB-231-*BCRP* cells, and **24** against CCRF-CEM cells. Furthermore, good cytotoxic activities were obtained with pterocarpans **15** and **19** against CCRF-CEM cells, **15** and **24** against CEM/ADR5000 cells, **15** against MDA-MB-231 cells, U87 cells, MDA-MB-231-*BCRP* cells, and HepG2 cells, **19** and **24** against HCT116 $p53^{-/-}$ cells, **24** against HCT116 $p53^{-/-}$ cells, **20** against SPC212 cells, and **23** against MCF-7 cells. Collateral sensitivity of MDA-MB-231-*BCRP* cells *vs* MDA-MB-231 cells was noted for **24** (D.R. 0.90), and of HCT116 $p53^{-/-}$ cells *vs* HCT116 $p53^{+/+}$ cells for **15** (D.R. 0.81) and **19** (D.R. 0.78). Normal sensitivity of MDA-MB-231-*BCRP* cells *vs* MDA-MB-231 cells (D.R. 1.03) and U87MG.Δ*EGFR* cells *vs* U87MG cells for **15** (D.R. 1.00), and HCT116 $p53^{-/-}$ cells *vs* HCT116 $p53^{+/+}$ cells for **24** (D.R. 0.95) were determined.

5. Modes of action of the cytotoxic isoflavonoids from the African medicinal plants

Several potent cytotoxic isoflavonoids isolated from African medicinal induced apoptotic cell death. In effect, the pterocarpan isoflavonoids 6α-hydroxyphaseollidin (**15**) and sophorapterocarpan A (**24**) as well as the isoflavanone sigmoidin I (**14**) and bidwillon A displayed cytotoxic effects against CCRF-CEM, CEM/ADR5000, HCT116 $p53^{+/+}$, HCT116 $p53^{-/-}$, U87. MG, U87. MGΔ*EGFR*, MDA-MB-231-pcDNA3, MDA-MB-231-*BCRP*, and HepG2 cells (Kuete, Sandjo, Djeussi, et al., 2014). Compound **15** had the best antiproliferative activity with IC_{50} values below 10 μM against the nine tested cancer cell lines. The IC_{50} values below 50 μM were also recorded for compounds **24** and **14** against the nine cancer cell lines whilst bidwillon A showed selective activities. Compound **15** induced apoptosis in CCRF-CEM cells mediated by the activation of caspases 3/7, 8, and 9, breakdown of mitochondrial membrane potential (MMP), and increase in reactive oxygen species (ROS) production, whereas the apoptotic process induced by compounds **24** and **14** was mediated by the loss of MMP as well as an increased ROS production (Kuete, Sandjo, Djeussi, et al., 2014).

The cytotoxicity of the isoflavone 5,7-dihydroxy-4′-methoxy-6,8-diprenylisoflavone (**3**) was also determined toward the above-mentioned cancer cell lines (Adem, Mbaveng, et al., 2019). It was found that the compound exhibited antiproliferative activity against all the nine tested cancer cell lines with the IC_{50} values ranging from 7.85 μM in U87MG.Δ*EGFR* cells to 14.44 μM in resistant MDA-MB231/*BCRP* cells. Compound **3** induced apoptosis in CCRF-CEM cells mediated by MMP alteration and increased ROS production (Adem, Mbaveng, et al., 2019).

The cytotoxicity and the modes of action of neobavaisoflavone (**8**) and a pterocarpan isoneorautenol (**19**) were evaluated against a panel of nine cancer cell lines, including CCRF-CEM, CEM/ADR5000, HCT116 $p53^{+/+}$, HCT116 $p53^{-/-}$, U87. MG, U87. MGΔ*EGFR*, MDA-MB-231-pcDNA3, MDA-MB-231-*BCRP*, and HepG2 using RRA (Kuete, Sandjo, Kwamou, et al., 2014). Kuete and his team found that compound **19** had significant cytotoxicity toward sensitive and drug-resistant cancer cell lines. Compound **8** was selectively active. IC_{50} values ranging from 2.67 μM (against MDA-MB 237*BCRP* cells) to 21.84 (toward U87MG) were recorded for compound **19**. They also demonstrated that compound **19** induced apoptosis in CCRF-CEM cells *via* activation of caspases 3/7, 8, and 9 as well as the loss of MMP and increased ROS production (Kuete, Sandjo, Kwamou, et al., 2014). Compound **8** also markedly augmented TRAIL-mediated apoptosis in LNCaP prostate cancer (Szliszka, Czuba, Sedek, Paradysz, & Krol, 2011).

The cytotoxicity and the mode of action of alpinumisoflavone (**5**) were investigated against a panel of nine cancer cell lines, including CCRF-CEM, CEM/ADR5000, HCT116 $p53^{+/+}$, HCT116 $p53^{-/-}$, U87. MG, U87. MGΔ*EGFR*, MDA-MB-231-pcDNA3, MDA-MB-231-*BCRP*, and HepG2 using RRA (Kuete et al., 2016). Compound **5** demonstrated good growth inhibitory effects towards the nine tested cancer cell lines with the recorded IC_{50} values ranging from 5.91 μM (towards CEM/ADR5000 cells) to 65.65 μM (MDA-MB-231-*BCRP* cells); compound **5** induced apoptosis in CCRF-CEM cells, mediated by loss of MMP and increased ROS production (Kuete et al., 2016).

The antiproliferative activity and the mode of action of the isoflavanone glyasperin F (**11**) and the pterocarpan isoflavonoid seputhecarpan D (**23**) were evaluated against MCF7, DLD 1, Caco-2, and HepG2 cell lines using NRUA (Kuete et al., 2018). Compounds **11** and **23** had IC_{50} values below 10 μM in 4/4 and 3/4 tested cell lines, respectively. The two isoflavonoids induced apoptosis in MCF-7 cells through the activation of caspases 3/7 and 9 as well as enhanced ROS production (Kuete et al., 2018).

6. Conclusion

In the present chapter, we have identified 24 isoflavonoids isolated from African medicinal plants with cytotoxic effects on various human cancer cell lines. They include nine isoflavones, five isoflavanones, and ten pterocarpans. The most potent cytotoxic isoflavones were 5,7-dihydroxy-4′-methoxy-6,8-diprenylisoflavone (**3**), 6,7,3′-Trimethoxy-4′,5′-methylenedioxyisoflavone (**4**), alpinumisoflavone (**5**), and osajin (**9**). The best isoflavanones include seputheisoflavone (**12**) and sigmoidin I (**14**). The most potent pterocarpans were 6α-hydroxyphaseollidin (**15**), isoneorautenol (**19**), seputhecarpan A (**20**), seputhecarpan D (**23**), and sophorapterocarpan A (**24**). Isoflavonoids that can combat the multidrug resistance of cancer were identified as compounds **3, 5, 9, 15, 19,** and **24.** Some of the active isoflavonoids displayed apoptotic cell death, mainly mediated by caspase activation, alteration of MMP, and enhanced ROS production in cancer cells. The identified isoflavonoids deserve further in-depth studies to develop novel cytotoxic products to fight cancer.

References

Adem, F. A., Kuete, V., Mbaveng, A. T., Heydenreich, M., Koch, A., Ndakala, A., et al. (2019). Cytotoxic flavonoids from two *Lonchocarpus* species. *Natural Product Research, 33*(18), 2609–2617.

Adem, F. A., Mbaveng, A. T., Kuete, V., Heydenreich, M., Ndakala, A., Irungu, B., et al. (2019). Cytotoxicity of isoflavones and biflavonoids from *Ormocarpum kirkii* towards multi-factorial drug resistant cancer. *Phytomedicine, 58*, 152853.

Argenta, D. F., Silva, I. T., Bassani, V. L., Koester, L. S., Teixeira, H. F., & Simões, C. M. (2015). Antiherpes evaluation of soybean isoflavonoids. *Archives of Virology, 160*(9), 2335–2342.

Bellou, S., Karali, E., Bagli, E., Al-Maharik, N., Morbidelli, L., Ziche, M., et al. (2012). The isoflavone metabolite 6-methoxyequol inhibits angiogenesis and suppresses tumor growth. *Molecular Cancer, 11*, 35.

Dixon, R. A., & Ferreira, D. (2002). Genistein. *Phytochemistry, 60*(3), 205–211.

Djeussi, D. E., Sandjo, L. P., Noumedem, J. A., Omosa, L. K., B, T. N., & Kuete, V. (2015). Antibacterial activities of the methanol extracts and compounds from *Erythrina sigmoidea* against Gram-negative multi-drug resistant phenotypes. *BMC Complementary and Alternative Medicine, 15*(1), 453.

Efferth, T., Kadioglu, O., Saeed, M. E. M., Seo, E. J., Mbaveng, A. T., & Kuete, V. (2021). Medicinal plants and phytochemicals against multidrug-resistant tumor cells expressing ABCB1, ABCG2, or ABCB5: A synopsis of 2 decades. *Phytochemistry Reviews, 20*(1), 7–53.

Efferth, T., Saeed, M. E. M., Kadioglu, O., Seo, E. J., Shirooie, S., Mbaveng, A. T., et al. (2020). Collateral sensitivity of natural products in drug-resistant cancer cells. *Biotechnology Advances, 38*, 107342.

He, X., Blount, J. W., Ge, S., Tang, Y., & Dixon, R. A. (2011). A genomic approach to isoflavone biosynthesis in kudzu (*Pueraria lobata*). *Planta, 233*(4), 843–855.

Ko, K. P. (2014). Isoflavones: Chemistry, analysis, functions and effects on health and cancer. *Asian Pacific Journal of Cancer Prevention, 15*(17), 7001–7010.

Křížová, L., Dadáková, K., Kašparovská, J., & Kašparovský, T. (2019). Isoflavones. *Molecules (Basel, Switzerland), 24*(6), 1076.
Kuete, V. (2010). Potential of Cameroonian plants and derived products against microbial infections: A review. *Planta Medica, 76*(14), 1479–1491.
Kuete, V. (2013). Medicinal plant research in Africa. In V. Kuete (Ed.). *Pharmacology and chemistry*. Oxford: Elsevier.
Kuete, V. (2025). Chapter Four-African medicinal plants and their derivative as the source of potent anti-leukemic products: rationale classification of naturally occurring anticancer agents. *Advances in Botanical research, 113*, https://doi.org/10.1016/bs.abr.2023.12.010.
Kuete, V., & Efferth, T. (2015). African flora has the potential to fight multidrug resistance of cancer. *BioMed Research International, 2015*, 914813.
Kuete, V., Mbaveng, A. T., Nono, E. C., Simo, C. C., Zeino, M., Nkengfack, A. E., et al. (2016). Cytotoxicity of seven naturally occurring phenolic compounds towards multi-factorial drug-resistant cancer cells. *Phytomedicine, 23*(8), 856–863.
Kuete, V., Ngameni, B., Simo, C. C., Tankeu, R. K., Ngadjui, B. T., Meyer, J. J., et al. (2008). Antimicrobial activity of the crude extracts and compounds from *Ficus chlamydocarpa* and *Ficus cordata* (Moraceae). *Journal of Ethnopharmacology, 120*(1), 17–24.
Kuete, V., Ngnintedo, D., Fotso, G. W., Karaosmanoğlu, O., Ngadjui, B. T., Keumedjio, F., et al. (2018). Cytotoxicity of seputhecarpan D, thonningiol and 12 other phytochemicals from African flora towards human carcinoma cells. *BMC Complementary and Alternative Medicine, 18*(1), 36.
Kuete, V., Sandjo, L. P., Djeussi, D. E., Zeino, M., Kwamou, G. M., Ngadjui, B., et al. (2014). Cytotoxic flavonoids and isoflavonoids from *Erythrina sigmoidea* towards multi-factorial drug resistant cancer cells. *Investigational New Drugs, 32*, 1053–1062.
Kuete, V., Sandjo, L. P., Kwamou, G. M., Wiench, B., Nkengfack, A. E., & Efferth, T. (2014). Activity of three cytotoxic isoflavonoids from Erythrina excelsa and *Erythrina senegalensis* (neobavaisoflavone, sigmoidin H and isoneorautenol) toward multi-factorial drug resistant cancer cells. *Phytomedicine, 21*(5), 682–688.
Kwamou, G. M., Sandjo, L. P., Kuete, V., Wandja, A. A., Tankeo, S. B., Efferth, T., et al. (2014). Unprecedented new nonadecyl para-hydroperoxycinnamate isolated from *Erythrina excelsa* and its cytotoxic activity. *Natural Product Research, 29*(10), 921–925.
Li, Y., Li, G., Yu, H., Jiao, X., & Gao, K. (2018). Antifungal activities of isoflavonoids from *Uromyces striatus* infected Alfalfa. *Chemistry & Biodiversity, 15*(12), e1800407.
Lorent, K., Gong, W., Koo, K. A., Waisbourd-Zinman, O., Karjoo, S., Zhao, X., et al. (2015). Identification of a plant isoflavonoid that causes biliary atresia. *Science Translational Medicine*, 7(286), 286ra267.
Mbaveng, A. T., Fotso, G. W., Ngnintedo, D., Kuete, V., Ngadjui, B. T., Keumedjio, F., et al. (2018). Cytotoxicity of epunctanone and four other phytochemicals isolated from the medicinal plants *Garcinia epunctata* and *Ptycholobium contortum* towards multi-factorial drug resistant cancer cells. *Phytomedicine, 48*, 112–119.
Mbaveng, A. T., Kuete, V., & Efferth, T. (2017). Potential of Central, Eastern and Western Africa medicinal plants for cancer therapy: spotlight on resistant cells and molecular targets. *Frontiers in Pharmacology, 8*, 343.
Mbaveng, A. T., Noulala, C. G. T., Samba, A. R. M., Tankeo, S. B., Fotso, G. W., Happi, E. N., et al. (2020). Cytotoxicity of botanicals and isolated phytochemicals from *Araliopsis soyauxii* Engl. (Rutaceae) towards a panel of human cancer cells. *Journal of Ethnopharmacology, 267*, 113535.
Ngameni, B., Fotso, G. W., Kamga, J., Ambassa, P., Abdou, T., Fankam, A. G., et al. (2013). 9—Flavonoids and related compounds from the medicinal plants of Africa. In V. Kuete (Ed.). *Medicinal plant research in Africa* (pp. 301–350). Oxford: Elsevier.

Ngnintedo, D., Fotso, G. W., Kuete, V., Nana, F., Sandjo, L. P., Karaosmanoğlu, O., et al. (2016). Two new pterocarpans and a new pyrone derivative with cytotoxic activities from *Ptycholobium contortum* (N.E.Br.) Brummitt (Leguminosae): revised NMR assignment of mundulea lactone. *Chemistry Central The Journal, 10*, 58.

Pandey, A., Misra, P., Khan, M. P., Swarnkar, G., Tewari, M. C., Bhambhani, S., et al. (2014). Co-expression of Arabidopsis transcription factor, AtMYB12, and soybean isoflavone synthase, GmIFS1, genes in tobacco leads to enhanced biosynthesis of isoflavones and flavonols resulting in osteoprotective activity. *Plant Biotechnology Journal, 12*(1), 69–80.

Ralston, L., Subramanian, S., Matsuno, M., & Yu, O. (2005). Partial reconstruction of flavonoid and isoflavonoid biosynthesis in yeast using soybean type I and type II chalcone isomerases. *Plant Physiology, 137*(4), 1375–1388.

Szliszka, E., Czuba, Z. P., Sedek, L., Paradysz, A., & Krol, W. (2011). Enhanced TRAIL-mediated apoptosis in prostate cancer cells by the bioactive compounds neobavaisoflavone and psoralidin isolated from *Psoralea corylifolia. Pharmacological Reports, 63*(1), 139–148.

Wang, X. (2011). Structure, function, and engineering of enzymes in isoflavonoid biosynthesis. *Functional & Integrative Genomics, 11*(1), 13–22.

CHAPTER SEVEN

Cytotoxic lignans, neolignans, and stilbenes from African medicinal plants

Victor Kuete[a,*], Ibrahim Hashim[b], and Leonidah K. Omosa[c]
[a]Department of Biochemistry, Faculty of Science, University of Dschang, Dschang, Cameroon
[b]Department of Soil Science, College of Agriculture and Life Sciences, University of Wisconsin-Madison, Madison, United States
[c]Department of Chemistry, Faculty of Science and Technology, University of Nairobi, Nairobi, Kenya
*Corresponding author. e-mail address: kuetevictor@yahoo.fr

Contents

Abstract

Lignans and stilbenes are two groups of related secondary metabolites widely distributed in the plant kingdom. Both lignans and stilbenes possess anticancer, antioxidant, antimicrobial, anti-inflammatory, and immunosuppressive activities. In the present chapter, we have reported the cytotoxicity of 13 lignans, 16 stilbenes, and 2 neolignans isolated from African medicinal plants on various human cancer cell lines. The most potent cytotoxic lignans were diphyllin (**2**), futokadsurin B (**3**), justicidins A (**4**) and B (**5**), kusunokinin (**6**), pycnanthulignenes A (**10**) and B (**11**), and tuberculatin (**13**). The best stilbenes include combretastatin A-1 (**14**), combretastatin B-1 (**15**), combretastatin B-3 (**16**), combretastatin B-4 (**17**), longistylins A (**18**) and (**19**), mundulea lactone (**20**), schweinfurthins A (**23**), B (**24**), E (**25**), F (**26**), and G (**27**), and vedelianin (**29**). The two reported cytotoxic neolignans were licarins A (**30**) and B (**31**).

Advances in Botanical Research, Volume 115
ISSN 0065-2296, https://doi.org/10.1016/bs.abr.2024.02.008

Amongst these compounds, those that can combat the multidrug resistance of cancer were identified as **3**, **10, 11**, and **20**. Some of the active lignans and stilbenes displayed apoptotic cell death in cancer cells, mainly mediated by caspase activation, alteration of MMP, and enhanced ROS production in cancer cells. The above lignans, neolignans, and stilbenes deserve further in-depth studies to develop novel cytotoxic products to fight cancer.

Abbreviations

D.R	degree of resistance
IC_{50}	inhibitory concentration 50
MDR	multidrug-resistant
MMP	mitochondrial membrane potential
PI	propidium iodide
ROS	reactive oxygen species
RRA	resazurin reduction assay

1. Introduction

Lignans and stilbenes (Fig. 1) are two groups of related secondary metabolites widely distributed in the plant kingdom (Tsopmo, Awah, & Kuete, 2013; Valletta, Iozia, & Leonelli, 2021). Lignans are a class of secondary plant metabolites produced by oxidative dimerization of two phenylpropanoid units with a backbone consisting generally consist of two phenylpropane (C6–C3) units (Saleem, Kim, Ali, & Lee, 2005). They are dimers that often contain two phenylpropane units (C6–C3) that are linked by their carbon 8 (β-β′ link) (Tsopmo et al., 2013). However, they show an enormous structural diversity. They are found in plants, particularly roots, rhizomes, stems, bark, leaves, seeds, fruits, and vegetables, and they are precursors to phytoestrogens (Korkina, Kostyuk, De Luca, & Pastore, 2011; Tsopmo et al., 2013). They may play a role as antifeedants in the defense of seeds and plants against herbivores (Saleem et al., 2005). High levels are found in flax seeds and sesame seeds (Landete, 2012). Secoisolariciresinol diglucoside is the principal lignan precursor found in flaxseeds

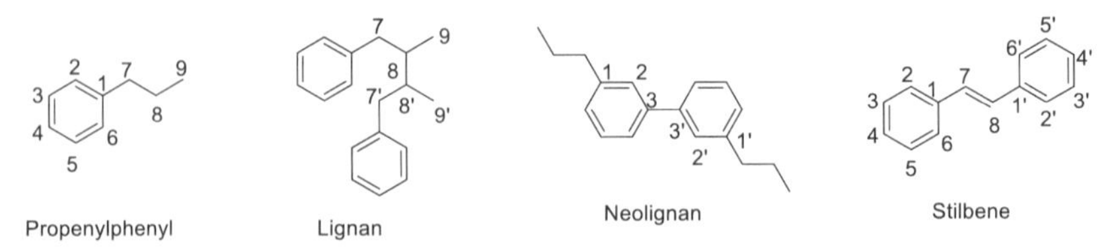

Fig. 1 Basic skeletons of penylpronane (C6–C3), lignans, neolignans and stilbene.

(Landete, 2012). They are also found in cereals (rye, wheat, oat, and barley), soybeans, tofu, cruciferous vegetables, such as broccoli and cabbage, and some fruits, such as apricots and strawberries. Stilbenes are also biosynthesized from phenylpropanoids that can further be oxidized to form oligomers (Tsopmo et al., 2013). They have a C6–C2–C6 structure and share most of their biosynthesis pathway with chalcones (Sobolev, Horn, Potter, Deyrup, & Gloer, 2006; Valletta et al., 2021). Most of them are produced by plants (Dubrovina & Kiselev, 2017; Valletta et al., 2021), and the only known exception is the anthelminthic and antimicrobial stilbene, 2-isopropyl-5-[*(E)*-2-phenylvinyl]benzene-1,3-diol, biosynthesized by the Gram-negative bacterium *Photorhabdus luminescens* (Eleftherianos et al., 2007; Mori et al., 2016; Richardson, Schmidt, & Nealson, 1988). Stilbenes act as phytoalexins, playing a crucial role in plant defense against phytopathogens, as well as being involved in the adaptation of plants to abiotic environmental factors (Valletta et al., 2021). Among stilbenes, trans-resveratrol is certainly the most popular and extensively studied for its health properties (Valletta et al., 2021). Most plant stilbenes are derivatives of the basic unit *trans*-resveratrol (3,5,4′-trihydroxy-*trans*-stilbene), although other structures are found in particular plant families (Tsopmo et al., 2013). Both lignans and stilbenes possess anticancer, antioxidant, antimicrobial, anti-inflammatory, and immunosuppressive activities (Kuete, 2013; Kuete, Kamga, et al., 2011; Nono et al., 2010; Saeed et al., 2014; Saeed et al., 2019; Saleem et al., 2005; Seukep, Sandjo, Ngadjui, & Kuete, 2016b). There is a growing interest in lignans as well as stilbenes due to applications in cancer chemotherapy and various other pharmacological effects (Mbaveng, Kuete, & Efferth, 2017; Saleem et al., 2005; Tsopmo et al., 2013). In this chapter, an overview of the cytotoxic potential of lignans, neolignans, and stilbenes isolated from African medicinal plants against human cancer cell lines including the multidrug-resistant (MDR) phenotypes is provided.

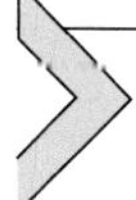

2. Biosynthesis of lignans, neolignans, and stilbenes

2.1 Biosynthesis of lignans and neolignans

The shikimic acid biosynthetic pathway is the source of plants' secondary metabolites, lignans and neolignans, and the first step involves the formation of monolignols (C_6–C_3 units). Monolignols are formed by converting chorismic acid to prephenic acid through a claisen rearrangement. The side chain derived from phosphoenolpyruvate is transferred and directly bonded to the carbocycle, constructing the fundamental carbon structure of phenylalanine. Prephenic acid undergoes decarboxylative aromatization to produce phenylpyruvic acid, and pyridoxal phosphate-dependent transamination produces *L*-phenylalanine. The phenylalanine ammonialyse enzyme deaminates *L*-phenylalanine to yield cinnamic acid, which is then further hydroxylated by P450 enzymes to yield pcoumaric acid and/or its polyhydroxylated analogs, such as caffeic acid. These can then go through methylation of the added phenolic functionalities *via* *O*-methyltransferase to produce the basic monomeric phenylpropanoid derivatives known as cinnamic acid derivatives. The cinnamic acid-derived carboxylic acids, which include substances like ferulic acid or sinapic acid, among others, can be further reduced by the corresponding coenzyme A ester to an intermediate aldehyde (a class of substances that includes, for example, coniferyl aldehyde), which can then be further reduced in the presence of NADPH to the corresponding alcohol, such as *p*-coumaryl and sinapyl alcohols (Scheme 1) (Teponno, Kusari, & Spiteller, 2016; Zálešák, Bon, & Pospíšil, 2019).

Lignans and neolignans share the common biosynthesis pathway (phenylpropanoid metabolism) and are formed by the stereo-selective dimerization of two units of monolignol, mostly coniferyl alcohol. The

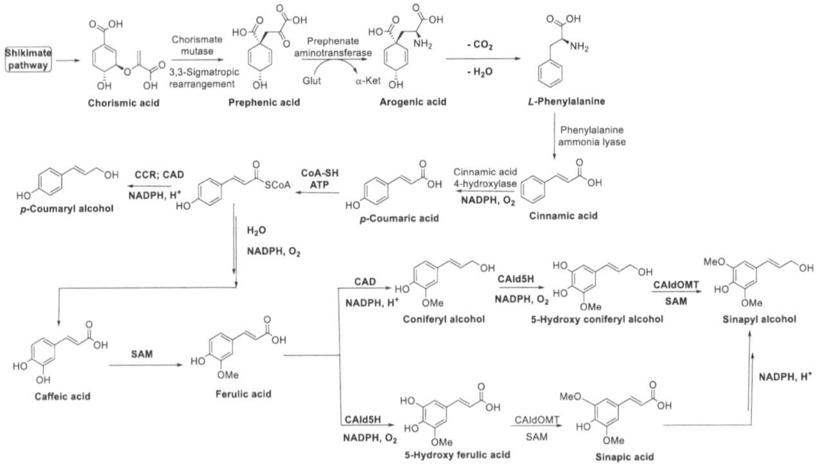

Scheme 1 Biogenesis of phenylpropanoid monomers (monolignols). *CCR,* cinnamoyl CoA reductase; *CAD,* cinnamyl alcohol dehydrogenase; *CAld5H,* coniferaldehyde-5-hydroxylase; *CAldOMT,* 5-hydroxyconiferylaldehyde *O*-methyltransferase; *Glut,* glutamate; *Ket,* ketoglutarate; *Me,* CH_3 (methyl) (Teponno et al., 2016; Zálešák et al., 2019).

initial step in the biosynthesis of lignans and neolignans is the formation of monolignols (C_6–C_3 units), as described above. This review, therefore, highlights the biosynthetic pathways of lignans and neolignans, as the detailed description was reported in our previous review (Tsopmo, Awah, & Kuete, 2013). Many lignans are formed from coniferyl alcohol, and the coupling of two coniferyl alcohol radicals occurs under the control of a unique asymmetric inducer, dirigent protein (DIP), yielding an optically active product, pinoresinol, which is further converted to a variety of lignans. Compounds of the neolignans class are frequently chiral, and the fact that naturally occurring neolignans are usually optically active suggests that DIPs are involved in the biosynthesis of neolignans (Scheme 2) (Anjum, Abbasi, Doussot, Favre-Réguillon, & Hano, 2017; Umezawa, Yamamura, Nakatsubo, Suzuki, & Hattori, 2011).

2.2 Biosynthesis of stilbenes

Natural stilbenes are present in many plant species, having a 1,2-diphenylethylene skeleton that is differently decorated with phenolic hydroxyl groups and exhibit substitution patterns specific to each species and indicative of the various biosynthetic pathways (Giacomini, Rupiani, Guidotti, Recanatini, & Roberti, 2016). The first steps of the stilbene biosynthesis follow the general phenylpropanoid pathway, which starts from the non-oxidative deamination of phenylalanine to trans-cinnamic acid catalyzed by phenylalanine ammonia lyase; oxidation catalyzed by cinnamate-4- hydroxylase followed by esterification carried out by coumaroyl CoA-ligase (4CL), leads to the central metabolite 4-coumaroyl CoA, a branch point from which all different metabolites are synthesized (Vogt, 2010).

The following steps involve the stilbene synthases (STSs), distinctive enzymes of the stilbene biosynthetic pathways (Austin, Bowman, Ferrer, Schröder, & Noel, 2004; Gehlert, Schöppner, & Kindl, 1990): A linear tetraketide intermediate is first formed by the condensation of the CoA ester and three units of malonyl CoA catalyzed by STSs. Next, the cyclization to produce the stilbene backbone is accomplished through intramolecular aldol condensation and a decarboxylative elimination involving the loss of one unit of CO_2 (Chong, Poutaraud, & Hugueney, 2009). The flavonoid family is produced by the same stepwise condensing reactions that are catalyzed by chalcone synthases (CHSs) (Kodan, Kuroda, & Sakai, 2002). Because these enzymes are related to one another and have similar substrates, there is evidence that STSs independently evolved from

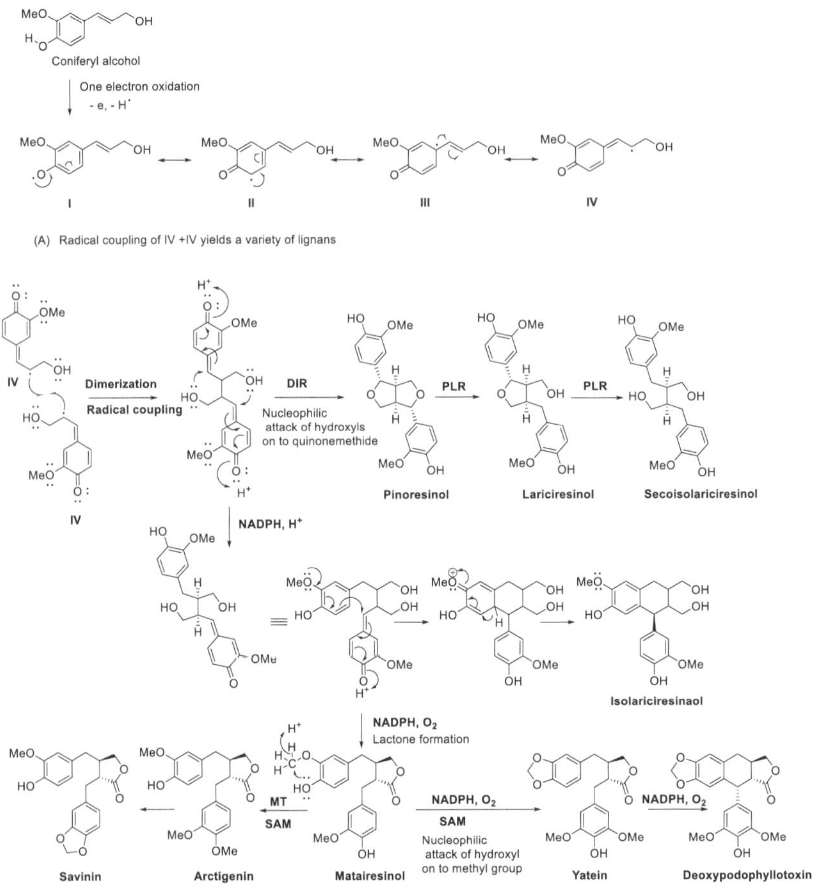

Scheme 2 Biosynthetic pathways of lignans and neolignans. (**A**) Biosynthesis pathway of two major lignans. *DIP,* dirigent protein; *PLR,* pinoresinol lariciresinol reductase; *SLD,* secoisolariciresinol dehydrogenase; *MT, O* -methyltransferase (**B**) Biosynthesis of two neolignans (dehydrodiconiferyl alcohol glucoside and guaiacylglycerol-β-coniferyl alcohol ether glucoside) by simple dimerization of two units of coniferyl alcohol; *Me,* CH_3 (methyl); *NADPH,* Nicotine adenine dinucleotide; *ATP,* Adenosine triphosphate; *SAM,* S-Adenosyl methionine (Anjum et al., 2017; Teponno et al., 2016; Tsopmo et al., 2013).

CHSs on multiple occasions (Parage et al., 2012; Vannozzi, Dry, Fasoli, Zenoni, & Lucchin, 2012). While CHSs are found everywhere, STSs are only expressed in plants that produce stilbenes, so the distribution of natural stilbenes in the plant kingdom is both constrained and diverse (Rivière, Pawlus, & Mérillon, 2012). Scheme 3 shows the biosynthesis pathway for stilbenes (particularly resveratrol).

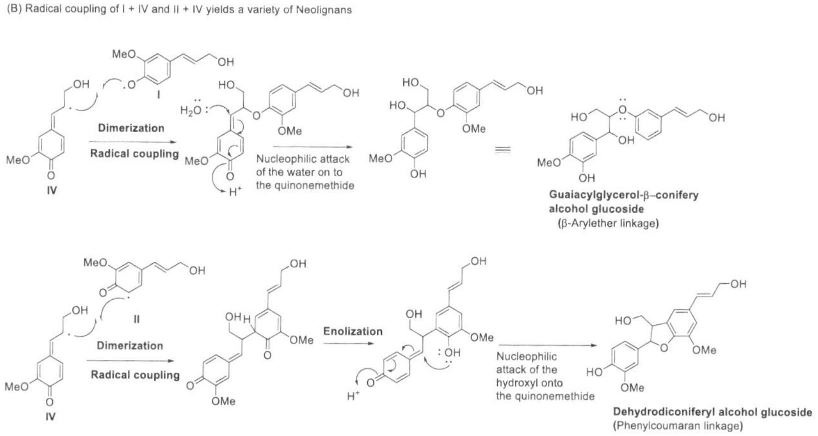

(B) Radical coupling of I + IV and II + IV yields a variety of Neolignans

Scheme 2 (*continued*)

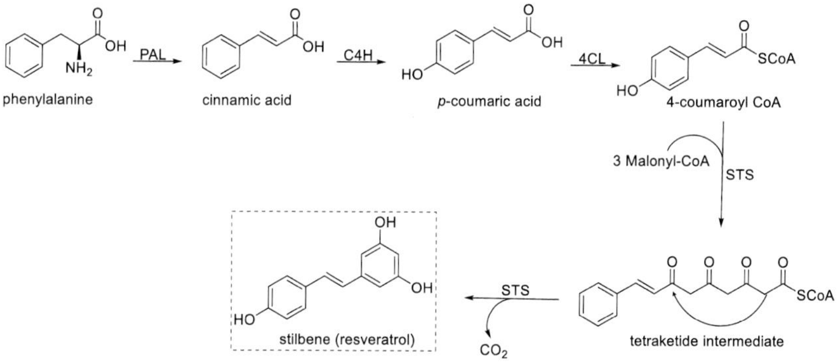

Scheme 3 Biosynthetic pathways of stilbenes (resveratrol). *C4H*, cinnamate-4-hydroxylase; *PAL*, phenylalanine ammonia lyase; *STS*, stilbene synthases.

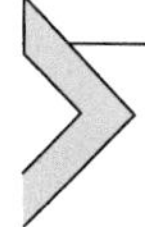

3. Cancer cell lines used in the screening of the cytotoxicity of lignans, neolignans, and stilbenes isolated from African medicinal plants

Several cancer cell lines were used to assess the cytotoxicity of lignans, neolignans, and stilbenes isolated from African medicinal plants. They include cell lines from breast cancer (MCF-7, MCF-7-ras, MDA-MB-231-*pcDNA*, and MDA-MB-231-*BCRP*, and MDA-MB-468), cervical cancer (CaSKi, HeLa, and SiHa), Cholangiocarcinoma (KKU-M213 and KKU-K100), colorectal cancer (HT-29, HCT116, HCT116 $p53^{+/+}$, HCT116 $p53^{-/-}$, T24, and

TSGH8301), glioblastoma (SF-295, U87MG, and U87MG.Δ*EGFR*), hepatocarcinoma (Hep3B and HepG2), esophageal cancer (ECA-109 and TE-1), fibrosarcoma (L-929 (mouse)), human gastric cancer (SGC7901), leukemia (CCRF-CEM, CEM/ADR5000, HL60, PBMC, RPMI-8226, U-266, and L1210 murine leukemia), lung cancer (A549), melanoma (A375), ovarian cancer (A2780, A2780cis, SKOV-3, SK-OV-3-MDR-1–6/6, and OVCAR-3), pancreatic cancer (Panc-1), prostate cancer (DU 145, PC-3, and LNCaP), and tongue cancer (SCC-4) (Table 1).

4. Cytotoxic lignans, neolignans, and stilbenes from the African medicinal plants towards drug sensitive and multidrug-resistant cancer cells

A total of 31 cytotoxic compounds belonging to lignans, neolignans, and stilbenes have been isolated from African medicinal plants. Their effects on various human cancer cell lines are summarized in Table 1. The degree of resistance (D.R.) determined as the ratio of the IC_{50} value in resistant cells *versus* the IC_{50} values in the corresponding sensitive cell line: CEM/ADR5000 cells *vs* CCRF-CEM cells, MDA-MB-231-*BCRP* *vs* MDA-MB-231-pcDNA3, HCT116 *p53*$^{-/-}$ cells *vs* HCT116 *p53*$^{+/+}$ cells, and U87MG.Δ*EGFR* *vs* U87MG, is also depicted in Table 1 (Mbaveng et al., 2020). In the laboratory scale, collateral sensitivity is achieved when the D.R., defined as the ratio of the IC_{50} values of the cytotoxic agent towards the resistant cell line *versus* that in its sensitive counterpart, is below 1 whilst the D.R. above 1 defines normal sensitivity. It has been established that the D.R. < 0.9 defines collateral sensitivity, whilst D.R. between 0.9 and 1.2 defines normal sensitivity. The cross-resistance is noted if the cytotoxic agent is more active in the sensitive cell line than its resistant subline, with D.R above 1.2 (Efferth et al., 2020; Efferth et al., 2021; Mbaveng et al., 2017). Collateral or normal sensitivities should be achieved for samples with the ability to combat drug resistant cancer cells (Efferth et al., 2020; Kuete & Efferth, 2015). In this section, the activity of these flavonoids will be discussed according to the cut-off points of cytotoxic phytochemicals defined as follows: outstanding activity ($IC_{50} \leq 0.5\,\mu M$), excellent activity ($0.5 < IC_{50} \leq 2\,\mu M$), very good activity ($2 < IC_{50} \leq 5\,\mu M$), good activity ($5 < IC_{50} \leq 10\,\mu M$), average activity ($10 < IC_{50} \leq 20\,\mu M$), weak activity ($20 < IC_{50} \leq 60\,\mu M$), very weak activity ($60 < IC_{50} \leq 150\,\mu M$), and not active ($IC_{50} > 150\,\mu M$) (Kuete, 2025).

Table 1 Cytotoxic lignans, neo ignans, and stilbenes isolated from African medicinal plants and their effects on sensitive and drug-resistant cancer cell lines.

Compound names	Source	Country	Cancer Cell Lines and IC_{50} values (μM) and degree of resistance in bracket	References
Lignans				
(3,4-Dimethoxybenzyl)-2-(3,4-methylenedioxybenzyl) butyrolactone (1)	*Bupleurum marginatum*	Egypt	41.30 (Hela); 52.70 (HepG2)	Ashour, El-Readi, Tahrani, Eid, and Wink (2012)
Diphyllin (2)	*Haplophyllum tuberculatum*	Libya	0.2792 (ECA-109); 0.2058 (TE-1), 7.8 (SGC7901), 11.5 (LNCaP); 7.2 (PC-3); 9.47 (Hep3B); 1.05 (HepG2); 6.58 (HT-29)	Chen et al. (2018); Day et al. (2002); Jiang et al. (2007); Shen et al. (2011); Sheriha et al. (1987); Sheriha and Abouamer (1984)
Futokadsurin B (3)	*Uapaca togoensis*	Cameroon	22.44 (CCRF-CEM) *vs* 8.16 (CEM/ADR5000) [0.36]; 17.53 (MDA-MB-231-*pcDNA*) *vs* 20.03 (MDA-MB-231-*BCRP*) [1.14]; 41.03 (HCT116 *p53*$^{-/-}$); 10.85 (HepG2)	Kuete et al. (2015); Seukep, Sandjo, Ngadjui, and Kuete (2016a)

(continued)

Table 1 Cytotoxic lignans, neolignans, and stilbenes isolated from African medicinal plants and their effects on sensitive and drug-resistant cancer cell lines. (*cont'd*)

Compound names	Source	Country	Cancer Cell Lines and IC_{50} values (μM) and degree of resistance in bracket	References
Justicidin A (4)	*Haplophyllum tuberculatum*	Libya, Sudan	0.11 (HT-29); 0.40 (HCT116); 0.02 (SiHa); 1.54 (MCF-7); 0.004 (T24); 25 (PBMC); 0.44 (TSGH8301); 0.01 (CaSKi); 0.07 (Hep3B); 0.05 (HepG2); 0.99 (MCF-7); 18.78 (MCF-7-ras)	Sheriha et al. (1987); Sheriha and Abouamer (1984); Lee et al. (2005); Wang et al. (2015)
Justicidin B (5)	*Haplophyllum tuberculatum*	Libya, Sudan	0.17 (RPMI-8226); 183 μM (U-266); 0.9 (HL60); 1.70 (A375)	Gertsch, Tobler, Brun, Sticher, and Heilmann, (2003); Ilieva et al. (2014); Momekov et al. (2014); Sheriha et al. (1987); Sheriha and Abouamer (1984)
Kusunokinin (6)	*Haplophyllum tuberculatum*	Libya	3.25 (A2780cis); 8.75 (A2780); 14.43 (SKOV-3); 14.26 (OVCAR-3); 3.85 (MCF-7); 5.19 (MDA-MB-468); 5.29 (MDA-MB-231); 4.47 (KKU-M213); 4.46 (KKU-K100); 5.34 (HT-29); 4.52 (A2780)	Mad-Adam et al. (2022); Rattanaburee, Sermmai, Tangthana-Umrung, Thongpanchang, and Graidist (2022); Rattanaburee, Tipmanee, Tedasen, Thongpanchang, and Graidist (2020); Sartorelli, Carvalho, Reimão, Lorenzi, and Tempone (2010); Sheriha et al. (1987)

Magnolin (7)	*Physalis peruviana*	Egypt	0.51 (Panc-1)	Sayed, El-Hawary, Abdelmohsen, and Ghareeb (2022)
Marginatoxin (8)	*Bupleurum marginatum*	Egypt	16.90 (Hela); 12.14 (HepG2)	Ashour et al. (2012)
Polygamain (9)	*Haplophyllum tuberculatum*	Libya	26.3 (MDA-MB-435); 54.9 (DU 145); 70.6 (PC-3); 36.4 (SCC-4); 65.8 (A549); 78.2 (MDA-MB-231); 37.9 (HeLa); 51.3 (SK-OV-3); 102.1 (SK-OV-3-MDR-1-6/6)	Hartley et al. (2012); Peng, Hartley, Fest, and Mooberry (2012); Sheriha et al. (1987); Sheriha and Abouamer (1984)
Pycnanthulignene A (10)	*Pycnanthus angolensis*	Cameroon	7.52 (CCRF-CEM) *vs* 5.84 (CEM/ADR5000) [0.78]; 37.58 (MDA-MB-231-*pcDNA*) *vs* 33.23 (MDA-MB-231-*BCRP*) [0.88]; 45.82 (HCT116 *p53*$^{+/+}$) *vs* 65.32 (HCT116 *p53*$^{-/-}$) [1.43]; 45.82 (U87MG) *vs* 46.92 (U87MG.Δ*EGFR*) [1.02]; 35.76 (HepG2)	Kuete et al. (2016); Kuete, Nono, et al. (2011); Nono et al. (2010)

(continued)

Table 1 Cytotoxic lignans, neolignans, and stilbenes isolated from African medicinal plants and their effects on sensitive and drug-resistant cancer cell lines. (*cont'd*)

Compound names	Source	Country	Cancer Cell Lines and IC_{50} values (μM) and degree of resistance in bracket	References
Pycnanthulignene B (11)	*Pycnanthus angolensis*	Cameroon	17.04 (CEM/ADR5000); 28.4 (U87MG) *vs* 9.5 (U87MG.Δ*EGFR*) [0.3]	Kuete et al. (2016); Kuete, Nono, et al. (2011); Nono et al. (2010)
Syringaresinol (12)	*Allanblackia floribunda*	Cameroon	14.59 (CCRF-CEM) *vs* 35.66 (CEM/ADR5000) [2.44]; 58.18 (MDA-MB-231-*pcDNA*)	Kuete, Azebaze, et al. (2011); Kuete et al. (2016)
Tuberculatin (13)	*Haplophyllum tuberculatum*	Libya	0.03 (Hep3B); 0.23 (SiHa); 0.08 (HepG2); 0.57 (HT-29); 0.55 (HCT116); 1.89 (MCF-7); 1.76 (MCF-7-ras)	Day et al. (2002); Sheriha et al. (1987); Sheriha and Abouamer (1984)
Stilbenes				
Combretastatin A-1 (14)	*Combretum caffrum*	South Africa	0.6 (L1210)	Pettit, Singh, Niven, Hamel, and Schmidt (1987)
Combretastatin B-1 (15)	*Combretum caffrum*	South Africa	3 (L1210)	Pettit et al. (1987)

Combretastatin B-3 (16)	*Combretum caffrum*	South Africa 3	3 (L1210)	Pettit et al. (1988)
Combretastatin B-4 (17)	*Combretum caffrum*	South Africa 4	1 (L1210)	Pettit et al. (1988)
Longistylin A (18)	*Cajanus cajan*	Nigeria	33.33 (CCRF-CEM) *vs* 35.03 (CEM/ADR5000) [1.05]; 17.69 (MCF-7); 2.38 (HepG2)	Ashidi, Houghton, Hylands, and Efferth (2010)
Longistylin C (19)	*Cajanus cajan*	Nigeria	34.01 (CCRF-CEM) *vs* 35.71 (CEM/ADR5000) [1.02]; 14.97 (MCF-7); 5.44 (HepG2)	Ashidi et al. (2010)
Mundulea lactone (20)	*Ptycholobium contortum*	Cameroon	24.13 (CCRF-CEM) *vs* 8.84 (CEM/ADR5000) [0.37]; 48.99 (MDA-MB-231-*pcDNA*) *vs* 22.42 (MDA-MB-231-*BCRP*) [0.46]; 15.25 (HCT116 $p53^{+/+}$) *vs* 14.73 (HCT116 $p53^{-/-}$) [0.97]; 11.39 (U87MG) *vs* 13.04 (U87MG.Δ*EGFR*) [1.15]; 30.49 (HepG2)	Mbaveng et al. (2018)

(continued)

Table 1 Cytotoxic lignans, neolignans, and stilbenes isolated from African medicinal plants and their effects on sensitive and drug-resistant cancer cell lines. (*cont'd*)

Compound names	Source	Country	Cancer Cell Lines and IC_{50} values (μM) and degree of resistance in bracket	References
Resveratrol (21)	*Nauclea pobeguinii*	Cameroon	28.15 (CCRF-CEM) *vs* 57.43 (CEM/ADR5000) [2.04]; 19.90 (MDA-MB-231-*pcDNA*) *vs* 22.93 (MDA-MB-231-*BCRP*) [1.15]; 73.65 (HCT116 $p53^{+/+}$); 76.59 (U87MG) *vs* 28.45 (U87MG.Δ*EGFR*) [0.37]; 88.60 (HepG2)	Kuete et al. (2015); Seukep et al. (2016b)
Resveratrol β-D-glucopyranoside (22)	*Nauclea pobeguinii*	Cameroon	25.08 (CCRF-CEM) *vs* 39.87 (CEM/ADR5000) [1.59]; 97.64 (MDA-MB-231-*pcDNA*) *vs* 95.59 (MDA-MB-231-*BCRP*) [0.98]; 63.77 (HCT116 $p53^{+/+}$) *vs* 47.03 (HCT116 $p53^{-/-}$) [0.74]; 55.64 (U87MG) *vs* 47.59 (U87MG.Δ*EGFR*) [0.86]; 60.08 (HepG2)	Kuete et al. (2015); Seukep et al. (2016b)

Schweinfurthin A (23)	*Macaranga schweinfurthii*	Cameroon	6.75 (A549); < 0.00002 (SF-295)	Beutler, Shoemaker, Johnson, and Boyd (1998); Klausmeyer, Van, Jato, McCloud, and Beutler (2010)
Schweinfurthin B (24)	*Macaranga schweinfurthii*	Cameroon	7.1 (A549); 0.02 (SF-295)	Beutler et al. (1998); Klausmeyer et al. (2010)
Schweinfurthin E (25)	*Macaranga schweinfurthii*	Madagascar	0.26 (A2780)	Yoder et al. (2007)
Schweinfurthin F (26)	*Macaranga schweinfurthii*	Madagascar	5.0 (A2780)	Yoder et al. (2007)
Schweinfurthin G (27)	*Macaranga schweinfurthii*	Madagascar	0.39 (A2780)	Yoder et al. (2007)
Schweinfurthin H (28)	*Macaranga schweinfurthii*	Madagascar	4.5 (A2780)	Yoder et al. (2007)
Vedelianin (29)	*Macaranga alnifolia*	Madagascar	0.13 (A2780)	Beutler et al. (1998); Klausmeyer et al. (2010); Yoder et al. (2007)

(*continued*)

Table 1 Cytotoxic lignans, neolignans, and stilbenes isolated from African medicinal plants and their effects on sensitive and drug-resistant cancer cell lines. (*cont'd*)

Compound names	Source	Country	Cancer Cell Lines and IC_{50} values (μM) and degree of resistance in bracket	References
Neolignans				
Licarin A (30)	*Piper capense*	Cameroon	4.3 (CCRF-CEM) *vs* 7.3 (CEM/ADR5000) [1.7]; 17.9 (MDA-MB-231-*pcDNA*) *vs* 17.9 (MDA-MB-231-*BCRP*) [1.0]; 14.5 (HCT116 *p53*$^{+/+}$) *vs* 21.8 (HCT116 *p53*$^{-/-}$) [1.5]; 10.7 (U87MG) *vs* 8.4 (U87MG.Δ*EGFR*) [0.8]; 10.0 (HepG2)	Mbaveng et al. (2021)
Licarin B (31)	*Piper capense*	Cameroon	10.90 (CCRF-CEM); 29.1 (MDA-MB-231-*pcDNA*) *vs* 33.5 (MDA-MB-231-*BCRP*) [1.2]; 11.3 (HCT116 *p53*$^{+/+}$) *vs* 12.6 (HCT116 *p53*$^{-/-}$) [1.1]; 10.3 (HepG2)	Mbaveng et al. (2021)

4.1 Cytotoxic lignans

Thirteen cytotoxic lignans have been isolated from African medicinal plants. They include (3,4-dimethoxybenzyl)-2-(3,4-methylenedioxybenzyl) butyrolactone (**1**), diphyllin (**2**), futokadsurin B (**3**), justicidins A (**4**) and B (**5**), kusunokinin (**6**), magnolin (**7**), marginatoxin (**8**), polygamain (**9**), pycnanthulignenes A (**10**) and B (**11**), syringaresinol (**12**), and tuberculatin (**13**). Their chemical structures are shown in Fig. 2. Their cytotoxicity in human cancer cell lines is recorded in Table 1. Amongst them, outstanding cytotoxic activities ($IC_{50} \leq 0.5\ \mu M$) were obtained with lignans **2** against ECA-109 cells

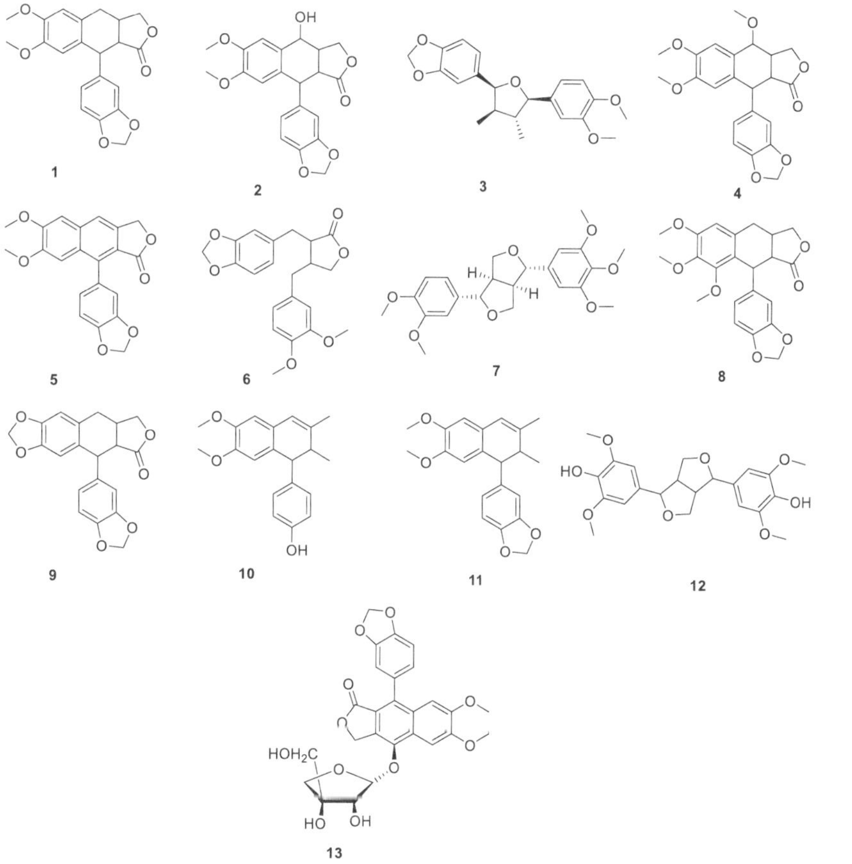

Fig. 2 Cytotoxic lignans isolated from African medicinal plants. **1:** (3,4-dimethoxybenzyl)-2-(3,4-methylenedioxybenzyl) butyrolactone; **2:** diphyllin; **3:** futokadsurin B; **4:** justicidin A; **5:** justicidin B; **6:** kusunokinin; **7:** magnolin; **8:** marginatoxin; **9:** polygamain; **10:** pycnanthulignene A; **11:** pycnanthulignene B; **12:** syringaresinol; **13:** tuberculatin.

and TE-1 cells, **4** and **13** against HT-29 cells, **4** against HCT116 cells, SiHa cells, T24 cells, TSGH8301 cells, MCF-7 cells and CaSKi cells, **4** and **13** against Hep3B cells and HepG2 cells, **5** against RPMI-8226 cells, and **13** against SiHa cells meanwhile excellent cytotoxic activities ($0.5 < IC_{50} \leq 2\,\mu M$) were obtained with lignans **2** against HepG2 cells, **5** against HL60 cells, A375 cells, and Panc-1 cells, and **13** against HT-29 cells, HCT116 cells, MCF-7 cells, and MCF-7-ras cells. Also, very good antiproliferative activities ($2 < IC_{50} \leq 5\,\mu M$) were obtained with lignans **6** against A2780cis cells, MCF-7 cells, KKU-M213 cells, KKU-K100 cells, A2780 cells, and **10** against CCRF-CEM cells and CEM/ADR5000 cells as well as the good antiproliferative activities ($5 < IC_{50} \leq 10\,\mu M$) of lignans **2** against SGC7901 cells, PC-3 cells, Hep3B cells, **2** and **6** against HT-29 cells, **3** against CEM/ADR5000 cells, **6** against MDA-MB-468 cells and MDA-MB-231 cells, and **11** against U87MG.Δ*EGFR* cells. Interestingly, collateral sensitivities of CEM/ADR5000 cells *vs* CCRF-CEM cancer cells to compounds **3** (D.R. 0.36) and **10** (D.R. 0.78), and of U87MG.Δ*EGFR* cells *vs* U87MG cells (D.R. 0.3) to **11** were also obtained.

4.2 Cytotoxic stilbenes

The stilbenes combretastatin A-1 (**14**), combretastatin B-1 (**15**), combretastatin B-3 (**16**), combretastatin B-4 (**17**), longistylins A (**18**) and (**19**), mundulea lactone (**20**), resveratrol (**21**), resveratrol *β*-D-glucopyranoside (**22**), schweinfurthins A (**23**), B (**24**), E (**25**), F (**26**), G (**27**), and H (**28**), and vedelianin (**29**) isolated from African medicinal plants also displayed cytotoxic effect in several cancer cell lines. Their chemical structures are shown in Fig. 3 meanwhile their cytotoxicity is summarized in Table 1. Amongst them, outstanding cytotoxic activities were obtained with stilbenes **23** and **24** against SF-295 cells and **25, 27**, and **29** against A2780 cells. Excellent cytotoxic activities were also obtained with stilbenes **14** against L1210 cells and **17** against L1210 cells, meanwhile very good antiproliferative activities were obtained with stilbenes **15** and **16** against L1210 cells, **18** and **19** against HepG2 cells, and **26** and **28** against A2780 cells. Good antiproliferative activities were also reported with stilbenes **20** against CEM/ADR5000 cells, **23** against A549 cells, and **24** against A549 cells. Importantly, collateral sensitivity of CEM/ADR5000 cells *vs* CCRF-CEM cancer cells to compounds **20** (D.R. 0.37) was achieved.

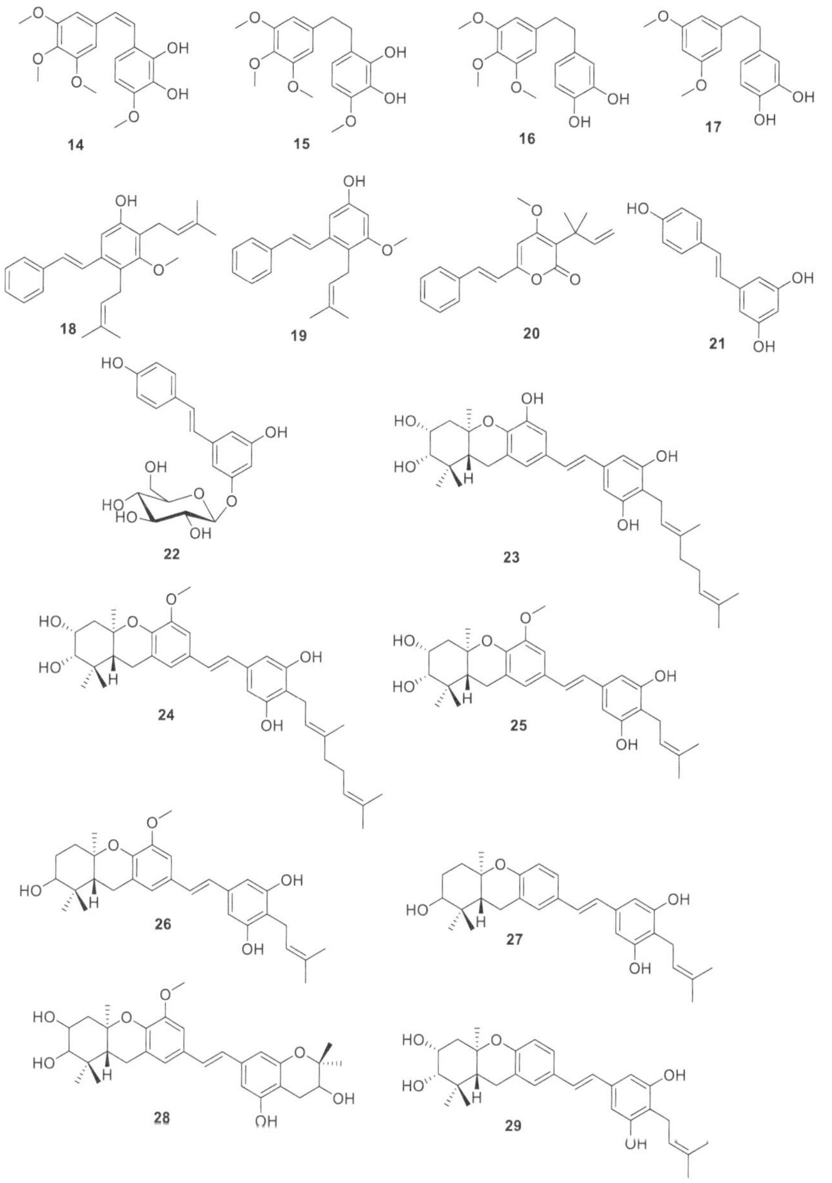

Fig. 3 Cytotoxic stilbenes isolated from African medicinal plants. **14:** combretastatin A-1; **15:** combretastatin B-1; **16:** combretastatin B-3; **17:** combretastatin B-4; **18:** longistylin A; **19:** longistylin C; **20:** mundulea lactone; **21:** resveratrol; **22:** resveratrol β-D-glucopyranoside; **23:** schweinfurthin A; **24:** schweinfurthin B; **25:** schweinfurthin E; **26:** schweinfurthin F; **27:** schweinfurthin G; **28:** schweinfurthin H; **29:** vedelianin.

4.3 Cytotoxic neolignans

Two cytotoxic neolignans licarins A (**30**) and B (**31**) have also been isolated from African medicinal plants. Their chemical structures are shown in Fig. 4. Their cytotoxicity in human cancer cell lines is recorded in Table 1. Neolignan **31** rather displayed cytotoxic effects ranging from average ($10 < IC_{50} \leq 20\ \mu M$) to weak ($20 < IC_{50} \leq 60\ \mu M$). In contrast, very good antiproliferative activities were obtained with neolignan **30** against CCRF-CEM cells and good activities against CEM/ADR5000 cells, U87MG.Δ*EGFR* cells, and HepG2 cells. Collateral sensitivity of U87MG.Δ*EGFR* cells *vs* U87MG cells (D.R. 0.8) to **30** was also observed.

5. Modes of action cytotoxic lignans and neolignans isolated from African medicinal plants

Chen et al. have demonstrated that diphyllin (**2**) inhibited proliferation and induced S arrest in TE-1 cells and ECA-109 cells, and reduced the formation of new blood vessels (Chen et al., 2018); they also showed that lignan **2** inhibited blood metastasis by regulating the mTORC1/HIF-1α-/VEGF pathway, therefore it could be considered as a new V-ATPase inhibitor in esophageal cancer (Chen et al., 2018). Shen and collaborators have shown that lignan **2** can inhibit the expression of V-ATPases in SGC7901 cells and that diphyllin treatment caused a decrease in phospho-LRP6, but not in LRP6. They also showed that β-catenin in Wnt/β-catenin signaling and its target genes, c-myc, and cyclin-D1, were also decreased with the inhibition of V-ATPases. Therefore, lignan **2** is a V-ATPase inhibitor in gastric cancer that inhibits the phosphorylation of LRP6 in Wnt/β-catenin signaling (Shen et al., 2011).

The study of the mechanism of action of justicidin A (**4**) in HT-29 cells and HCT 116 cells was performed by the team of Lee and collaborators (Lee et al., 2005). They demonstrated that lignan **4** induced apoptosis in

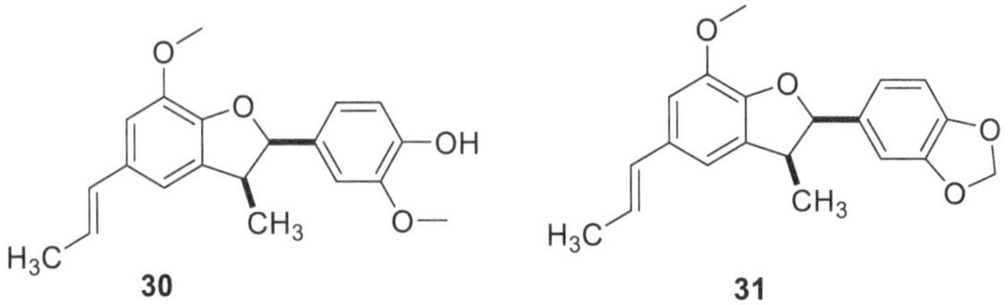

Fig. 4 Cytotoxic neolignans licarins A (**30**) and B(**31**) isolated from African medicinal plants.

the two cancer cell lines. lignan **4** treatment caused DNA fragmentation and an increase in phosphatidylserine exposure in the cells. Caspase-9 but not caspase-8 was activated, suggesting that justicidin A treatment damaged mitochondria. The mitochondrial membrane potential (MMP) was altered and cytochrome c and Smac were released from mitochondria to the cytoplasm upon justicidin A treatment. They also found that the level of Ku70 in the cytoplasm was decreased, but that of Bax in mitochondria was increased by justicidin A. Since Ku70 normally binds and sequesters Bax, they suggested that compound **4** decreases the level of Ku70 leading to translocation of Bax from the cytosol to mitochondria to induce apoptosis (Lee et al., 2005). Wang et al. also demonstrated that compound **4** had antiangiogenic effects *in vitro* and *in vivo* through pleiotropic positive and negative regulators of angiogenesis molecules (Wang et al., 2015). lignan **4** also induced autophagy flux enhances apoptosis in HT-29 cells *via* class III PI3K and Atg5 pathway (Won et al., 2015).

Justicidin B (**5**) was found to possess potent selective antineoplastic activity, related to its ability to induce programmed cell death in NHL-derived human cell lines (Ilieva et al., 2014). Momekov et al. also demonstrated that lignan **5** is a potent cytotoxic and proapoptotic agent against HL-60 and that the induction of apoptosis proceeds *via* activation of the intrinsic mitochondrial cell-death signaling pathways (Momekov et al., 2014). Compound **5** caused both early and late apoptosis in A375 melanoma cells mediated by the activation of caspase-3/7 and enhancing Bax expression (Al-Qathama, Gibbons, & Prieto, 2017).

It was shown that kusunokinin (**6**) induced apoptosis and increased multi-caspase activity in A2780 and A2780cis cells. Compound **6** significantly downregulated topoisomerase II, cyclin D1 and CDK1 expression, but upregulated Bax and PUMA expression in both A2780 and A2780cis cells (Mad-Adam, Rattanaburee, Tanawattanasuntorn, & Graidist, 2022).

Polygamain (**9**) was also reported as a microtubule destabilizer that seems to occupy a unique pharmacophore within the colchicine site of tubulin (Hartley et al., 2012).

The cytotoxicity and the modes of action of pycnanthulignene A (**10**) were determined against a panel of nine cancer cell lines, including CCRF-CEM, CEM/ADR5000, HCT116 *p53*$^{+/+}$, HCT116 *p53*$^{-/-}$, U87. MG, U87. MGΔ*EGFR*, MDA-MB-231-pcDNA3, MDA-MB-231-*BCRP*, and HepG2 using RRA (Kuete et al., 2016). Lignan **50** had anti-proliferative effects towards the nine tested cancer cell lines with the recorded IC_{50} values ranging from 5.84 μM (towards CEM/ADR5000 cells) to 65.32 μM (towards colon carcinoma HCT116 (*p53*$^{-/-}$) cells);

Compound **10** induced apoptosis in CCRF-CEM cells, mediated by loss of MMP, and increased reactive oxygen species (ROS) production (Kuete et al., 2016).

The cytotoxicity and the mode of action of the neolignan licarin A (**30**) were also determined in a broad panel of animal and human cancer cell lines by the team of Mbaveng and collaborators (Mbaveng et al., 2021). The tested cancer cell line included CCRF-CEM, CEM/ADR5000, HCT116 $p53^{+/+}$, HCT116 $p53^{-/-}$, U87. MG, U87. MGΔ*EGFR*, MDA-MB-231-pcDNA3, MDA-MB-231-*BCRP*, HepG2, CC531, B16-F1, B16-F10, A2058, SK-Mel505, MaMel-80a, MV3, SkMel-28 and Mel-2A. The authors have applied the resazurin reduction assay to determine the cytotoxicity of the neolignane **30**, and its mode of induction of apoptosis using propidium iodide (PI) staining for cell cycle distribution, annexin V/PI staining for apoptosis, 5,5′,6,6′-tetrachloro-1,1′,3,3′-tetraethylbenzimidazolylcarbocyanine iodide staining for MMP analyzes, and 2′,7′-dichlorodihydrofluoresceine diacetate for the quantification of ROS by flow cytometry. They found that compound **30** had cytotoxic effects on the 18 tested cancer cell lines with the IC_{50} values ranging from 4.3 μM (against CCRF-CEM cells) to 21.8 μM (against HCT116 $p53^{-/-}$). Compound **30** also induced apoptosis in CCRF-CEM cells mediated by activation of caspase 3/7, 8, and 9, MMP alteration, and increased ROS production.

6. Conclusion

In this review, we have identified 13 lignans, 16 stilbenes, and 2 neolignans isolated from African medicinal plants with cytotoxic effects on various human cancer cell lines. The most potent cytotoxic lignans were diphyllin (**2**), futokadsurin B (**3**), justicidins A (**4**) and B (**5**), kusunokinin (**6**), pycnanthulignenes A (**10**) and B (**11**), and tuberculatin (**13**). The best stilbenes include combretastatin A-1 (**14**), combretastatin B-1 (**15**), combretastatin B-3 (**16**), combretastatin B-4 (**17**), longistylins A (**18**) and (**19**), mundulea lactone (**20**), schweinfurthins A (**23**), B (**24**), E (**25**), F (**26**), and G (**27**), and vedelianin (**29**). The two reported cytotoxic neolignans were licarins A (**30**) and B (**31**). Amongst these compounds, those that can combat the multidrug resistance of cancer were identified as **3**, **10, 11**, and **20**. Some of the active lignans and stilbenes displayed apoptotic cell death in cancer cells, mainly mediated by caspase activation, alteration of MMP, and enhanced ROS production in cancer cells. The above lignans, neolignans, and stilbenes deserve further in-depth studies to develop novel cytotoxic products to fight cancer.

References

Al-Qathama, A., Gibbons, S., & Prieto, J. M. (2017). Differential modulation of Bax/Bcl-2 ratio and onset of caspase-3/7 activation induced by derivatives of Justicidin B in human melanoma cells A375. *Oncotarget, 8*(56), 95999–96012.

Anjum, S., Abbasi, B. H., Doussot, J., Favre-Réguillon, A., & Hano, C. (2017). Effects of photoperiod regimes and ultraviolet-C radiations on biosynthesis of industrially important lignans and neolignans in cell cultures of Linum usitatissimum L. (Flax). *Journal of Photochemistry and Photobiology B: Biology, 167*, 216–227.

Ashidi, J. S., Houghton, P. J., Hylands, P. J., & Efferth, T. (2010). Ethnobotanical survey and cytotoxicity testing of plants of South-western Nigeria used to treat cancer, with isolation of cytotoxic constituents from *Cajanus cajan* Millsp. leaves. *Journal of Ethnopharmacology, 128*(2), 501–512.

Ashour, M. L., El-Readi, M. Z., Tahrani, A., Eid, S. Y., & Wink, M. (2012). A novel cytotoxic aryltetraline lactone from *Bupleurum marginatum* (Apiaceae). *Phytochemistry Letters, 5*(2), 387–392.

Austin, M. B., Bowman, M. E., Ferrer, J. L., Schröder, J., & Noel, J. P. (2004). An aldol switch discovered in stilbene synthases mediates cyclization specificity of type III polyketide synthases. *Chemistry & Biology, 11*(9), 1179–1194.

Beutler, J. A., Shoemaker, R. H., Johnson, T., & Boyd, M. R. (1998). Cytotoxic geranyl stilbenes from *Macaranga schweinfurthii*. *Journal of Natural Products, 61*(12), 1509–1512.

Chen, H., Liu, P., Zhang, T., Gao, Y., Zhang, Y., Shen, X., et al. (2018). Effects of diphyllin as a novel V-ATPase inhibitor on TE-1 and ECA-109 cells. *Oncology Reports, 39*(3), 921–928.

Chong, J., Poutaraud, A., & Hugueney, P. (2009). Metabolism and roles of stilbenes in plants. *Plant Science, 177*(3), 143–155.

Day, S. H., Lin, Y. C., Tsai, M. L., Tsao, L. T., Ko, H. H., Chung, M. I., et al. (2002). Potent cytotoxic lignans from *Justicia procumbens* and their effects on nitric oxide and tumor necrosis factor-alpha production in mouse macrophages. *Journal of Natural Products, 65*(3), 379–381.

Dubrovina, A. S., & Kiselev, K. V. (2017). Regulation of stilbene biosynthesis in plants. *Planta, 246*(4), 597–623.

Efferth, T., Kadioglu, O., Saeed, M. E. M., Seo, E. J., Mbaveng, A. T., & Kuete, V. (2021). Medicinal plants and phytochemicals against multidrug-resistant tumor cells expressing ABCB1, ABCG2, or ABCB5: A synopsis of 2 decades. *Phytochemistry Reviews, 20*(1), 7–53.

Efferth, T., Saeed, M. E. M., Kadioglu, O., Seo, E. J., Shirooie, S., Mbaveng, A. T., et al. (2020). Collateral sensitivity of natural products in drug-resistant cancer cells. *Biotechnology Advances, 38*, 107342.

Eleftherianos, I., Boundy, S., Joyce, S. A., Aslam, S., Marshall, J. W., Cox, R. J., et al. (2007). An antibiotic produced by an insect-pathogenic bacterium suppresses host defenses through phenoloxidase inhibition. *Proceedings of the National Academy of Sciences, 104*(7), 2419–2424.

Gertsch, J., Tobler, R. T., Brun, R., Sticher, O., & Heilmann, J. (2003). Antifungal, antiprotozoal, cytotoxic and piscicidal properties of Justicidin B and a new arylnaphthalide lignan from *Phyllanthus piscatorum*. *Planta Medica, 69*(5), 420–424.

Gehlert, R., Schöppner, A., & Kindl, H. (1990). Stilbene synthase from seedlings of *Pinus sylvestris*: Purification and induction in response to fungal infection. *Molecular Plant-Microbe Interactions, 3*, 444–449.

Giacomini, E., Rupiani, S., Guidotti, L., Recanatini, M., & Roberti, M. (2016). The use of stilbene scaffold in medicinal chemistry and multi- target drug design. *Current Medicinal Chemistry, 23*(23), 2439–2489.

Hartley, R. M., Peng, J., Fest, G. A., Dakshanamurthy, S., Frantz, D. E., Brown, M. L., et al. (2012). Polygamain, a new microtubule depolymerizing agent that occupies a unique pharmacophore in the colchicine site. *Molecular Pharmacology, 81*(3), 431–439.

Ilieva, Y., Zhelezova, I., Atanasova, T., Zaharieva, M. M., Sasheva, P., Ionkova, I., et al. (2014). Cytotoxic effect of the biotechnologically-derived justicidin B on human lymphoma cells. *Biotechnology Letters, 36*(11), 2177–2183.

Jiang, R. W., Zhou, J. R., Hon, P. M., Li, S. L., Zhou, Y., Li, L. L., et al. (2007). Lignans from *Dysosma versipellis* with inhibitory effects on prostate cancer cell lines. *Journal of Natural Products, 70*(2), 283–286.

Klausmeyer, P., Van, Q. N., Jato, J., McCloud, T. G., & Beutler, J. A. (2010). Schweinfurthins I and J from *Macaranga schweinfurthii*. *Journal of Natural Products, 73*(3), 479–481.

Kodan, A., Kuroda, H., & Sakai, F. (2002). A stilbene synthase from Japanese red pine (*Pinus densiflora*): Implications for phytoalexin accumulation and down-regulation of flavonoid biosynthesis. *Proceedings of the National Academy of Sciences of the United States of America, 99*(5), 3335–3339.

Korkina, L., Kostyuk, V., De Luca, C., & Pastore, S. (2011). Plant phenylpropanoids as emerging anti-inflammatory agents. *Mini-Reviews in Medicinal Chemistry, 11*(10), 823–835.

Kuete, V. (2013). Medicinal plant research in Africa. In Kuete (Ed.). *Pharmacology and chemistry*. Oxford: Elsevier.

Kuete, V. (2025). Chapter Four-African medicinal plants and their derivative as the source of potent anti-leukemic products: Rationale classification of naturally occurring anticancer agents. *Advances in botanical research, 113*. https://doi.org/10.1016/bs.abr.2023.12.010.

Kuete, V., Azebaze, A. G., Mbaveng, A., Nguemfo, E. L., Tshikalange, E. T., Chalard, P., et al. (2011). Antioxidant, antitumor and antimicrobial activities of the crude extract and compounds of the root bark of *Allanblackia floribunda*. *Pharmaceutical Biology, 49*(1), 57–65.

Kuete, V., & Efferth, T. (2015). African flora has the potential to fight multidrug resistance of cancer. *BioMed Research International, 2015*, 914813.

Kuete, V., Kamga, J., Sandjo, L. P., Ngameni, B., Poumale, H. M., Ambassa, P., et al. (2011). Antimicrobial activities of the methanol extract, fractions and compounds from *Ficus polita* Vahl. (Moraceae). *BMC Complementary and Alternative Medicine, 11*, 6.

Kuete, V., Mbaveng, A. T., Nono, E. C., Simo, C. C., Zeino, M., Nkengfack, A. E., et al. (2016). Cytotoxicity of seven naturally occurring phenolic compounds towards multi-factorial drug-resistant cancer cells. *Phytomedicine, 23*(8), 856–863.

Kuete, V., Nono, E. C., Mkounga, P., Marat, K., Hultin, P. G., & Nkengfack, A. E. (2011). Antimicrobial activities of the CH_2Cl_2-CH_3OH (1:1) extracts and compounds from the roots and fruits of *Pycnanthus angolensis* (Myristicaceae). *Natural Product Research, 25*(4), 432–443.

Kuete, V., Sandjo, L., Seukep, J., Maen, Z., Ngadjui, B., & Efferth, T. (2015). Cytotoxic compounds from the fruits of *Uapaca togoensis* towards multi-factorial drug-resistant cancer cells. *Planta Medica, 81*(1), 32–38.

Kuete, V., Sandjo, L. P., Mbaveng, A. T., Seukep, J. A., Ngadjui, B. T., & Efferth, T. (2015). Cytotoxicity of selected Cameroonian medicinal plants and *Nauclea pobeguinii* towards multi-factorial drug-resistant cancer cells. *BMC Complementary and Alternative Medicine, 15*, 309.

Landete, J. M. (2012). Plant and mammalian lignans: A review of source, intake, metabolism, intestinal bacteria and health. *Food Research International, 46*(1), 410–424.

Lee, J. C., Lee, C. H., Su, C. L., Huang, C. W., Liu, H. S., Lin, C. N., et al. (2005). Justicidin A decreases the level of cytosolic Ku70 leading to apoptosis in human colorectal cancer cells. *Carcinogenesis, 26*(10), 1716–1730.

Mad-Adam, N., Rattanaburee, T., Tanawattanasuntorn, T., & Graidist, P. (2022). Effects of trans-(±)-kusunokinin on chemosensitive and chemoresistant ovarian cancer cells. *Oncology Letters, 23*(2), 59.

Mbaveng, A. T., Fotso, G. W., Ngnintedo, D., Kuete, V., Ngadjui, B. T., Keumedjio, F., et al. (2018). Cytotoxicity of epunctanone and four other phytochemicals isolated from the medicinal plants *Garcinia epunctata* and *Ptycholobium contortum* towards multi-factorial drug resistant cancer cells. *Phytomedicine, 48*, 112–119.

Mbaveng, A. T., Kuete, V., & Efferth, T. (2017). Potential of Central, Eastern and Western Africa medicinal plants for cancer therapy: Spotlight on resistant cells and molecular targets. *Frontiers in Pharmacology, 8*, 343.

Mbaveng, A. T., Noulala, C. G. T., Samba, A. R. M., Tankeo, S. B., Fotso, G. W., Happi, E. N., et al. (2020). Cytotoxicity of botanicals and isolated phytochemicals from *Araliopsis soyauxii* Engl. (Rutaceae) towards a panel of human cancer cells. *Journal of Ethnopharmacology, 267*, 113535.

Mbaveng, A. T., Wamba, B. E. N., Bitchagno, G. T. M., Tankeo, S. B., Çelik, İ., Atontsa, B. C. K., et al. (2021). Bioactivity of fractions and constituents of *Piper capense* fruits towards a broad panel of cancer cells. *Journal of Ethnopharmacology, 271*, 113884.

Momekov, G., Yossifov, D., Guenova, M., Michova, A., Stoyanov, N., Konstantinov, S., et al. (2014). Apoptotic mechanisms of the biotechnologically produced arylnaphtalene lignan justicidin B in the acute myeloid leukemia-derived cell line HL-60. *Pharmacological Reports, 66*(6), 1073–1076.

Mori, T., Awakawa, T., Shimomura, K., Saito, Y., Yang, D., Morita, H., et al. (2016). Structural Insight into the enzymatic formation of bacterial stilbene. *Cell Chemical Biology, 23*(12), 1468–1479.

Nono, E. C., Mkounga, P., Kuete, V., Marat, K., Hultin, P. G., & Nkengfack, A. E. (2010). Pycnanthulignenes A-D, antimicrobial cyclolignene derivatives from the roots of *Pycnanthus angolensis*. *Journal of Natural Products, 73*(2), 213–216.

Parage, C., Tavares, R., Réty, S., Baltenweck-Guyot, R., Poutaraud, A., Renault, L., et al. (2012). Structural, functional, and evolutionary analysis of the unusually large stilbene synthase gene family in grapevine. *Plant Physiology, 160*(3), 1407–1419.

Peng, J., Hartley, R. M., Fest, G. A., & Mooberry, S. L. (2012). Amyrisins A-C, O-prenylated flavonoids from *Amyris madrensis*. *Journal of Natural Products, 75*(3), 494–496.

Pettit, G. R., Singh, S. B., Niven, M. L., Hamel, E., & Schmidt, J. M. (1987). Isolation, structure, and synthesis of combretastatins A-1 and B-1, potent new inhibitors of microtubule assembly, derived from *Combretum caffrum*. *Journal of Natural Products, 50*(1), 119–131.

Pettit, G. R., Singh, S. B., Schmidt, J. M., Niven, M. L., Hamel, E., & Lin, C. M. (1988). Isolation, structure, synthesis, and antimitotic properties of combretastatins B-3 and B-4 from *Combretum caffrum*. *Journal of Natural Products, 51*(3), 517–527.

Rattanaburee, T., Sermmai, P., Tangthana-Umrung, K., Thongpanchang, T., & Graidist, P. (2022). Anticancer activity of (±)-kusunokinin derivatives towards cholangiocarcinoma cells. *Molecules (Basel, Switzerland), 27*(23).

Rattanaburee, T., Tipmanee, V., Tedasen, A., Thongpanchang, T., & Graidist, P. (2020). Inhibition of CSF1R and AKT by (±)-kusunokinin hinders breast cancer cell proliferation. *Biomedicine & Pharmacotherapy, 129*, 110361.

Richardson, W. H., Schmidt, T. M., & Nealson, K. H. (1988). Identification of an anthraquinone pigment and a hydroxystilbene antibiotic from *Xenorhabdus luminescens*. *Applied and Environmental Microbiology, 54*(6), 1602–1605.

Rivière, C., Pawlus, A. D., & Mérillon, J. M. (2012). Natural stilbenoids: Distribution in the plant kingdom and chemotaxonomic interest in Vitaceae. *Natural Product Reports, 29*(11), 1317–1333.

Saeed, M., Kuete, V., Kadioglu, O., Börtzler, J., Khalid, H., Greten, H. J., et al. (2014). Cytotoxicity of the bisphenolic honokiol from Magnolia officinalis against multiple drug-resistant tumor cells as determined by pharmacogenomics and molecular docking. *Phytomedicine, 21*(12), 1525–1533.

Saeed, M. E. M., Rahama, M., Kuete, V., Dawood, M., Elbadawi, M., Sugimoto, Y., et al. (2019). Collateral sensitivity of drug-resistant ABCB5- and mutation-activated EGFR overexpressing cells towards resveratrol due to modulation of SIRT1 expression. *Phytomedicine, 59*, 152890.

Saleem, M., Kim, H. J., Ali, M. S., & Lee, Y. S. (2005). An update on bioactive plant lignans. *Natural Product Reports, 22*(6), 696–716.

Sartorelli, P., Carvalho, C. S., Reimão, J. Q., Lorenzi, H., & Tempone, A. G. (2010). Antitrypanosomal activity of a diterpene and lignans isolated from *Aristolochia cymbifera*. *Planta Medica, 76*(13), 1454–1456.

Sayed, A. M., El-Hawary, S. S., Abdelmohsen, U. R., & Ghareeb, M. A. (2022). Antiproliferative potential of *Physalis peruviana*-derived magnolin against pancreatic cancer: A comprehensive in vitro and in silico study. *Food & Function, 13*(22), 11733–11743.

Seukep, J. A., Sandjo, L. P., Ngadjui, B. T., & Kuete, V. (2016a). Antibacterial activities of the methanol extracts and compounds from *Uapaca togoensis* against Gram-negative multi-drug resistant phenotypes. *South African Journal of Botany, 103*, 1–5.

Seukep, J. A., Sandjo, L. P., Ngadjui, B. T., & Kuete, V. (2016b). Antibacterial and antibiotic-resistance modifying activity of the extracts and compounds from *Nauclea pobeguinii* against Gram-negative multi-drug resistant phenotypes. *BMC Complementary and Alternative Medicine, 16*, 193.

Shen, W., Zou, X., Chen, M., Liu, P., Shen, Y., Huang, S., et al. (2011). Effects of diphyllin as a novel V-ATPase inhibitor on gastric adenocarcinoma. *European Journal of Pharmacology, 667*(1-3), 330–338.

Sheriha, G. M., Abouamer, K., Elshtaiwi, B. Z., Ashour, A. S., Abed, F., & Alhallao, H. H. (1987). Quinoline alkaloids and cytotoxic lignans from *Haplophyllum tuberculatum*. *Phytochemistry, 26*(12), 3339–3341.

Sheriha, G. M., & Abouamer, K. M. (1984). Lignans of *Haplophyllum tuberculatum*. *Phytochemistry, 23*(1), 151–153.

Sobolev, V. S., Horn, B. W., Potter, T. L., Deyrup, S. T., & Gloer, J. B. (2006). Production of stilbenoids and phenolic acids by the peanut plant at early stages of growth. *Journal of Agricultural and Food Chemistry, 54*(10), 3505–3511.

Teponno, R. B., Kusari, S., & Spiteller, M. (2016). Recent advances in research on lignans and neolignans. *Natural Product Reports, 33*(9), 1044–1092.

Tsopmo, A., Awah, F. M., & Kuete, V. (2013). 12—Lignans and stilbenes from African medicinal plants. In Kuete (Ed.). *Medicinal plant research in Africa* (pp. 435–478). Oxford: Elsevier.

Umezawa, T., Yamamura, M., Nakatsubo, T., Suzuki, S., & Hattori, T. (2011). Stereoselectivity of the biosynthesis of norlignans and related compounds. In D. Gang (Ed.). *The biological activity of phytochemicals* (pp. 179–197). (41st ed.,). New York: Springer.

Vannozzi, A., Dry, I. B., Fasoli, M., Zenoni, S., & Lucchin, M. (2012). Genome-wide analysis of the grapevine stilbene synthase multigenic family: Genomic organization and expression profiles upon biotic and abiotic stresses. *BMC Plant Biology, 12*, 130.

Valletta, A., Iozia, L. M., & Leonelli, F. (2021). Impact of environmental factors on stilbene biosynthesis. *Plants (Basel), 10*(1).

Vogt, T. (2010). Phenylpropanoid biosynthesis. *Mol Plant, 3*(1), 2–20.

Wang, Y. W., Chuang, J. J., Chang, T. Y., Won, S. J., Tsai, H. W., Lee, C. T., et al. (2015). Antiangiogenesis as the novel mechanism for justicidin A in the anticancer effect on human bladder cancer. *Anti-cancer Drugs, 26*(4), 428–436.

Won, S. J., Yen, C. H., Liu, H. S., Wu, S. Y., Lan, S. H., Jiang-Shieh, Y. F., et al. (2015). Justicidin A-induced autophagy flux enhances apoptosis of human colorectal cancer cells via class III PI3K and Atg5 pathway. *Journal of Cellular Physiology, 230*(4), 930–946.

Yoder, B. J., Cao, S., Norris, A., Miller, J. S., Ratovoson, F., Razafitsalama, J., et al. (2007). Antiproliferative prenylated stilbenes and flavonoids from *Macaranga alnifolia* from the Madagascar rainforest. *Journal of Natural Products, 70*(3), 342–346.

Zálešák, F., Bon, D. J. D., & Pospíšil, J. (2019). Lignans and neolignans: Plant secondary metabolites as a reservoir of biologically active substances. *Pharmacological Research, 146*, 104284.

CHAPTER EIGHT

Pharmaceutical xanthones from African medicinal plants to fight cancers and their recalcitrant phenotypes

Hugues Fouotsa[a,*], Julio Issah Mawouma Pagna[b], and Victor Kuete[c]
[a]Department of Process Engineering, National Higher Polytechnic School of Douala, University of Douala, Douala, Cameroon
[b]Department of Organic Chemistry, Faculty of Science, University of Yaoundé I, Yaoundé, Cameroon
[c]Department of Biochemistry, Faculty of Science, University of Dschang, Dschang, Cameroon
*Corresponding author. e-mail address: fhugosuns@yahoo.fr

Contents

Abstract

Xanthones are one of the biggest classes of compounds in natural product chemistry. Several xanthones have been isolated from natural sources of higher plants, fungi, ferns, and lichens. They have gradually risen to great importance because of their

Advances in Botanical Research, Volume 115
ISSN 0065-2296, https://doi.org/10.1016/bs.abr.2024.02.003

medicinal properties such as antitumor, antidiabetic, and antimicrobial depending on their diverse structures, which are modified by substituents on the ring system. Although several reviews have already been published on xanthone compounds, few of them have focused on the anticancer activity of xanthone derivatives.

In this chapter, we summarized the phytochemistry, the biosynthetic pathway, and cytotoxic potential of xanthones derived from plants native to Africa towards both drug sensitive and multidrug-resistant cancer cells.

Abbreviations

^{13}C NMR Carbon-13 nuclear magnetic resonance.
1D One dimensional (RMN).
^{1}H NMR Proton nuclear magnetic resonance.
2D Two dimensional (RMN).
AcOEt Ethyl acetate.
AcOH Acetic acid.
AH6809 isopropoxy-9-oxoxanthene-2-carboxylic acid.
COSY Correlation spectroscopy.
DEPT Distortionless enhancement by polarisation transfer.
DMXAA 5,6-dimethylxanthenone-4-acetic acid.
GC Gas chromatography.
GLC Gas-liquid chromatography.
H_2O Water.
HIV/AIDS Human immunodeficiency virus/acquired immune deficiency syndrome.
HMBC Heteronuclear multiple bond correlation.
HPLC High performance liquid chromatography.
HROESY High-rotating-frame overhauser effect spectroscopy.
HSQC Heteronuclear single quantum coherence.
IC_{50} Inhibitory concentration 50 (concentration of a compound required for 50% inhibition).
IR Infrared.
MAO A Monoamine oxidase A.
MDR Multidrug-resistant.
MeOH Methanol.
MIC Minimum inhibitory concentration.
MS Mass spectrometry.
NADPH Nicotinamide adenine dinucleotide phosphate.
NMR Nuclear magnetic resonance.
NOESY Nuclear overhauser effect spectroscopy.
PKSs polyketide synthases.
THPA 2,3′,4,6-tetrahydroxybenzophenone.
TLC Thin-layer chromatography.
TMS Tetramethylsilane.
TOCSY Total Correlation Spectroscopy.
TTR Transthyretin.
UV Ultravioletsss.

1. Introduction

According to the United States "National Cancer Institute", cancer is a disease in which some of the body's cells grow uncontrollably and spread to other parts of the body. Nowadays, cancer is considered the leading cause of death in modern society with estimated 9.6 million deaths representing one in six deaths and with an estimated five-year prevalence of 43.8 million people (IARC, 2020). There is an urgent need to find better cures for cancer-related diseases. Although enormous efforts have been dedicated to developing novel drugs for cancer treatment, cancer remains a major life-threatening disease. To develop more potent anticancer drugs, many tools have been employed, including chemical and biological methods. Using chemical tools, numerous structural scaffolds have been created for disease treatments. Among these scaffolds, xanthones are simple three-membered heterocyclic ring compounds mainly found as secondary metabolites in higher plants and microorganisms. Xanthones are secondary metabolites found in a few higher plants, fungi, and lichens. The xanthone skeleton (the word "xanthon" is derived from the Greek word xanthos, meaning yellow) is a planar, conjugated ring system composed of carbons 1–4 (aromatic ring A) and carbons 5–8 (aromatic ring B), fused through a carbonyl group and an oxygen atom (Fig. 1). The simplest member of the class, 9H-xanthen-9-one, is a symmetrical compound with a dibenzo-γ-pyrone skeleton (El-Seedi et al., 2009). The numbering starts from ring A, while ring B is given prime locants or consecutively numbered from ring A. According to literature reports, around 650 xanthones are known from natural sources. These xanthones have been isolated from 62 families of higher plants, fungi, and lichens (El-Seedi et al., 2009; Masters & Brase, 2012; Peres, Nagem, & De Oliveira, 2000; Viera & Kijjoa, 2005; Yang, Ma, Wie, Han, & Gao, 2012). Xanthones from higher plants are distributed in Gentianaceae, Moraceae, Guttiferae, Polygalaceae, and Leguminosae families. The

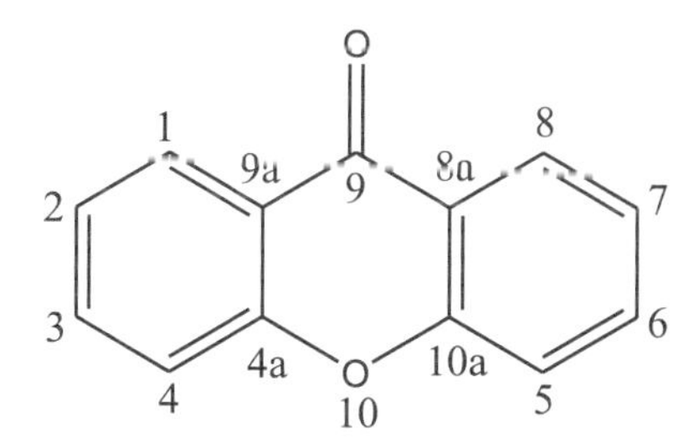

Fig. 1 Xanthone basic skeleton (9H-xanthen-9-one).

families with higher proportions are Clusiaceae or Guttiferae (55 species in 12 genera) and Gentianaceae (121 species in 21 genera) (Jensen & Schripsema, 2002; Masters & Brase, 2012; Viera & Kijjoa, 2005; Yang et al., 2012).

Xanthones have diverse pharmacological properties, mainly due to their oxygenation nature and diversity of functional groups. The biological activities discussed in some review articles on xanthones during 2000–12 include antibacterial, antiviral, antioxidative, anti-inflammatory, antiproliferative, antihypertensive, antithrombotic, *in vitro* and *in vivo* antitumor, cytotoxic, coagulant, monoamine oxidase (MAO) inhibition, gastro-protective effects, antiatherosclerosis activity, inhibition of hypotension, cardioprotection, inhibition of cholinesterase, cyclooxygenase activity, immunosuppression, and binding to transthyretin (El-Seedi et al., 2010; Masters & Brase, 2012; Peres et al., 2000; Pinto, Sousa, & Nascimento, 2005; Pouli & Marakos, 2009; Viera & Kijjoa, 2005; Wang, Liu, & Zhang, 2010). The α-glucosidase inhibitory activity may lead to xanthones, and their sources being used against diabetes and HIV/AIDS (Li et al., 2011). Xanthones have shown the potential to prevent disease development by their concerted action of protecting cells from oxidative stress damage and acting as phytoalexin to impair pathogen growth (Franklin, Conceic, Kombrink, & Dias, 2009). A biological activity that is of significance for tropical Africa is the antimalarial property (Alaribe et al., 2012; Pontius, Krick, Kehraus., Brun, & Konig, 2008). The African continent has a rich treasure of plant natural resources. Several plants are in traditional use as remedies for various diseases, including life-threatening diseases such as malaria, tuberculosis, HIV/AIDS, diabetes, and hypertension. Over 90 xanthones with different functionalities are known from the plants occurring on the African continent. Some African plant sources of xanthones are *Garcinia polyantha* and *Garcinia nobilis* (Guttiferae) (Fouotsa et al., 2012; Lannang et al., 2005), *Symphonia globulifera* (Guttiferae) (Ngouela et al., 2005), *Securidaca longepedunculata* (Polygalaceae) (Meyer, Rakuambo, & Hussein, 2008), *Allanblackia floribunda* and *Allanblackia monticola* (Guttiferae) (Azebaze, Meyer, Bodo, & Nkengfack, 2004; Nkengfack, Mkounga, Meyer, Fomum, & Bodo, 2002), *Anthocleista vogelii* (Loganiaceae) (Alaribe et al., 2012), *Garcinia gerrardii*, *Garcinia livingstonei*, *Hypericum roeperanum* (Guttiferae), *Polygala nyikensis* (Polygalaceae), and *Swertia calycina* (Gentianaceae) (Hostettmann, Marston, Ndjoko, & Wolfender, 2000).

Xanthones have gradually risen to great importance because of their medicinal properties. This review focuses on the types, isolation, characterization, biological applications, and biosynthesis of naturally occurring xanthones isolated so far. Different physicochemical and instrumental methods such as liquid-solid and liquid-liquid extraction, Thin-layer chromatography (TLC), flash chromatography, column chromatography, infrared (IR), Proton nuclear magnetic resonance (^{1}H NMR), and Carbon-13 nuclear magnetic resonance (^{13}C NMR) spectroscopy, Gas-liquid chromatography, high performance liquid chromatography (HPLC), gas chromatography, and Liquid chromatography–mass spectrometry have been widely used for isolation and structural elucidation of xanthones. In the present Chapter, a synopsis of the cytotoxic potential of xanthones isolated from African medicinal plants will be presented.

2. Classification

Xanthones isolated from natural sources are classified into six main groups, namely, simple oxygenated xanthones, prenylated xanthones, xanthone glycosides, xanthonolignoids, bis-xanthones, and miscellaneous xanthones, which include caged xanthones (Jensen & Schripsema, 2002; Pouli & Marakos, 2009; Viera & Kijjoa, 2005). The polyphenolic xanthones are further subdivided according to the degree of oxygenation into non-, mono-, di-, tri-, tetra-, penta-, and hexa-oxygenated substances (Jamwal, 2012; Velisek, Davidek, & Cejpek, 2008; Viera & Kijjoa, 2005). The other xanthone subclasses are based on the level of oxidation of ring A, which can occur either as completely aromatic or as dihydro-, tetrahydro-, and hexahydro derivatives, either in monomeric or dimeric form (Masters & Brase, 2012). There are several xanthones that are hydroxylated xanthones with prenyl or geranyl units (El-Seedi et al., 2010). The Clusiaceae family mainly has prenylated xanthones, while the Gentianaceae family has oxygenated xanthones (Scheme 1).

2.1 Simple oxygenated xanthones

Simple oxygenated xanthones are subdivided according to the degree of oxygenation into non-, mono-, di-, tri-, tetra-, penta-, and hexaoxygenated substances (Viera & Kijjoa, 2005; Mandal, Das, & Joshi, 1992; Sultanbawa, 1980). In these xanthones, the substituents are simple hydroxy, methoxy, or methyl groups. About 150 simple oxygenated xanthones have been reported.

Non-oxygenated simple xanthones The non-oxygenated xanthones, namely, methylxanthones (1-,2-,3-,4-methylxanthone), were reported in crude oils from off-shore Norway (Oldenburg et al., 2002). This was the first description of xanthones in fossil organic matter. These xanthones might have been generated as diagenetic products, formed by oxidation of xanthenes in the reservoir, or might have originated by biosynthesis from aromatic precursors.

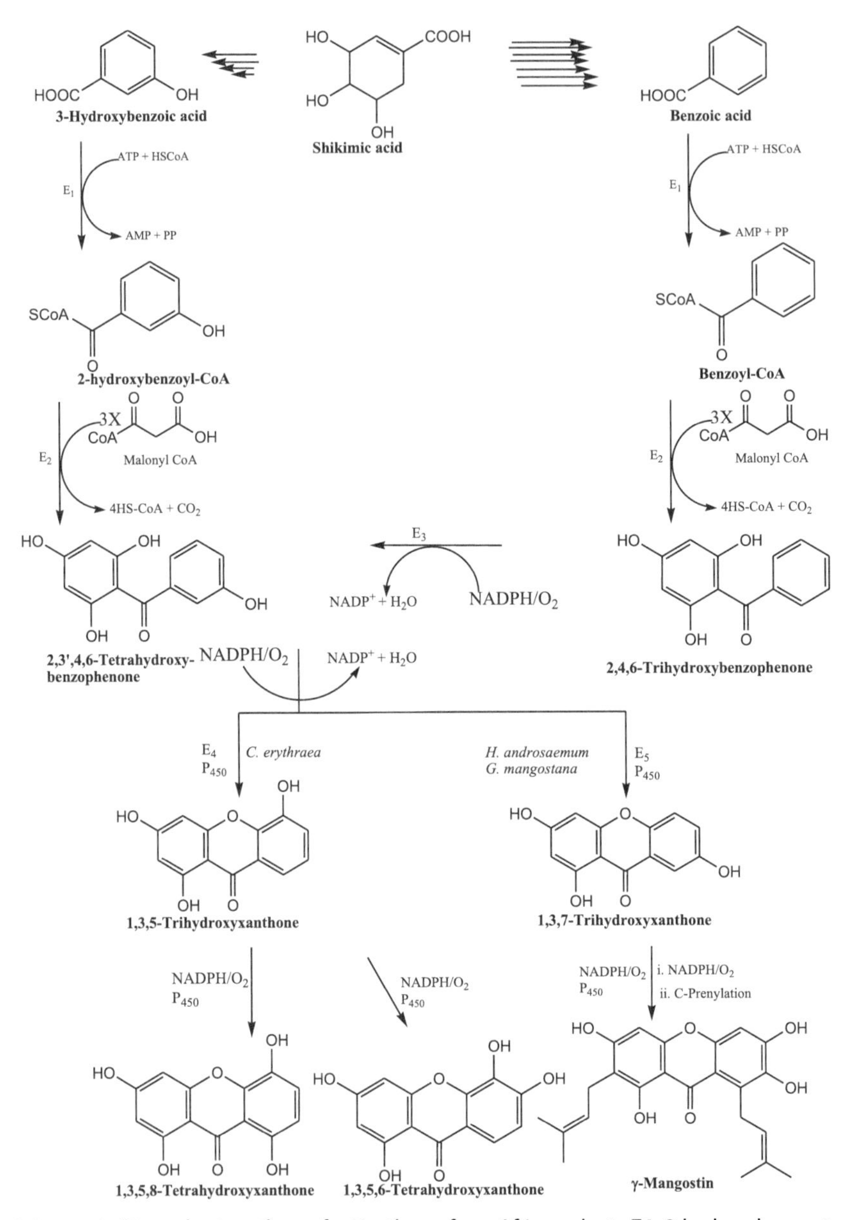

Scheme 1 Biosynthesis pathway for Xanthone from African plants **E1**: 3-hydroxybenzoate-CoA ligase, **E2**: Benzophenone synthase, **E3**: Benzophenone 3′hydroxylase, **E4=E5**: xanthone synthase.

Mono-oxygenated xanthones. Besides, six mono-oxygenated xanthones from *Swertia*, 2-hydroxyxanthone, 4-hydroxyxanthone, and 2-methoxyxanthone have been isolated from four genera, namely, *Calophyllum*, *Kielmeyera*, *Mesua,* and *Ochrocarpus*.

Di-oxygenated xanthones. More than fifteen deoxygenated xanthones were reported from plants of the families Clusiaceae and Euphorbiaceae. 1,5-Dihydroxyxanthone, 1,7-dihydroxyxanthone, and 2,6-dihydroxyxanthone are found extensively. Other deoxygenated xanthones such as 1-hydroxy-5-methoxyxanthone, 1-hydroxy-7-methoxyxanthone, 2-hydroxy-1-methoxy-xanthone, 3-hydroxy-2-methoxyxanthone, 3-hydroxy-4-methoxyxanthone, 5-hydroxy-1-methoxyxanthone, and 1,2-methylenedioxyxanthone have been reported from 11 plants genera.

Tri-oxygenated xanthones. Forty-five trioxygenated xanthones have been reported; out of these fifteen were described as new. Among these, only two natural sulfonated xanthones, namely, 1,3-dihydroxy-5-methoxyx-anthone-4-sulfonate and 5-O-β-D-glucopyranosyl-1,3-dihydroxyxanthone-4-sulfonate, are reported from *Hypericum sampsonii*. *Tetra-oxygenated xanthones.* Among the 53 tetra-oxygenated xanthones identified so far, 21 were found to be new natural products. These xanthones were mainly reported from plants of the families Gentianaceae, Clusiaceae, and Polygalaceae. Interestingly, 7-chloro-1,2,3-trihydroxy-6-methoxyxanthone isolated from *Polygala vulgaris* (Dall'Acqua et al., 2002) appeared to be the first chloroxanthone of the family Polygalaceae. This compound exhibited antiproliferative activity against the human intestinal adenocarcinoma cell line. The free hydro-xyxanthones are 1,3,5,6-, 1,3,5,7-, and 1,3,6,7-tetrahydroxyxanthone (Chen, Wang, Wu, & Qin, 2013).

Penta-oxygenated xanthones. Twenty-seven penta-oxygenated xanthones have been identified. Four partially methylated penta-oxygenated xanthones, namely, 1,8-dihydroxy-2,3,7-trimethoxyxanthone, 5,6-dihydroxy-1,3,7-trimethoxyxanthone, 1,7-dihydroxy-2,3,8-trimethoxyxanthone, 3,8-dihy-droxy-1,2,6-trimethoxyxanthone (Xue, Li, Zuo, Yang, & Zhang, 2009), and 3,7-dihydroxy-1,5,6-trimethoxyxanthone, have been isolated from three plants genera.

Hexa-oxygenated xanthones. Two hexa-oxygenated xanthones, 8-hydroxy 1,2,3,4,6-pentamethoxyxanthone (Jankovic, Krstic, Savikin-Fodulovic, Menkovic, & Grubisic, 2002; Valentao et al., 2002) and 1,8-dihydroxy-2,3,4,6-tetramethoxyxanthone (Krstic, Jankovic, Savikin-Fodulovic, Menkovic, & Grubisic, 2003), are isolated from two *Centaurium* species and 3-hydroxy-1,2,5,6,7-pentamethoxyxanthone was isolated from the roots of *Polygala japonica*. The natural occurrence of penta-oxygenated, hexa-oxygenated, and dimeric xanthones has been reviewed by Peres and Nagem (Peres & Nagem, 1997).

2.2 Xanthone glycosides

Sixty-one naturally occurring glycosylated xanthones, 39 of which are new compounds, have been reported predominantly in the families Gentianaceae and Polygalaceae as C- or O-glycosides. The details of naturally occurring xanthone glycosides have been reviewed (Hostettmann & Miura, 1977) and a distinction between C-glycosides and Oglycosides has also been made. In C-glycosides, C–C bond links the sugar moiety to the xanthone nucleus and they are resistant to acidic and enzymatic hydrolysis whereas the *O*-glycosides have typical glycosidic linkage.

C-Glycosides. C-glycosides are rare; thus, only seven C-glycosides were mentioned in Sultanbawa's review (Sultanbawa, 1980) and 17 in Al-Hazimi's review (Al-Hazimi & Miana, 1990). Mangiferin and isomangiferin are the most common C-glycosides. Mangiferin (2,-C-β-D-glucopyranosyl-1,3,6,7-tetrahydroxyxanthone) is of widespread occurrence in angiosperms and ferns and was first isolated from *Mangifera indica* (Bhatia, Ramanathan, & Seshadri, 1967; Haynes & Taylor, 1966; Iseda, 1957). An isomer, isomangiferin (4-C-β-D-glucopyranosyl-1,3,6,7-tetrahydroxyxanthone), has been isolated from the aerial parts of *Anemarrhena asphodeloides* (Aritomi & T. Kawasaki, 1970a). Homomangiferin (2-C-β-D-glucopyranosyl-3-methoxy-1,6,7-trihydroxyxanthone) has also been isolated from the bark of *M. indica* (Aritomi & Kawasaki, 1970b). In 1973, another glycoxanthone (2-C-β-D-glucopyranosyl-1,3,5,6-tetrahydroxyxanthone) with an oxidation pattern other than that of mangiferin was found in *Canscora decussate* (Ghosal & Chaudhuri, 1973). Arisawa and Morita (1976) isolated tetraoxygenated xanthone glycoside 2-C-β-D-glucopyranosyl-5-methoxy-1,3,6-trihydroxyxanthone from *Iris florentina*.

O-Glycosides. More than 20 xanthone O-glycosides are known. A few are from natural sources, namely, gentiacauloside from *Gentiana acaulis*, gentioside from *Gentiana lutea*, and swertianolin from *Swertia japonica* (Arisawa & Morita, 1976). Their natural occurrence is restricted to the family Gentianaceae. The first xanthone O-glycoside, norswertianin-1-O-glucosyl-3-O-glucoside, was isolated from *Swertia perennis* (Hostettmann & Miura, 1977). A tetraoxygenated xanthone O-glycoside (3,7,8-trihydroxyxanthone-1-O-β-laminaribioside) was isolated from the fern species (Imperato, 1980). 1-Hydroxy-7-methoxy-3-O-primeverosylxanthone

(Verney & Debelmas, 1973) and 1-methoxy-5-hydroxyxanthone-3-O-rutinoside (Ghosal, Chauhan, Biswas, & Chaudhuri, 1976) have been isolated from *Gentiana* species and *C. decussata*.

2.3 Prenylated and related xanthones

Among 285 prenylated xanthones, 173 were described as new compounds. The occurrence of prenylated xanthones is restricted to the plant species of the family Guttiferae. The major C5 unit of the substituents included the commonly found 3-methylbut-2-enyl or isoprenyl group as in isoemericellin and the less frequent 3-hydroxy-3-methylbutyl as in nigrolineaxanthone P and 1,1-dimethylprop-2-enyl as in globuxanthone, respectively (Bringmann, Lang, Steffens, Gunther, & Schaumann, 2003; Nguyen & Harrison, 2000; Rukachaisirikul et al., 2003). Prenylated xanthones, caloxanthone O and caloxanthone P were isolated from *Calophyllum inophyllum* (Dai et al., 2010), and polyprenylated xanthones and benzophenones from *Garcinia oblongifolia* (Shan, Lin, Yu, Chen, & Zhan, 2012).

2.4 Xanthonolignoids

Naturally occurring xanthonolignoids are rare, so only five compounds are known. The first xanthonolignoid was isolated from Kielmeyera species by Castelao et al. (Castelao et al., 1977). They also isolated two other xanthonolignoids named cadensins A and B from *Caraipa densiflora*. A xanthonolignoid Kielcorin was obtained from *Hypericum* species (Nielsen & Arends, 1978). Recently, kielcorin was also isolated from *Vismia guaramirangae* (Monache, Mac-Quhae., Monache, Bettolo, & De Lima, 1983), *Kielmeyera variabilis* (Pinheiro et al., 2003), and *Hypericum canariensis* (Cardona, Fernandez, Pedro, Seoane, & Vidal, 1986), whereas cadensin C and cadensin D from *V. guaramirangae* and *H. canariensis* have been reported (Nikolaeva, Glyzin, Mladentseva, Sheichenko, & Patudin, 1983).

2.5 Bis-xanthones

A total of twelve bis-xanthones, five from higher plants, one from lichen, and six from fungi, have been reported to date. These include jacarelhyperols A and B (Ishiguro, Nagata, Oku, & Masae, 2002), from the aerial parts of *Hypericum japonicum* and dimeric xanthone, and globulixanthone E, from the roots of *Symphonia globulifera* (Nkengfack, Azebaze, Vardamides, Fomum, & Heerden, 2002; Nkengfack et al., 2002). Three C2-C2′ dimeric tetrahydroxyxanthones dicerandrols A, B, and C, are also isolated from the fungus *Phomopsis longicolla* (Wagenaar & Clardy, 2001).

2.6 Miscellaneous

Xanthones with substituents other than those mentioned above are included in this group. Xanthofulvin and vinaxanthone were isolated from *Penicillium* species (Kumagai, Hosotani, Kikuchi, Kimura, & Saji, 2003). A polycyclic substance (xanthopterin) with the ability to inhibit the HSP47 (heat shock protein) gene expression was isolated from the culture broth of a *Streptomyces* species (Terui et al., 2003). Xantholiptin is a potent inhibitor of collagen production induced by treatment with TGF-b in human dermal fibroblasts. Xanthones have been synthesized by different methods. The elements of synthetic methods such as building blocks, Diels-Alder reaction, and heterogeneous catalysts have also been reviewed (Yang et al., 2012).

3. Methods for isolation and characterization of xanthones

3.1 Isolation of xanthones

Plants xanthones are commonly isolated by column chromatography on silica gel using different solvent mixtures with increasing polarity (Dua et al., 2004; Khan, Rahman, & Islam, 2010; Purev et al., 2002). Xanthone glycosides are usually crystallized from methanol (MeOH). They may also be separated and identified using TLC (Chawla, Chibber, & Khera, 1975) and HPLC (Negi & Singh, 2011; Negi, Singh, Pant, & Rawat, 2010; Song, Hu, Lin, Lu, & Shi, 2004; Suryawanshi, Mehrotra, Asthana, & Gupta, 2006; Zhang, Wang, Dong, Yang, & Li, 2009) by comparison with authentic samples. The structure of xanthones has been established on the basis of UV, IR, MS, and NMR data (Chakravarty, Mukhopadhyay, Moitra, & Das, 1994; Chakravarty et al., 2001; Hajimehdipour, Amanzadeh, Sadat, Ebrahimi, & Mozaffarian, 2003; Khetwal & Bisht, 1988; Kikuchi & Kikuchi, 2004; Koolen et al., 2013; Mandal & Chatterjee, 1994; Sun et al., 2013; Verma & Khetwal, 1985; Wang, Zhao, Han, & Liang, 2005; Yang, Ding, Duan, & Liu, 2005). Preparative TLC on silica gel using ethyl acetate (AcOEt), MeOH, and H_2O (21:4:3) as mobile phase has been used in instances of difficult separation. Frequently used solvents in TLC are on polyamide, MeOH-H_2O (9:1) and MeOH-H_2O-acetic acid (AcOH; 90: 5: 5); on cellulose, AcHO (5%–30%); on silica gel, Py-H_2O-AcOEt-MeOH (12: 10: 80: 5) and AcOEt-MeOH-H_2O (21: 4: 3) and chromatoplates are viewed in UV light. In certain cases, spraying with 5% KOH in MeOH or 5% aqueous H_2SO_4 has been advantageous (Verney & Debelmas, 1973). Polyamide columns are frequently applied for the separation of xanthone glycosides.

Purification of xanthones on Sephadex LH-20 column has also been carried out (Hostettmann & Miura, 1977). Xanthones are also isolated from resin of *Garcinia hanburyi* (Deng, Guo, Xie Shao, & Pan, 2013) and from the fermentation products of an endophytic fungus *Phomopsis* (Yang et al., 2013). HPLC has been proven as the best technique for the separation, identification, and quantification of xanthones. Several HPLC methods have been developed for naturally occurring xanthones using microporous chemically bonded silica gel (Micropak CN column), solvent hexane-chloroform (13: 7, v/v), isooctane-$CHCl_3$ (3: 17, v/v), or dioxane-dichloromethane (1: 9) detected at 254 nm by UV detector (Negi & Singh, 2011). Polar aglycones, as well as glycosides of xanthones, are also resolved on a reversed-phase column (C8 and C18) using acetonitrile-water as mobile phase (Aberham, Schwaiger, Stuppner, & Ganzera, 2007; Dai, Gao, & Bi, 2004). High-speed counter-current chromatography and high-performance centrifugal partition chromatography were also used for the separation and isolation of mangiferin and neomangiferin from an extract of *A. asphodeloides* (Zhou et al., 2007) and α-mangostins and γ-mangostins from mangosteen pericarp, respectively (Shan & Zhang, 2010).

3.2 Characterization of xanthones

Ultraviolet visible spectroscopy (UV). The ultraviolet visible spectroscopy technique is useful for locating free hydroxyl groups in xanthones. In particular, the OH group at position 3 is easily detected by the addition of NaOAc which results in a bathochromic shift of the 300–330 nm bands with increased intensity. Three or four bands of maximum absorption are always found in the region 220–410 nm and it is noteworthy that all bands show high intensity. Most of the substances show a marked absorption in the 400 nm regions, which accounts for their yellow colour (Van & Nessler, 1891).

Infrared spectroscopy (IR). The carbonyl group in xanthones is always easily detectable in IR spectra as a strong band (stretching frequency) in the region of 1657 cm^{-1} (Purev et al., 2002). The presence of a hydroxyl group in the 1 or 8 position lowers the frequency to about 1650 cm^{-1} by hydrogen bonding. Substituents in the 3 or 6 position of the xanthone nucleus may have a marked effect upon the carbonyl stretching frequency (Whitman & Wiles, 1956).

Proton nuclear magnetic resonance spectroscopy (^1H NMR). 1D and 2D NMR spectra (^1H, ^{13}C, DEPT, COSY, TOCSY, HROESY, HSQC, HMBC, and NOESY) have been used for the characterization of the xanthones. The ^1H NMR spectrum appears predominantly in the range of 0–12 ppm downfield from the reference signal of TMS. The integral of the signals is proportional to the number of protons present. ^1H NMR gives information about the substitution pattern on each ring. Acetylated derivatives have been utilized in the structure determination of glycosides (Barraclough, Locksley, Scheinmann, Magalhaes, & Gottlieb, 1970). The number and relative position of acetyl and methoxy groups can be determined by observing the shift in the position of absorption for the aromatic protons which occurs upon replacing methoxy group by an acetyl group. Signals between δ 2.40–2.50 are indicative of acetylation at peri-position to the carbonyl group (1 or 8 position) since for other positions the acetyl signals fall between δ 2.30 and 2.35. In nonacetylated xanthones, the presence of hydrogen bonded OH at δ 12–13 also confirms hydroxyl substitution at 1 or 8. But when these positions are unsubstituted, then absorption for the aromatic protons appears at δ 7.70–8.05 (Kaldas, Hostettmann, & Jacot-Guillarmod, 1974). Tetraoxygenated xanthones, namely, 1,3,7,8- and 1,3,5,8-, showed two meta and two ortho-coupled protons in the ^1H NMR spectrum. They can also be distinguished by the fact that the presence of the ortho-coupled proton in the 1,3,7,8- system appears at lower field (Hostettmann, Tabacchi, & Jacot-Guillarmod, 1974) than that for 1,3,5,8- (bellidifolin) system (Gentili & Horowitz, 1968). The signals of 2″-O-acetyl methyl protons of 8-Cglucosyl flavone acetate are found at higher field than those of corresponding 6-C-glucosyl flavone acetate (Holdsworth, 1973). In a similar manner, 2-C and 4-C isomeric glycosyl xanthones can be distinguished.

Carbon nuclear magnetic resonance spectroscopy (^{13}C NMR). The number of signals in the ^{13}C NMR spectrum indicates the number of different types of C atoms. It gives information about the total number of the C atoms present in the molecule. It is particularly diagnostic for determining the sugar linkage in di- or polysaccharides; the signal of the carbon carrying the primary alcohols appears at δ 62 in glucose. This signal is shifted to δ 67 in disaccharides possessing a 1–6 linkage (Negi & Singh, 2011; Negi et al., 2010). The chemical shift for carbonyl carbon is δ 184.5 when positions 1 and 8 are substituted by hydroxyl groups. But when one of these positions is occupied either by a methoxy or a sugar moiety, the carbonyl signal is shifted upfield by about 4 ppm. If both positions are occupied by a methoxy group or sugar moieties, the upfield shift is about 10 ppm. When methoxy groups are located in position 1 or 8, the corresponding absorption appears at δ 60–61, whereas they appear at about δ 56 when the methoxy group is located in the remaining positions on the xanthone nucleus (Purev et al., 2002).

Mass spectrometry (MS). Mass spectrometry is also a useful tool in the structure elucidation of xanthone glycosides. Prox established the fragmentation pattern of mangiferin and related C-glycosides (Prox, 1968). Aritomi & Kawasaki obtained satisfactory results using peracetylated derivatives of the same and analogous compounds (Aritomi & Kawasaki, 1970a, 1970b). In the mass spectrum of *O*-glycosides, no discernible molecular ion peak can be observed, but an important fragment ion peak due to the aglycone moiety appears, followed by further fragmentation. Significant fragment ions from the loss of OH, H_2O, and CHO are typical for xanthones and related compounds with a methoxy substituent peri to the carbonyl group (Arends, Helboe, & Moller, 1973; Purev et al., 2002).

4. Biosynthesis and biological activities

4.1 Biosynthesis of xanthones

The structural diversity of xanthones lies in their mixed shikimate and acetate biogenic origin. The biosynthetic pathway defines xanthones as cyclized 2,3-dihydroxybenzophenone derivatives. Based on the biogenic pathway, 9H-xanthen-9-one carbons 1–4 are assigned to the acetate-derived ring A, and carbons 5–8 to the shikimate-derived ring B. The biosynthesis of xanthones is based on the synthesis of 2,3′,4,6-Tetrahydroxybenzophenone (Beerhues, 1996) (Scheme 1). The benzophenone undergoes regioselective intramolecular cyclization through one-electron oxidation steps to form the simplest xanthone. The xanthone biosynthetic pathway in higher plants has been proposed to involve the condensation of shikimate acid derivatives and malonyl-CoA as extension units (Beerhues, 1996; Masters & Brase, 2012). Some xanthones are entirely derived from the acetate pathway by folding a C16-polyketide in lower plants Helminthosporium ravenielii and Helminthosporium turcicum (Birch, Baldas, Hlubucek, Simpson, & Westerman, 1976; Raistrick, Robinson, & White, 1936). The use of shikimate acid derivatives as entry compounds for 2,3′,4,6-tetrahydroxybenzophenone (THPA) biosynthesis depends on the individual plants. The enzyme used is a benzophenone synthase (Beerhues, 1996), which belongs to the family of type III polyketide synthases (PKSs). Type III PKSs generate a diverse variety of secondary metabolites by varying the starter substrate, the number of condensation reactions, and cyclization reactions (Austin & Noel, 2003) to afford scaffolds such as chalcones, pyrones, chromones, and stilbenes (Abe, 2008). In *Centaurium erythraea*, 2,3′,4,6-tetrahydroxybenzophenone is obtained from the condensation of 3-hydroxybenzoyl-CoA and three malonyl-CoA units (Beerhues, 1996; Peters, Schmidt, & Beerhues, 1998). The benzophenone synthase from *Hypericum androsaemum* and *Garcinia mangostana* is substrate specific on benzoyl-CoA, which is rarely used as a starter substrate by PKSs (Beerhues & Liu, 2009). The 2,4,6-trihydroxybenzophenone intermediate is further hydroxylated by benzophenone 3′-hydroxylase and converted to 2,3′,4,6-tetrahydroxy benzophenone (Nualkaew et al., 2012; Schmidt & Beerhues, 1997). The 3-hydroxybenzoyl-CoA substrate appears to originate directly from the shikimic acid degradation (*C. erythraea*) or from 3-hydroxybenzoic acid (Abd El-Mawla, Schmidt, & Beerhues, 2001; Wang et al., 2003).

In *H. androsaemum*, benzoyl-CoA is derived from cinnamoyl-CoA degradation (Abd El-Mawla & Beerhues, 2002). The m-hydroxybenzoic acid in Gentiana lutea is derived from phenylalanine (El-Seedi et al., 2010). In *Swertia chirata*, phenylalanine, cinnamic acid, and benzoic acid were ruled out as intermediates for 3-hydroxybenzoyl-CoA, which was rather derived from an early shikimate pathway intermediate, [carboxy-^{13}C] shikimate (Wang et al., 2003).

In xanthone biosynthesis in plants, the central step is the regioselective cyclization of the 2,3′,4,6-tetrahydroxybenzophenone (THPA). The xanthone synthase from *C. erythraea* cyclizes THPA via the ortho position with respect to the 3-hydroxy group to give 1,3,5-trihydroxyxanthone. A para-position cyclization occurs in *H. androsaemum* and *G. mangostana* to produce 1,3,7-trihydroxyxanthone (Peters et al., 1998; Schmidt & Beerhues, 1997). The last step is hydroxylation to form 1,3,6,7-tetrahydroxyxanthone. The xanthone synthases are cytochrome P450 oxidases and require NADPH and O_2. The cyclization reaction mechanisms follow the oxidative phenol coupling reactions that are strongly favoured by the presence of the ortho-para-directing 32-hydroxyl group (El-Seedi et al., 2010; Peters et al., 1998; Velisek et al., 2008).

4.2 Bioactivities of xanthones

The study of xanthones is interesting not only from a chemosystematic viewpoint but also from a pharmacological point of view. Xanthones possess antidepressant action and antitubercular activity, while xanthone glycosides exhibit depressant action. The choleretic, diuretic, antimicrobial, antiviral, and cardiotonic action of some xanthones has also been established (Kitanov & Blinova, 1987; Suzuki et al., 1980, 1981). The inhibition of type A and type B MAO by a number of xanthones has also been observed (Suzuki et al., 1980, 1981). Despite their restricted occurrence in the plant kingdom, some xanthones are reported to possess antileukemic, antitumour, antiulcer, antimicrobial, antihepatotoxic, and CNS-depressant activities (Banerji et al., 1994). For example, bellidifolin (1,5,8-trihydroxy-3-methoxy-xanthone) was found to be a selective inhibitor of MAO A, whereas psorospermin, a dihydrofuranoxanthone epoxide, exhibits significant activity against leukemia and in colon and mammary tumour models (Beerhues & Berger, 1994). In general, xanthones and their derivatives have also been shown to be effective as allergy inhibitors and bronchodilators in the treatment of asthma (Balasubramanian & Rajagopalan, 1988). A series of isoprenylated xanthones isolated from moraceous plants showed interesting biological activities such as hypotensive effect, anti-rhinoviral activity, inhibition of the formation of some prostanoids, and anti-tumour promoting activity (Hano, Matsumoto, Sun, & Nomura, 1990). In a study of structure-activity relationships, it was found that 1,3,5,6-, 1,3,6,7-, 2,3,6,7- and 3,4,6,7-tetraoxygenated xanthones possess antiplatelet effects and that the mechanism of action of 1,3,6,7-tetraoxygenated xanthones is due to both inhibition of thromboxane formation and phosphoinositide breakdown (Lin, Liou, Ko, & Teng, 1993). Some xanthone dicarboxylic acids have shown potent inhibition of the binding of leukotriene B4 to receptors on intact neutrophils. Norathyriol showed anti-inflammatory effects mediated partly through the suppression of mast cell degranulation, and partly through (at least at higher doses) a non-selective blockade of the increase in vascular plasma exudation caused by various mediators (Lin, Chung, Liou, Lee, & Wang, 1996). Bellidifolin, isolated from *S. japonica*, was found to be a potent hypoglycemic agent in STZ-induced diabetic rats by both oral and intraperitoneal administration (Basnet et al., 1995). Recently, various bioactivities of xanthones including cytotoxic and antitumour activities, anti-inflammatory activity, antifungal activity, enhancement of choline acetyltransferase activity, and inhibition of lipid peroxidase have been described (Iinuma, Tosa, Tanaka, & Yonemori, 1994). In Mehta, Shah, and Venkateswarlu (1994), reported the total synthesis of novel xanthone antibiotics, cervinomycins A1 and A2, with promising activity against anaerobic bacteria, mycoplasma, and some Gram-positive bacteria.

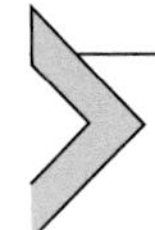

5. Cancer cell lines used to assess the cytotoxicity of xanthones isolated from African medicinal plants

Several cancer and normal cell lines were used to evaluate the cytoxoxicity of xanthones isolated African medicinal plants. They include cell lines from breast cancer (MCF-7, MDAMB-231-*pcDNA*, MDA-MB-231-*BCRP* clone 23, and SKBR3), cervical cancer (HeLa and KB-3–1), leukemia (BC-1, CCRF-CEM, CEM/ADR5000, PF-382, THP-1, HL-60, K562, K562, K-562/Adr, and L5178Y and P-388 (from mouse)), normal cells (AML12, CRL2120), hepatocarcinoma (HepG2 and SMMC-7721), glioblastoma (SF-268, U87MG, and U87MG.ΔEGFR), squamous cell carcinoma (KB and KB-C2), colon cancer (Caco-2, COLO205, HT-29, HCT116 $p53^{+/+}$, HCT116 $p^{53-/-}$, and SW-680), gastric cancer (BGC-823 and SGC-7901), lung cancer (786–0, A549, and NCIH187), melanoma (Colo-38 and UACC-62), ovarian cancer (A2780 and A2780CisR), pancreatic cancer (Capan-1 and MiaPaCa-2), protate cancer (PC-3), and renal cancer (TK-10) (Table 1).

6. Cytotoxic xanthones isolated from the African medicinal plants towards drug sensitive and MDR cancer cells

The global cancer burden remains a serious concern with the alarming incidence of one in eight men and one in eleven women dying in developing countries. This situation is aggravated by the multidrug resistance of cancer cells that hampers chemotherapy. It is responsible for many therapeutic failures and high burdens globally, in patients suffering from cancer (Fitzmaurice et al., 2015; Vorobiof & Abratt, 2007). Any modern protocol for new cytotoxic drug discovery today should integrate the ability of neoplastic cells to rapidly develop resistant phenotypes. thus, resistant cell lines should be integrated into the cell panel used for the discovery of more efficient substances. Some clinically established cytotoxic drugs such as camptothecin, paclitaxel, vinblastine, or vincristine are naturally occurring compounds (Farnsworth & Soejarto, 1991; Gullett et al., 2010; Ludueña, 1997). In addition, numerous botanicals and phytochemicals derived from African medicinal plants have been found active against MDR cancer cell lines (Kuete, Efferth, & 2015, 2015; Mbaveng, Kuete, & Efferth, 2017). Some of such prominent phytochemicals include xanthones (Mbaveng et al., 2018). The potential of xanthones as anticancer

Table 1 Cytotoxic xanthones isolated from African medicinal plants and their effects on drug sensitive and multidrug-resistant cancer cell lines.

Compounds names	African plant source	Country	Cancer cell lines and IC_{50} values (μM) and degree of resistance in bracket	References
3′-Hydroxymethyl-2′-(4″-hydroxy-3″,5″-dimethoxyphenyl)-5′,6′:5,6-(6,8-dihydroxyxanthone)-1′,4′-dioxane (**34**)	*Hypericum roeperianum*	Cameroon	23.28 (CCRF-CEM) vs 54.04 (CEM/ADR5000) [D.R.: 2.32]; 20.73 (MDA-MB-231-*pcDNA*) vs 22.16 (MDA-MB-231-*BCRP*) [D.R.: 1.07]; >91.74 (HCT116 $p53^{+/+}$) vs >91.74 (HCT116 $p53^{-/-}$); 29.70 (U87MG); 12.72 (U87MG.Δ*EGFR*) [D.R.: 0.43]; 25.19 (HepG2) vs 20.89 (AML12) [S.I.: 0.83]	Guefack et al. (2020)
1,3,5,6-Tetrahydroxyxanthone (**4**)	*Garcinia polyantha; Hypericum roeperianum*	Cameroon	98.44 (KB); 101.91 (COLO205); 72.68 (MCF-7); 97.68 (K-562); 94.98 (K-562/Adr); 38.58 (CCRF-CEM) vs 38.46 (CEM/ADR5000) [D.R.: 1.00]; 75.15 (MDA-MB-231-*pcDNA*) vs 62.94 (MDA-MB-231-*BCRP*) [D.R.: 0.84]; 75.48 (HCT116 $p53^{+/+}$) vs 112.27 (HCT116 $p53^{-/-}$)(D.R.: 1.49); 61.42 (U87MG) vs 59.04 (U87MG.Δ*EGFR*) [D.R.: 0.96]; 64.73 (HepG2) vs 150.02 (AML12) [S.I.: 2.32]	Tanaka et al. (2009); Guefack et al. (2020)
1,3,6-Trihydroxy-7-methoxy-2,8-diprenylxanthone (**20**)	*Pentadesma butyracea*	Cameroon	8.53 (MCF-7)	Zelefack et al. (2009)
1,3,6-trihydroxy-8-methylxanthone (**6**)	*Ledebouria graminifolia*	Botswana	1.16 (L5178Y); 253.5 (K562); 68.2 (A2780); 74 (A2780CisR)	Jiang et al. (2003)
1,4-Dihydroxy-7-methoxyxanthone (**31**)	*Securidaca longipedunculata*	Ethiopia	52 (KB-3-1)	Wedajo et al. (2022)

(*continued*)

Table 1 Cytotoxic xanthones isolated from African medicinal plants and their effects on drug sensitive and multidrug-resistant cancer cell lines. *(cont'd)*

Compounds names	African plant source	Country	Cancer cell lines and IC_{50} values (μM) and degree of resistance in bracket	References
1,5,6-trihydroxy-3,7-dimethoxyxanthone (**13**)	*Allablanckia floribunda*	Cameroon	92.08 (SF-268); 115.10 (MCF-7); 32.88 (HepG2)	Shen, Tian, and Yang (2007); Chen, Chen, and Duh (2004); Nkengfack, Azebaze, Vardamides, Fomum and Van Heerden (2002)
1,5,6-trihydroxy-3,7-dimethoxyxanthone (**13**)	*Ledebouria graminifolia*	Botswana	8.22 (KB)	Mutanyatta, Matapa, Shushu, and Abegaz (2003)
1,5-Dihydroxy-2-methoxyxanthone **(32)**	*Hypericum riparium*	Cameroon	18.50 (BGC-823)	Tala et al. (2015)
1,5-dihydroxy-3-methoxyxanthone (**7**)	*Centaurium spicatum*	Egypt	108.89 (KB); 171.28 (COLO205); 124.78 (MCF-7); 100.36 (KB-C2); 60.84 (K-562/Adr)	Tanaka et al. (2009)
1,5-dihydroxy-3-methoxyxanthone (**7**)	*Centaurium spicatum*	Egypt	29.87 (HT-29); 10.69 (P-388)	Gnerre et al. (2001)
1,5-Dihydroxyxanthone (**2**)	*Pentadesma butyracea*; *Allanblackia floribunda*; *Garcinia afzelii, Garcinia polyantha*	Cameroon	21.96 (HT-29); 20.65 (P-388); 14.47 (KB); 438.51 (MDCK)	Wabo, Kikuchi, Katou, Tane and Oshima (2010); Nkengfack, Azebaze, Vardamides, Fomum and Van Heerden (2002); Lannang, Tatsimo, Fouotsa, Dzoyem, Saxena and Sewald (2014)

1,6-dihydroxy-3,5-dimethoxyxanthone (**8**)	*Centaurium spicatum*	Egypt	25.27 (HT-29); 16.45 (P-388)	Chen et al. (2004); El-Shanawany et al. (2011)
1,6-dihydroxy-5-methylxanthone (**5**)	*Garcinia afzelii; Garcinia polyantha*	Cameroon	3.25 (HT-29); 1.04 (P-388 cells)	Shen et al. (2007); Lannang et al. (2010); Lannang et al. (2014)
1,7-Dihydroxy-4-methoxyxanthone (**30**)	*Securidaca longipedunculata*	Ethiopia	0.38 (KB-3-1)	Wedajo et al. (2022)
2-(3,3-dimethylallyl)-7-methoxy-1,5,6-trihydroxy-2″,2″-dimethylpyrano (6″,5″:3,4) xanthone (**10**)	*Symphonia pauciflora*	Madagascar	3.8 (A2780)	Pana et al. (2010)
2-hydroxy-3-methoxyxanthone (**9**)	*Calophyllum flavoramulum*	Algeria or malasia	83.86 (KB); 97.49 (KB-C2); K-562/Adr (237.54); 360.65 (COLO205); 360.65 (MCF-7)	Ferchichi et al. (2012)
2-hydroxy-5-methoxyxanthone (**33**)	*Hypericum roeperianum*	Cameroon	16.80 (CCRF-CEM) vs 52.95 (CEM/ADR5000) [D.R.: 3.15]; 43.80 (MDA-MB-231-*pcDNA*) vs 33.60 (MDA-MB-231-*BCRP*) [D.R.: 0.77]; 46.67 (HCT116 *p53*$^{+/+}$) vs 29 (HCT116 *p53*$^{-/-}$) (>165 μM) [D.R.: >3.54]; 74.44 (U87MG) vs 44.98 (U87MG.Δ*EGFR*) (D.R.: 0.60); 44.21 (HepG2) vs >165.29 (AML12) [S.I.: >3.74]	Guefack et al. (2020)
3′,4′-deoxy-4′-chloropsoroxanthin-3′, 5′-diol (**11**)	*Psorospermum molluscum*	Madagascar	0.042 (A2780); 0.068 (HCT-116); 2.0 (SKBR3)	Leet et al. (2008)

(*continued*)

Table 1 Cytotoxic xanthones isolated from African medicinal plants and their effects on drug sensitive and multidrug-resistant cancer cell lines. *(cont'd)*

Compounds names	African plant source	Country	Cancer cell lines and IC_{50} values (μM) and degree of resistance in bracket	References
3′-Hydroxymethyl-2′-(4″-hydroxy-3″,5″-dimethoxyphenyl)-5′,6′:5,6-(xanthone)-1′,4′-dioxane (**35**)	*Hypericum roeperianum*	Cameroon	16.31 (CCRF-CEM) vs 43.47 (CEM/ADR5000) [D.R.: 2.66]; 36.89 (MDA-MB-231-*pcDNA*) vs 30.50 (MDA-MB-231-*BCRP*) [D.R.: 0.83]; 37.79 (HCT116 $p53^{+/+}$) vs >85.47 (HCT116 $p53^{-/-}$) [D.R.: >2.26]; 35.50 (U87MG) vs 30.61 (U87MG.Δ*EGFR)* [D.R.: 0.86]; 31.29 (HepG2) vs >85.47 (AML12) [S.I.: >2.73]	Guefack et al. (2020)
4-[(2E)-3,7-dimethylocta-2,6-dien-1-yl]-1,5,8-trihydroxy-3-methoxy-9H-xanthen-9-one (**36**)	*Garcinia polyantha*	Cameroon	19.74 (HeLa); 53.66 (PC-3); 6.82 (THP-1)	Lannang, Tatsimo, Fouotsa, Dzoyem, Saxena and Sewald (2014)
8-hydroxycudraxanthone G (**39**)	*Garcinia nobilis*	Cameroon	7.15 (CCRF-CEM) vs 30.12 (CEM/ADR5000) (D.R.: 4.21); 30.00 (MDA-MB-231-*pcDNA*) vs 30.93 (MDA-MB-231-*BCRP*) [D.R.: 1.03]; 29.12 (HCT116 $p53^{+/+}$) vs 38.83 (HCT116 $p53^{-/-}$) [1.33]; 26.78 (U87MG) vs 53.85 (U87MG.Δ*EGFR)* [D.R.: 2.01]; 39.22 (HepG2) vs >97.56 (AML12) [S.I.: >2.49]	Kuete et al. (2014)
Allanxanthone A (**14**)	*Allablanckia floribunda*	Cameroon	3.94 (KB)	Abe (2008)
Allanxanthone A (**14**)	*Pentadesma butyracea*	Cameroon	155.93 (HeLa); 179.69 (BGC-823)	Tala et al. (2013)
Banganxanthone C (**37**)	*Garcinia polyantha*	Cameroon	84.12 (A549); 65.17 (THP-1); 52.01 (MCF-7)	Lannang et al. (2014)
Butyraxanthones A (**18**)	*Pentadesma butyracea*	Cameroon	7.31 (MCF-7)	Zelefack et al. (2009)

Butyraxanthones B (**19**)	*Pentadesma butyracea*	Cameroon	7.07 (MCF-7)	Zelefack et al. (2009)
Butyraxanthones D (**22**)	*Pentadesma butyracea*	Cameroon	3.03 (MCF-7)	Zelefack et al. (2009)
Caloxanthone A (**15**)	*Calophyllum inophyllum*	Cameroon	18.74 (KB)	Yimdjo et al. (2004)
Cudraxanthone I (**41**)	*Milicia excelsa*	Cameroon	10.63 (CCRF-CEM) vs 8.23 (CEM/ADR5000) [D.R.: 0.78]; 7.80 (MDA-MB-231-*pcDNA*) vs 2.78 (MDA-MB-231-*BCRP*) [D.R.: 0.36]; 9.79 (HCT116 $p53^{+/+}$) vs 13.28 (HCT116 $p53^{-/-}$) [D.R.: 1.36]; 22.49 (U87MG) vs 19.13 (U87MG.Δ*EGFR*) (D.R.: 0.85); 9.63 (HepG2) vs >105.82 (AML12) [S.I.: >10.65]	Kuete et al. (2014)
Decussatin (**16**)	*Anthocleista vogelli*	Nigeria	27.90 (HepG2); 25.09 (HL-60)	Ding et al. (2011); Okorie (1976)
Euxanthone (**1**)	*Cassia. obtusifolia; Garcinia nobilis; Pentadesma butyracea; Vismia laurentii; Vismia rubescens; Psorospermum aurantiacum; Psorospermum molluscum; Pentadesma butyracea; Oricia suaveolens*	Cameroon; Madagascar	66.21 (KB); 63.58 (KB-C2); 18.46 (K-562/Adr); 184.17 (COLO205); 206.10 (MCF-7); 17.27 (HT-29); 17.27 (P-388); 60.1 (TK-10); 20.2 (UACC-62 cells); 34.23 (CCRF-CEM) vs 49.34 (CEM/ADR5000) [D.R.: 1.44]; >175.44 (MDA-MB-231-*pcDNA*) vs 13.00 (MDA-MB-231-*BCRP*) [D.R.: <0.07]; 136.07 (HCT116 $p53^{+/+}$) vs 88.86 (HCT116 $p53^{-/-}$) [D.R.: 0.65]; 136.07 (U87MG) vs 159.36 (U87MG.Δ*EGFR*) [D.R.: 1.17]; 14.98 (HepG2) vs >175.44 (AML12) [S.I.: >11.71]; 32.91 (MCF-7)	Tanaka et al. (2009); Wabo, Kikuchi, Katou, Tane and Oshima (2010); Sob et al. (2010); Fouotsa et al. (2012); Nguemeving et al. (2006); Leet et al. (2008); Pedro, Cerqueira, Sousa, Nascimento, and Pinto (2002); Kuete et al. (2016); Fouotsa et al. (2020)
	Pentadesma butyracea	Cameroon	3.23 (MCF-7); 2.72 (NCIH187); 39.22 (BC-1); 39.45 (KB);	Zelefack et al. (2009)

(continued)

Table 1 Cytotoxic xanthones isolated from African medicinal plants and their effects on drug sensitive and multidrug-resistant cancer cell lines. (*cont'd*)

Compounds names	African plant source	Country	Cancer cell lines and IC_{50} values (μM) and degree of resistance in bracket	References
Gartanin (**17**)	*Pentadesma butyracea; Garcinia staudtii; Garcinia nobilis*		13.02 (MCF-7)	Zelefack et al. (2009); Fouotsa et al. (2012); Suksamrarn et al. (2006)
Globulixanthones A (**27**)	*Symphonia globulifera*	Cameroon	6.63 (KB)	Nkengfack et al. (2002)
Globulixanthones B (**26**)	*Symphonia globulifera*	Cameroon	4.70 (KB)	Nkengfack et al. (2002)
Globuxanthone (**42**)	*Pentadesma butyracea*	Cameroon	87.51 (A549); 58.18 (HeLa); 62.11 (BGC-823)	Tala et al. (2013)
Mangostanin (**25**)	*Pentadesma butyracea*	Cameroon	9.14 (MCF-7)	Zelefack et al. (2009)
Mboudiexanthone (**38**)	*Garcinia nobilis*	Cameroon	35.26 (MCF-7)	Fouotsa et al. (2020)
Morusignin I (**40**)	*Garcinia nobilis*	Cameroon	16.65 (CCRF-CEM) vs >101.52 (CEM/ADR5000) [D.R.: >6.10]; 39.26 (MDA-MB-231-*pcDNA*) vs 28.98 (MDA-MB-231-*BCRP*) [D.R.: 0.74]; 41.88 (HCT116 ($p53^{+/+}$)) vs >7.64 (HCT116 $p53^{-/-}$) [D.R.: 0.90]; 69.62 (U87MG) vs 38.53 (U87MG.Δ*EGFR*) [D.R.: 0.55]; 70.38 (HepG2) vs >101.52 (AML12) [S.I.: >1.44]	Kuete et al. (2014)

Norathyriol (**3**)	*Hypericum roeperianum; Garcinia polyantha*	Cameroo	19.94 (CCRF-CEM) vs 23.21 (CEM/ADR5000) [D.R.: 1.16]; >153.85 (MDA-MB-231-*pcDNA*) vs 20.38 (MDA-MB-231-*BCRP*) [D.R.: <0.13]; 40.17 (HCT116 $p53^{+/+}$) vs >153.85 (HCT116 $p53^{-/-}$) [D.R.: >6.15]; 106.00 (U87MG) vs 30.37 (U87MG.Δ*EGFR*) [D.R.: 0.29]; 32.40 (HepG2) vs 45.35 (AML12) [S.I.: 1.40]; 128.06 (KB); 109.60 (COLO205), 132.67 (K-562/Adr); 359.57 (KB-C2), 86.52 (K-562), 71.52 (MCF-7); 203.82 (SF-268); 42.30 (HepG2)	Guefack et al. (2020); Tanaka et al. (2009); Shen et al. (2007)
psoroxanthin (**12**)	*Psorospermum molluscum*	Madagascar	0.33 (A2780); 1.0 (HCT-116)	Leet et al. (2008)
Rheediaxanthone A (**43**)	*Garcinia epunctata*	Cameroon	3.15 (Caco-2); 14.60 (HepG2); 80.29 (CRL2120)	Kuete et al. (2018)
Rubraxanthone (**24**)	*Allanblackia monticola; Pentadesma butyracea*	Cameroon	6.33 (MCF-7)	Zelefack et al. (2009); Azebaze et al. (2004)
Tovophyllin (**23**)	*Allanblackia monticola Pentadesma butyracea*	Cameroon	5.62 (MCF-7)	Zelefack et al. (2009); Azebaze et al. (2004)

(continued)

Table 1 Cytotoxic xanthones isolated from African medicinal plants and their effects on drug sensitive and multidrug-resistant cancer cell lines. (*cont'd*)

Compounds names	African plant source	Country	Cancer cell lines and IC_{50} values (µM) and degree of resistance in bracket	References
Xanthone V1 (**28**)	*Symphonia globulifera; Vismia laurentii*	Cameroon	12.44 (CCRF-CEM); 48.07 (CEM/ADR5000); 11.57 (HL-60); 47.11 (MiaPaCa-2); 1.42 (MCF-7); 21.32 (SW-680); 9.62 (786-0); 9.64 (U87MG); 10.13 (A549); 3.02 (Colo-38); 0.58 (HeLa); 0.61 (Caski); >50.76 (AML12)	Kuete et al. (2011)
Xanthone V2 (**29**)	*Vismia guineensis*	Ivory Coast	8.01 (SGC-7901); 14.61 (SMMC-7721); 3.06 (HCT-116)	Zou et al. (2004)

KB: human epidermoid carcinoma; HCT116 $p53^{+/+}$: colon cancer cells and its knockout clone, HCT116 $p53^{-/-}$; A2780: human ovarian cancer cell line and its cisplatin resistant (A2780CisR); CCRF-CEM: drug sensitive leukemia and its (CEM/ADR5000) multidrug-resistant P-glycoprotein-over-expressing subline; MDAMB-231-pcDNA: breast cancer cells and its resistant subline (MDA-MB-231-BCRP clone 23), HepG2: liver cancer cells; A549: human alveolar basal epithelial adenocarcinoma cells; HeLa: human cervical cancer cell line; HT-29: human colorectal cancer; U87MG: glioblastoma cells and its resistant subline (U87MG.ΔEGFR); PC-3: prostate cancer; MCF-7: human breast cancer cell line; 786-0 renal carcinoma cells; THP-1: leukemia cancer cell line; AML12: Murine hepatocyte cell line; BC-1: lymphoma cell line; BGC-823: human gastric adenocarcinoma cell line; Caco-2: human colon carcinoma cell line; Capan-1: human pancreatic ductal adenocarcinoma cell line; COLO205: human colon adenocarcinoma cell line; UACC-62: melanoma cell line; Caski: human cervical epidermoid carcinoma cell line; DLD-1: colorectal adenocarcinoma cell line; HL-60: human leukemia cell line; K562: human chronic myelogenous leukemia cell line (K-562/Adr); SW-680: colon carcinoma cell line; MiaPaCa-2: pancreatic adenocarninoma cell line; Colo-38: skin melanoma cell line; PF-382: human leukemic T cell line; KB-C2: squamous cell carcinoma; SKBR3: human breast cancer cell line; NCIH187: lung cancer cell line; SMMC-7721: human hepatocellular carcinoma cell line; SF-268: human glioblastoma cell line; SGC-7901: Human gastric cancer cell line; TK-10: human renal carcinoma cell line; P-388: monocyte/macrophage-like cell line; L5178Y: mouse lymphoma cell line; CRL2120: human fibroblast cell line; KB-3-1: cervical cancer cell line; D.R. or the degree of resistance was determined as the ratio of IC50 value in the resistant divided by the IC50 in the sensitive cells for CEM/ADR5000 cells vs CCRF-CEM cells, MDA-MB-231-BCRP cells vs MDA-MB-231-pcDNA cells, HCT116 $p53^{-/-}$ cells vs HCT116 $p53^{+/+}$ cells, and U87MG.ΔEGFR cells vs U87MG cells. S.I. or the selectivity index was determined as the ratio of IC50 value in AML12 cells divided by the IC50 in HepG2 cells.

agents is well known. The hit compounds involved in cancer chemotherapy include 5,6-dimethylxanthenone-4-acetic acid (DMXAA), psorospermin, mangiferin, norathyriol, mangostins, and 6-isopropoxy-9-oxoxanthene-2-carboxylic acid (AH6809), a prostanoid receptor antagonist (Pinto et al., 2005). Prenylated caged xanthones, both naturally occurring and synthetic analogs, have been identified as promising anticancer agents (Anantachoke, Tuchinda, Kuhakarn, Pohmakotr, & Reutrakul, 2012). Gambogic acid was found to be a highly valuable lead compound for antitumor chemotherapy (Anantachoke et al., 2012). The cytotoxicity of several xanthones found in African plants (Fig. 2), including mangiferin and mangostin, has been documented on a panel of cancer cell lines, and the modes of action of some of these molecules have been provided. Herein, the cytotoxicity of euxanthone (**1**), 1,5-Dihydroxyxanthone (**2**), 1,3,6,7-tetrahydroxyxanthone or norathyriol (**3**), 1,3,5,6-tetrahydroxyxanthone (**4**), 1,6-dihydroxy-5-methoxyxanthone (**5**), 1,3,6-trihydroxy-8-methylxanthone (**6**), 1,5-dihydroxy-3-methoxyxanthone (**7**), 1,6-dihydroxy-3,5-dimethoxyxanthone (**8**), 2-hydroxy-3-methoxyxanthone (**9**), 2-(3,3-dimethylallyl)-7-methoxy-l,5,6-trihydroxy-2″,2″-dimethylpyrano (6″,5″:3,4) xanthone (**10**), 3′,4′-deoxy-4′-chloropsoroxanthin-3′,5′-diol (**11**), psoroxanthin(**12**), 1,5,6-trihydroxy-3,7-dimethoxyxanthone (**13**), allanxanthone A (**14**), caloxanthone A (**15**), decussatin (**16**), gartanin (**17**), butyraxanthones A (**18**), Butyraxanthones B (**19**), 1,3,6-trihydroxy-7-methoxy-2,8-diprenylxanthone (**20**), garcinone E (**21**), butyraxanthones D (**22**), tovophyllin (**23**), rubraxanthone (**24**), mangostanin (**25**), globulixanthones B (**26**), globulixanthones A (**27**), xanthone V1 (**28**), xanthone V2 (**29**), 1,7-dihydroxy-4-methoxyxanthone (**30**), 1,4-dihydroxy-7-methoxyxanthone (**31**), 1,5-dihydroxy-2-methoxyxanthone (**32**), 2-hydroxy-5-methoxyxanthone (**33**), 3′-hydroxymethyl-2′-(4″-hydroxy-3″,5″-dimethoxyphenyl)-5′,6′:5,6-(6,8-dihydroxyxanthone)-1′,4′ -dioxane (**34**), 3′-hydroxymethyl-2′-(4″-hydroxy-3″,5″-dimethoxyphenyl)-5′,6′:5,6-(xanthone)-1′,4′-dioxane (**35**), 4-[(2*E*)-3,7-dimethylocta-2,6-dien-1-yl]-1,5,8- trihydroxy-3-methoxy-9*H*-xanthen-9-one (**36**), banganxanthone C (**37**), mboudiexanthone (**38**), 8-hydroxycudraxanthone G (**39**), morusignin I (**40**), cudraxanthone I (**41**), globuxanthone (**42**), and rheediaxanthone A (**43**) isolated from African medicinal plants is shown in Table 1. Their chemical structures are shown in Fig. 2.

The cytotoxicity of phytochemicals will be appreciated according to the classification criteria established as follows: outstanding activity ($IC_{50} \leq 0.5\ \mu M$), excellent activity ($0.5 < IC_{50} \leq 2\ \mu M$), very good activity ($2 < IC_{50} \leq 5\ \mu M$), good activity ($5 < IC_{50} \leq 10\ \mu M$), average activity ($10 < IC_{50} \leq 20\ \mu M$), weak

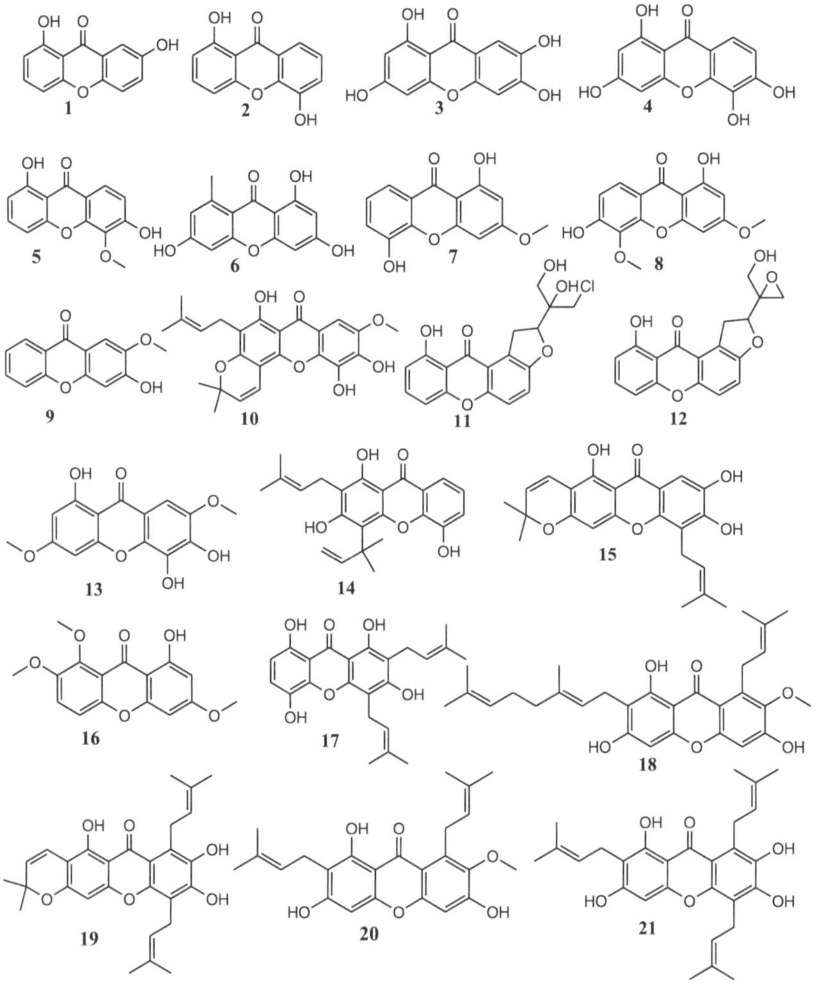

Fig. 2 Cytotoxic xanthones isolated from African medicinal plants. **1:** 1,7-dihydroxyxanthone; **2:** 1,5-dihydroxyxanthone; **3:** 1,3,6,7-tetrahydroxyxanthone or norathyriol; **4:** 1,3,5,6-tetrahydroxyxanthone; **5;** 1,6-dihydroxy-5-methoxyxanthone; **6:** 1,3,6-trihydroxy-8-methylxanthone; **7:** 1,5-dihydroxy-3-methoxyxanthone; **8:** 1,6-dihydroxy-3,5-dimethoxyxanthone; **9:** 2-hydroxy-3-methoxyxanthone; **10:** 2-(3,3-dimethylallyl)-7-methoxy-l,5,6-trihydroxy-2″,2″-dimethylpyrano (6″,5″:3,4) xanthone; **11:** 3′,4′-deoxy-4′-chloropsoroxanthin-3′,5′-diol; **12:** psoroxanthin; **13:** 1,5,6-trihydroxy-3,7-dimethoxyxanthone; 14**:** allanxanthone A; **15:** caloxanthone A; **16:** decussatin; **17:** gartanin; **18:** butyraxanthones A; **19:** Butyraxanthones B; **20:** 1,3,6-trihydroxy-7-methoxy-2,8-diprenylxanthone; **21:** garcinone E **22:** butyraxanthones D; **23:** tovophyllin; **24:** rubraxanthone; **25:** Mangostanin; **26:** globulixanthones B; **27:** globulixanthones A; **28:** xanthone V1; **29:** xanthone V2; **30:** 1,7-dihydroxy-4-methoxyxanthone; **31:** 1,4-dihydroxy-7-methoxyxanthone; **32:** 1,5-dihydroxy-2-methoxyxanthone; **33:** 2-hydroxy-5-methoxyxanthone; **34:** 3′-hydroxymethyl-2′-

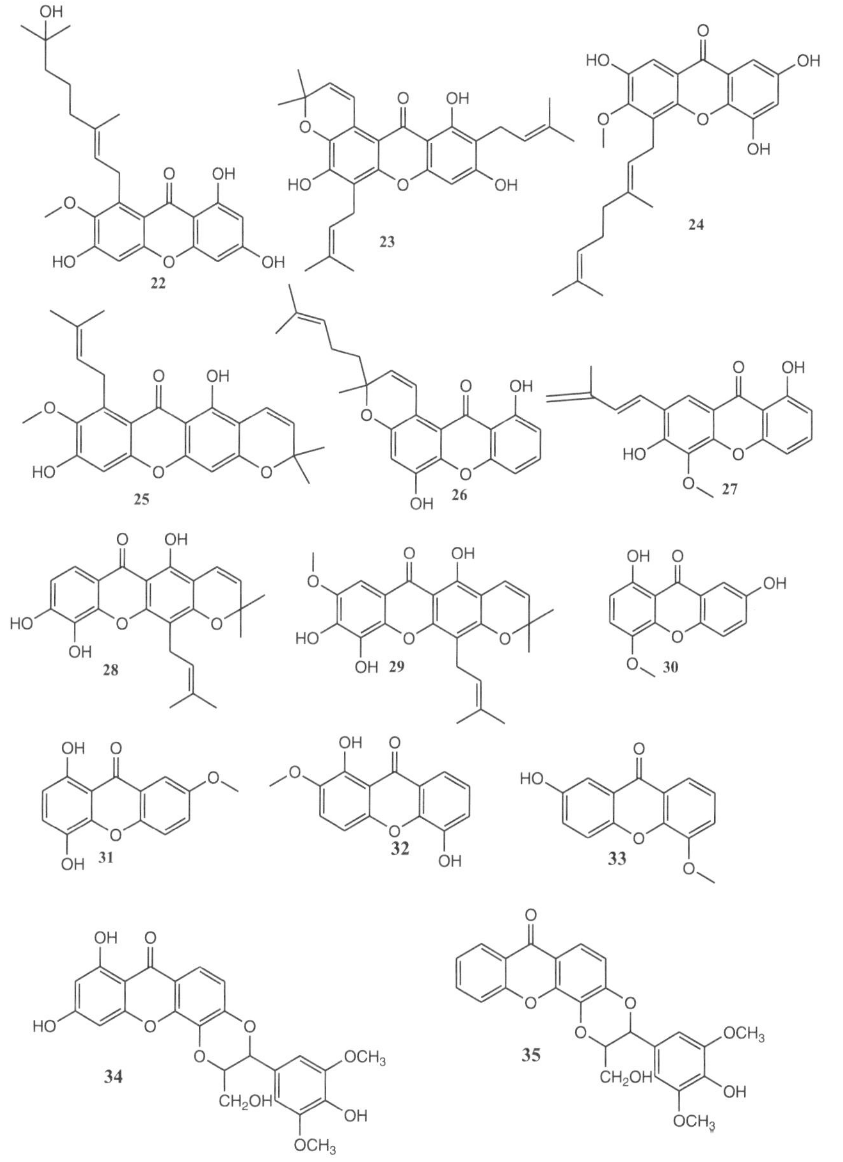

Fig. 2 (Continued)

(4″-hydroxy-3″,5″-dimethoxyphenyl)-5′,6′:5,6-(6,8-dihydroxyxanthone)-1′,4′-dioxane; **35:** 3′-hydroxymethyl-2′-(4″-hydroxy-3″,5″-dimethoxyphenyl)-5′,6′:5,6-(xanthone)-1′,4′-dioxane **36:** 4-[(2*E*)-3,7-dimethylocta-2,6-dien-1-yl]-1,5,8- trihydroxy-3-methoxy-9*H*-xanthen-9-one; **37:** banganxanthone C; **38:** mboudiexanthone; **39:** 8-hydroxycudraxanthone G; **40:** morusignin I; **41:** cudraxanthone I; **42:** globuxanthone; **43:** rheediaxanthone A.

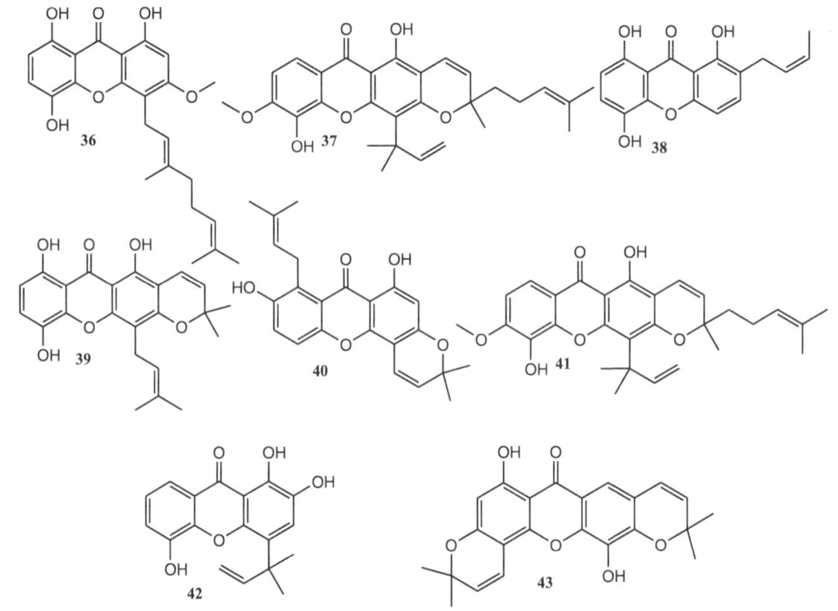

Fig. 2 (*Continued*)

activity ($20 < IC_{50} \leq 60\ \mu M$), very weak activity ($60 < IC_{50} \leq 150\ \mu M$), and not active ($IC_{50} > 150\ \mu M$) (Kuete, 2025). It has also been established that the degree of resistance (D.R.) < 0.9 defines collateral sensitivity, whilst D.R. between 0.9 and 1.2 defines normal sensitivity. The cross-resistance is noted if the cytotoxic agent is more active in the sensitive cell line than its resistant subline, with D.R above 1.2 (Efferth et al., 2020; Efferth et al., 2021; Mbaveng et al., 2017). Collateral sensitivity or normal sensitivity of resistant versus sensitive cancer cell lines should be achieved for compounds with ability to combat cancer drug resistance (Efferth et al., 2020; Efferth et al., 2021; Kuete et al., 2015; Mbaveng et al., 2017). This basis of classification will be used to discuss the cytotoxicity of xanthones isolated from African medicinal plants.

From the data summarized in Table 1, it appears that outstanding cytotoxic effects were obtained with compounds **11** and **12** against A2780 cells, **11** against HCT-116 cells, **28** against MCF-7 cells, and **30** against KB-3–1 cells; excellent cytotoxic were obtained with compounds **5** against P-388 cells, **6** against L5178Y cells, **12** against HCT-116 cells, and **28** against HeLa cells and Caski cells; very cytotoxic effects were obtained with compounds **5** against HT-29 cells, **10** against A2780 cells, **11** against SKBR3 cells, **14** and **26** against KB cells, **17** against NCIH187 cells, **21** and **22** against MCF-7 cells, **28** against

Colo-38 cells, **29** against HCT-116 cells, **41** against MDA-MB-231-*BCRP* cells, and **44** against Caco-2 cells; good cytotoxic effects were obtained with compounds **13** and **27** against KB cells, **18, 19, 20, 23, 24,** and **25** against MCF-7 cells, **28** against 786–0 cells and U87MG cells, **29** against SGC-7901 cells, **36** against THP-1 cells, **39** against CCRF-CEM cells, and **41** against CEM/ADR5000 cells, MDA-MB-231-*pcDNA* cells, HCT116 $p53^{+/+}$ cells, and HepG2 cells. These compounds are more suitable to fight malignant diseases. Interestingly, the collateral sensitivity resistant CEM/ADR5000 cells versus CCRF-CEM cells to compound **41** (D.R: 0.78), MDA-MB-231-*BCRP* cells versus MDA-MB-231-*pcDNA* cells to **41** (D.R: 0.36), and U87MG.Δ*EGFR* cells U87MG cells to **41** (D.R: 0.85) was achieved. This compound appears to be more suitable to combat cancer drug resistance.

7. Conclusion

The present chapter gives some insight into the state of the art and summarizes the most important data for xanthones isolated from African flora with a potential for anticancer drug discovery. According to literature reports, around 650 xanthones are known from natural sources. These xanthones have been isolated from 62 families of higher plants, fungi, and lichens. Xanthones from higher plants are distributed in Gentianaceae, Moraceae, Guttiferae, Polygalaceae, and Leguminosae families. The best cytotoxic xanthones were identified as **5, 6, 10–14, 17–30, 36, 39, 41** and **43**. Amongst these compounds, **41** was the most suitable to combat cancer drug resistance. They deserve further preclinical and clinical studies to develop new drugs to combat cancers and their drug resistance.

References

Abd El-Mawla, A. M., & Beerhues, L. (2002). Benzoic acid biosynthesis in cell cultures of *Hypericum androsaemum*. *Planta Medica, 214*(5), 727–733.

Abd El-Mawla, A. M., Schmidt, W., & Beerhues, L. (2001). Cinnamic acid is a precursor of benzoic acids in cell cultures of *Hypericum androsaemum* but not in cell cultures of *Centaurium erythraea* L. *Planta Medica, 212*(2), 288–293.

Abe, I. (2008). Engineering of plant polyketide biosynthesis. *Chemical and Pharmaceutical Bulletin, 56*(11), 1505–1514.

Aberham, A., Schwaiger, S., Stuppner, H., & Ganzera, M. (2007). Quantitative analysis of iridoids, secoiridoids, xanthones and xanthone glycosides in *Gentiana lutea* L. roots by RP-HPLC and LC-MS. *Journal of Pharmaceutical andBiomedical Analysis, 45*(3), 437–442.

Alaribe, C. S. A., Coker, H. A. B., Shode, F. O., Ayoola, G., Adesegun, S. A., Bamiro, J., et al. (2012). Antiplasmodial and phytochemical investigations of leaf extracts of *Anthocleista vogelii* (Planch). *Journal of Natural Products, 5*, 60–67.

Al-Hazimi, H. M. G., & Miana, G. A. (1990). Naturally occurring xanthones in higher plants and ferns. *Journal of the Chemical Society of Pakistan, 12*(2), 174–188.

Anantachoke, N., Tuchinda, P., Kuhakarn, C., Pohmakotr, M., & Reutrakul, V. (2012). Prenylated caged xanthones: chemistry and biology. *Pharmaceutical Biology, 50*(1), 78–91.

Arends, P., Helboe, P., & Moller, J. (1973). Mass spectrometry of xanthones. I. The electron impact-induced fragmentation of xanthone, monohydroxy- and Monomethoxyxanthones. *Organic Mass Spectrometry, 7*(6), 667–681.

Arisawa, M., & Morita, N. (1976). Studies on constituents of genus Iris. VII. The constituents of Iris unguicularis POIR. (I). *Chemical and Pharmaceutical Bulletin, 24*(4), 815–817.

Aritomi, M., & Kawasaki, T. (1970a). A new xanthone C-glucoside, position isomer of Mangiferin, from *Anemarrhena asphodeloides* Bunge. *Chemical and Pharmaceutical Bullelin, 18*, 2327–2333.

Aritomi, M., & Kawasaki, T. (1970b). A Mangiferin monomethyl ether from *Mangifera indica* L. *Chemical and Pharmaceutical Bullelin, 18*, 2224–2234.

Austin, M. B., & Noel, J. P. (2003). The chalcone synthase superfamily of type III polyketidensynthases. *Natural Product Reports, 20*(1), 79–110.

Azebaze, A. G. B., Meyer, M., Bodo, B., & Nkengfack, A. E. (2004). Allanxanthone B, a polyisoprenylated xanthone from the stem bark of *Allanblackia monticola* Staner L.C. *Phytochemistry, 65*(18), 2561–2564.

Balasubramanian, K., & Rajagopalan, K. (1988). Novel xanthones from *Garcinia mangostana*, structures of Br-xanthone-A and Br-xanthone-B. *Phytochemistry, 27*(5), 1552–1554.

Banerji, A., Deshpande, A, D., Prabhu, B, R., & Pradhan, P. (1994). Tomentonone, a new xanthonoid from the stem bark of *Calophyllum tomentosum*. *Journal of Natural Products, 57*, 396–399.

Barraclough, D., Locksley, H. D., Scheinmann, F., Magalhaes, T. M., & Gottlieb, O. R. (1970). Applications of proton magnetic resonance spectroscopy in the structural investigation of xanthones. *Journal of the Chemical Society B*, 603–612.

Basnet, P., Kadota, S., Shimizu, M., Takata, Y., Kobayashi, M., & Namba, T. (1995). Bellidifolin stimulates glucose uptake in rat 1 fibroblasts and ameliorates hyperglycemia in streptozotocin (STZ)-induced diabetic rats. *Planta Medica, 61*(5), 402–405.

Beerhues, L. (1996). Benzophenone synthase from cultured cells of *Centaurium erythraea*. *FEBS Letters, 383*(3), 264–266.

Beerhues, L., & Berger, U. (1994). Xanthones in cell suspension cultures of two centaurium species. *Phytochemistry, 35*(5), 1227–1231.

Beerhues, L., & Liu, B. (2009). Biosynthesis of biphenyls and benzophenones—evolution of benzoicbacid specific type III polyketide synthases in plants. *Phytochemistry, 70*(15-16), 1719–1727.

Bhatia, V. K., Ramanathan, J. D., & Seshadri, T. R. (1967). Constitution of mangiferin. *Tetrahedron, 23*(3), 1363–1368.

Birch, A. J., Baldas, J., Hlubucek, J. R., Simpson, T. J., & Westerman, P. W. (1976). Biosynthesis of fungal xanthone ravenelin. *Journal of the Chemical Society, Perkin Transactions I*, 898–904.

Bringmann, G., Lang, G., Steffens, S., Gunther, E., & Schaumann, K. (2003). Evariquinone, isoemericellin, and stromemycin from a sponge derived strain of the fungus *Emericella variecolor*. *Phytochemistry, 63*(4), 437–443.

Cardona, M. L., Fernandez, M. I., Pedro, J. R., Seoane, E., & Vidal, R. (1986). Additional new xanthones and xanthonolignoids from *Hypericum canariensis*. *Journal of Natural Products, 49*(1), 95–100.

Castelao, J. R., Gottlieb, O. R., Lima, R. A., Mesquita, A. A. L., Gottlieb, H. E., & Wenkert, E. (1977). Xanthonolignoids from Kielmeyera and Caraipa species 1-H and 13-C NMR spectroscopy of xanthones. *Phytochemistry, 16*(6), 735–740.

Chakravarty, A. K., Mukhopadhyay, S., Moitra, S. K., & Das, B. (1994). (–) Syringaresinol, a hepatoprotective agent and other constituents from *Swertia chirata. Indian Journal of Chemistry, 33*, 405–408.

Chakravarty, A. K., Sarkar, T., Masuda, K., Takey, T., Doi, H., Kotani, E., et al. (2001). Structure and synthesis of glycoborine, a new carbazole alkaloid from the roots of *Glycosmis arborea*: A note on the structure of glycozolicine. *Indian Journal of Chemistry B, (40)*, 484–489.

Chawla, H. M., Chibber, S. S., & Khera, U. (1975). Separation and identification of some hydroxyxanthones by thin-layer chromatography. *Journal of Chromatography A, 111*, 246–247.

Chen, J. J., Chen, I. S., & Duh, C. Y. (2004). Cytotoxic xanthones and biphenyls from the root of *Garcinia linii. Planta Medica, 70*(12), 1195–1200.

Chen, Y., Wang, G. K., Wu, C., & Qin, M. J. (2013). Chemical constituents of *Gentiana rhodantha. Zhongguo Zhong Yao Za Zhi = Zhongguo Zhongyao Zazhi = China Journal of Chinese Materia Medica, 38*, 362–365.

Dai, H.-F., Zeng, Y.-B., Xiao, Q., Han, Z., Zhao, Y.-X., & Mei, W.-L. (2010). Caloxanthones O and P: Two new prenylated xanthones from*Calophyllum inophyllum. Molecules (Basel, Switzerland), 15*, 606–612.

Dai, R., Gao, J., & Bi, K. (2004). High-performance liquid chromatographic method for the determination and pharmacokinetic study of mangiferin in plasma of rats having taken the traditional Chinese medicinal preparation Zi-Shen Pill. *Journal of Chromatographic Science, 42*, 88–90.

Dall'Acqua, S., Innocenti, G., Viola, G., Piovan, A., Caniato, R., & Cappelletti, E. M. (2002). Cytotoxic compounds from *Polygala vulgaris. Chemical and Pharmaceutical Bulletin, 50*, 1499–1501.

Deng, Y. X., Guo, T., Xie Shao, Z. Y. H., & Pan, S. L. (2013). Three new xanthones from the resin of *Garcinia hanburyi. PlantaMedica, 79*, 792–796.

Ding, L., Liu, B., Zhang, S. D., Hou, Q., Qi, L. L., & Zhou, Q. Y. (2011). Cytotoxicity, apoptosis-inducing effects and structure-activity relationships of four natural xanthones from Gentianopsis paludosa Main HepG2 and HL-60 cells. *Natural Product Research, 25*(7), 669–683.

Dua, V. K., Ojha, V. P., Roy, R., Joshi, B. C., Valecha, N., Usha, D. C., et al. (2004). Anti-malarial activity of some xanthones isolated from the roots of *Andrographis paniculate. Journal of Ethnopharmacology, 95*, 247–251.

Efferth, T., Saeed, M. E. M., Kadioglu, O., Seo, E. J., Shirooie, S., Mbaveng, A. T., et al. (2020). Collateral sensitivity of natural products in drug-resistant cancer cells. *Biotechnology Advances, 38*, 107342.

Efferth, T., Kadioglu, O., Saeed, M. E. M., Seo, E. J., Mbaveng, A. T., & Kuete, V. (2021). Medicinal plants and phytochemicals against multidrug-resistant tumor cells expressing ABCB1, ABCG2, or ABCB5: A synopsis of 2 decades. *Phytochemistry Reviews, 20*(1), 7–53.

El-Seedi, H. R., El-Ghorab, D. M. H., El-Barbary, M. A., Zayed, M. F., Goransson, U., Larsson, S., et al. (2009). Naturally occurring xanthones; Latest investigations: Isolation, structure elucidation and chemosystematic significance. *Current Medicinal Chemistry, 16*, 2581–2626.

El-Seedi, H. R., El-Barbary, M. A., El-Ghorab, D. M., Bohlin, L., Borg-Karlson, A. K., Goransson, U., et al. (2010). Recent insights into the biosynthesis and biological activities of natural xanthones. *Current Medicinal Chemistry, 17*(9), 854–901.

El-Shanawany, M. A., Mohamed, G. A., Nafady, A. M., Ibrahim, S. R. M., Radwan, M. M., & Ross, S. A. (2011). A new xanthone from the roots of *Centaurium spicatum*. *Phytochemistry Letters, 4*(2), 126–128.

Farnsworth, N. R., & Soejarto, D. D. (1991). *Global importance of medicinal plants. Conservation of medicinal plants*. Cambridge: Cambridge University Press, 25–51.

Ferchichi, L., Derbre, S., Mahmood, K., Toure, K., Guilet, D., Litaudon, M., et al. (2012). Bioguided fractionation and isolation of natural inhibitors of advanced glycation end-products (AGEs) from *Calophyllum flavoramulum*. *Phytochemistry, 78*, 98–106.

Fitzmaurice, C., Dicker, D., Pain, A., Hamavid, H., Moradi-Lakeh, M., MacIntyre, M. F., et al. (2015). The global burden of cancer 2013. *JAMA Oncology, 1*(4), 505–527.

Fouotsa, H., Lannang, A. M., Mbazoa, C. D., Rasheed, S., Marasini, B. P., Ali, M. Z., et al. (2012). Xanthones inhibitors of α-glucosidase and glycation from *Garcinia nobilis*. *Phytochemistry Letters, 5*(2), 236–239.

Fouotsa, H., Dzoyem, J. P., Lannang, A. M., Stammler, H.-G., Mbazoa, C. D., Luhmer, M., et al. (2020). Antiproliferative activity of a new xanthone derivative from leaves of *Garcinia nobilis* Engl. *Natural Product Research, 35*(24), 5604–5611.

Franklin, G., Conceic, L. F. R., Kombrink, E., & Dias, A. C. P. (2009). Xanthones biosynthesis in *Hypericum perforatum* cells provides antioxidant and antimicrobial protection upon biotic stress. *Phytochemistry, 70*(1), 60–68.

Gentili, B., & Horowitz, R. M. (1968). Flavonoids of Citrus. IX. Some new c-glycosylflavones and a nuclearmagnetic resonance method for differentiating 6- and 8-c-glycosyl isomers. *Journal of Organic Chemistry, 33*, 1571–1577.

Ghosal, S., & Chaudhuri, R. K. (1973). New tetraoxygenated xanthones of *Canscora decussata*. *Phytochemistry, 12*, 2035–2038.

Ghosal, S., Chauhan, R. B. P. S., Biswas, K., & Chaudhuri, R. K. (1976). New 1,3,5-trioxygenated xanthones in *Canscora decussata*. *Phytochemistry, 15*, 1041–1043.

Gnerre, C., Thull, U., Gaillard, P., Carrupt, P. A., Testa, B., Fernandes, E., et al. (2001). Natural and synthetic xanthones as monoamine oxidase inhibitors: Biological assay and 3D-QSAR. *Helvetica Chimica Acta, 84*(3), 552–570.

Guefack, M.-G. F., Damen, F., Mbaveng, A. T., Tankeo, S. B., Bitchagno, G. T. M., Çelik, I., et al. (2020). Cytotoxic constituents of the bark of *Hypericum roeperianum* towards multidrug-resistant cancer cells. *Evidence-Based Complementary and Alternative Medicine, 2020*, 4314807.

Gullett, N. P., Ruhul, A. A. R. M., Bayraktar, S., Pezzuto, J. M., Shin, D. M., Khuri, F. R., et al. (2010). Cancer prevention with natural compounds. *Seminars in Oncology, 37*, 258–281.

Hajimehdipour, H., Amanzadeh, Y., Sadat, Ebrahimi, S. E., & Mozaffarian, V. (2003). Three tetraoxygenated xanthones from *Swertia longifolia*. *Pharmaceutical Biology, 41*, 497–499.

Hano, Y., Matsumoto, Y., Sun, J. Y., & Nomura, T. (1990). Structures of three new isoprenylated xanthones, cudraxanthones E, F, and G[1,2]. *Planta Medica, 56*, 399–402.

Haynes, L. J., & Taylor, D. R. (1966). *C*-glycosyl compounds. Part V. Mangiferin; The nuclear magnetic resonance spectra of xanthones. *Journal of the Chemical Society C*, 1685–1687.

Holdsworth, D. K. (1973). A method to differentiate isomeric Cglucosyl chromones, isoflavones and xanthones. *Phytochemistry, 12*, 2011–2015.

Hostettmann, K., & Miura, I. (1977). A new xanthone diglucoside from *Swertia perennis* L. *Helvetica Chimica Acta, 60*, 262–264.

Hostettmann, K., Tabacchi, R., & Jacot-Guillarmod, A. (1974). Contribution à la phytochimie du genre *Gentiana*, VI. Etude des xanthones dans les feuilles de *Gentiana bavarica* L. *Helvetica Chimica Acta, 57*, 294–301.

Hostettmann, K., Marston, A., Ndjoko, K., & Wolfender, J.-L. (2000). The potential of African plants as a source of drugs. *Current Organic Chemistry, 4*(10), 973–1010.

IARC. (2020). *Latest global cancer data: Cancer burden rises to 18.1 million new cases and 9.6 million cancer deaths in 2018.* (Lyon, France: International Agency for Research on Cancer) ⟨https://www.who.int/cancer/PRGlobocanFinal.pdf⟩.

Iinuma, M., Tosa, H., Tanaka, T., & Yonemori, S. (1994). Two xanthones from roots of *Calophyllum inophyllum. Phytochemistry, 35*, 527–532.

Imperato, F. (1980). A xanthone-O-glycoside from *Asplenium adiantum-nigrum. Phytochemistry, 19*, 2030–2031.

Iseda, S. (1957). Isolation of 1,3,6,7-tetrahydroxyxanthone and the skeletal structure of Mangiferin. *Bulletin of the Chemical Society of Japan, 30*, 625–629.

Ishiguro, K., Nagata, S., Oku, H., & Masae, Y. (2002). Bisxanthones from *Hypericum japonicum*: Inhibitors of PAF-induced hypotension. *Planta Medica, 68*, 258–261.

Jamwal, A. (2012). Systematic review on xanthones and others isolates from genus *Swertia. International Journal of Chemical and Pharmaceutical Sciences, 1*(3), 1115–1133.

Jankovic, T., Krstic, D., Savikin-Fodulovic, K., Menkovic, N., & Grubisic, D. (2002). Xanthones and secoiridoids from hairy root cultures of *Centaurium erythraea* and *C. pulchellum. Planta Medica, 68*, 944–946.

Jensen, S. R., & Schripsema, J. (2002). Chemotaxonomy and pharmacology of Gentianaceae. In L. Struwe, & V. Albert (Eds.). *Gentianaceae—systematics and natural history* (pp. 573–631). Cambridge: Cambridge University Press.

Jiang, D. J., Hu, G. Y., Jiang, J. L., Xiang, H. L., Deng, H. W., & Li, Y. J. (2003). Relationship between protective effect of xanthone on endothelial cells and endogenous nitric oxide synthase inhibitors. *Bioorganic & Medicinal Chemistry, 11*(23), 5171–5177.

Kaldas, M., Hostettmann, K., & Jacot-Guillarmod, A. (1974). *Contribution a la phytochimie du genre Gentiana IX. Etude de composés flavoniques et xanthoniques dans les feuilles de Gentiana campestris L. 1`ere communication. Helvetica Chimica Acta, 57*, 2557–2561.

Khan, A., Rahman, M., & Islam, S. (2010). Isolation and bioactivity of a xanthone glycoside from *Peperomia pellucida. Life Sciences and Medicine Research: A Journal of Science and its Applications,* 1–15.

Khetwal, K. S., & Bisht, R. S. (1988). A xanthone glycoside from *Swertia speciosa. Phytochemistry, 27*, 1910–1911.

Kikuchi, M., & Kikuchi, M. (2004). Studies on the constituents of *Swertia japonica* MAKINO I. On the structures of new secoiridoid diglycosides. *Chemical and Pharmaceutical Bulletin, 52*, 1210–1214.

Kitanov, G. M., & Blinova, K. F. (1987). Modern state of the chemical study of species of the genus *Hypericum. Chemical Natural Compounds, 2*, 151–166.

Koolen, H. H. F., Menezes, L. S., Souza, M. P., Silva, F. M. A., Almeida, F. G. O., De Souza, A. Q. L., et al. (2013). Talaroxanthone, a novel xanthone dimer from the endophytic fungus *Talaromyces* sp. associated with *Duguetia stelechanthan* (Diels) R. E. Fries. *Journal of the Brazilian Chemical Society, 24*, 880–883.

Krstic, D., Jankovic, T., Savikin-Fodulovic, K., Menkovic, N., & Grubisic, A. (2003). Secoiridoids and xanthones in the shoots and roots of *Centaurium pulchellumcultured in-vitro. In Vitro Cellular and Developmental Biology, 39*, 203–207.

Kuete, V. (2025). Chapter Four-African medicinal plants and their derivative as the source of potent anti-leukemic products: rationale classification of naturally occurring anticancer agents. *Advances in Botanical Research, 113*. https://doi.org/10.1016/bs.abr.2023.12.010.

Kuete, V., & Efferth, T. (2015). African flora has the potential to fight multidrug resistance of cancer. *BioMed Research International, 2015*, 914813.

Kuete, V., Wabo, H. K., Eyong, K. O., Feussi, M. T., Wiench, B., Krusche, B., et al. (2011). Anticancer activities of six selected natural compounds of some Cameroonian medicinal plants. *PLoS One, 6*(8), e21762.

Kuete, V., Sandjo, L. P., Ouete, J. L. N., Fouotsa, H., Wiench, B., & Efferth, T. (2014). Cytotoxicity and modes of action of three naturally occurring xanthones (8-hydroxycudraxanthone G, morusignin I and cudraxanthone I) against sensitive and multidrug-resistant cancer cell lines. *Phytomedicine, 21*, 315–322.

Kuete, V., Ngnintedo, D., Fotso, G. W., Karaosmanoğlu, O., Ngadjui, B. T., Keumedjio, F., et al. (2018). Cytotoxicity of seputhecarpan D, thonningiol and 12 other phytochemicals from African flora towards human carcinoma cells. *BMC Complementary and Alternative Medicine, 18*, 36.

Kuete, V., Mbaveng, A. T., Nono, E. C. N., Simo, C. C., Zeino, M., Nkengfack, A. E., & Efferth, T. (2016). Cytotoxicity of seven naturally occurring phenolic compounds towards multi-factorial drug-resistant cancer cells. *Phytomedicine, 23*, 856–863.

Kumagai, K., Hosotani, N., Kikuchi, K., Kimura, T., & Saji, I. (2003). Xanthofulvin, a novel semaphorin inhibitor produced by a strain of *Penicillium*. *The Journal of Antibiotics, 56*, 610–616.

Lannang, A. M., Komguem, J., Ngninzeko, F. N., Tangmouo, J. G., Lontsi, D., Ajaz, A., et al. (2005). Bangangxanthone A and B, two xanthones from the stem bark of *Garcinia polyantha* Oliv. *Phytochemistry, 66*(19), 2351–2355.

Lannang, A. M., Tatsimoa, S. J. N., Fouotsa, H., Dzoyem, J. P., Saxena, A. K., & Sewald, N. (2014). Cytotoxic Compounds from the Leaves of *Garcinia polyantha*. *Chemistry & Biodiversity, 11*(6), 975–981.

Lannang, A. M., Louh, G. N., Biloa, B. M., Komguem, J., Mbazoa, C. D., Sondengam, B. L., ... El Ashry, E. H. S. (2010). Cytotoxicity of natural compounds isolated from the seeds of Garcinia afzelii. *Planta medica, 76*(07), 708–712.

Leet, J. E., Liu, X., Drexler, D. M., Cantone, J. L., Huang, S., Mamber, S. W., et al. (2008). Cytotoxic xanthones from Psorospermum molluscum from the Madagascar rain forest. *Journal of Natural Products, 71*(3), 460–463.

Li, G.-L., He, J.-Y., Zhang, A., Wan, Y., Wang, B., & Chen, W.-H. (2011). Toward potent α-glucosidaes inhibitors based on xanthones: A closer look into the structure activity correlations. *European Journal of Medicinal Chemistry, 46*(9), 4050–4055.

Lin, C. N., Liou, S. S., Ko, F. N., & Teng, C. M. (1993). Gamma-Pyrone compounds. IV: Synthesis and antiplatelet effects of mono- and dioxygenated xanthones and xanthonoxypropanolamine. *Journal of Pharmaceutical Science, 82*, 11–16.

Lin, C. N., Chung, M., Liou, S. J., Lee, T. H., & Wang, J. P. (1996). Synthesis and Anti-inflammatory Effects of Xanthone Derivatives. *Journal of Pharmaceutical Pharmacology, 48*, 532–538.

Ludueña, R. F. (1997). Immunsystem. *Klinische Biochemie, 178*, 207–230.

Mandal, S. & Chatterjee, A. (1994). *Seminar on research in ayurveda and siddha* (New Delhi: CCRAS).

Mandal, S., Das, P. C., & Joshi, P. C. (1992). Naturally occurring xanthones from terrestrial flora. *Journal of Indian Chemical Society, 69*, 611–636.

Masters, K.-S., & Brase, S. (2012). Xanthones from fungi, lichens, and bacteria: The natural products and their synthesis. *Chemical Reviews, 112*(7), 3717–3776.

Mbaveng, A. T., Kuete, V., & Efferth, T. (2017). Potential of Central, Eastern and Western Africa medicinal plants for cancer therapy: Spotlight on resistant cells and molecular targets. *Frontiers in Pharmacology, 8*, 343.

Mbaveng, A. T., Fotso, G. W., Ngnintedo, D., Kuete, V., Ngadjui, B. T., Keumedjio, F., et al. (2018). Cytotoxicity of epunctanone and four other phytochemicals isolated from the medicinal plants *Garcinia epunctata* and *Ptycholobium contortum* towards multi-factorial drug resistant cancer cells. *Phytomedicine, 48*, 112–119.

Mehta, G., Shah, S. R., & Venkateswarlu, Y. (1994). Total synthesis of novel xanthone antibiotics (±)-cervinomycins A_1, and A_2. *Tetrahedron, 50*, 11729–11742.

Meyer, J. J. M., Rakuambo, N. C., & Hussein, A. A. (2008). Novel xanthones from *Securidaca longepedunculata* with activity against erectile dysfunction. *Journal of Ethnopharmacol, 119*(3), 599–603.

Monache, F. D., Mac-Quhae, M. M., Monache, G. D., Bettolo, G. B. M., & De Lima, R. A. (1983). Xanthones, xanthonolignoids and other constituents of the roots of *Vismia guaramirangae*. *Phytochemistry, 22*(1), 227–232.

Mutanyatta, J., Matapa, B. G., Shushu, D. D., & Abegaz, B. M. (2003). Homoisoflavonoids and xanthones from the tubers of wild and in vitro regenerated *Ledebouria graminifolia* and cytotoxic activities of some of the homoisoflavonoids. *Phytochemistry, 62*(5), 797–804.

Negi, J. S., & Singh, P. (2011). HPLC fingerprinting of highly demanding medicinal plant *Swertia:* An overview. *International Journal of Medicinal and Aromatic Plants, 1*, 333–337.

Negi, J. S., Singh, P., Pant, G. J., & Rawat, M. S. M. (2010). RP-HPLC analysis and antidiabetic activity of *Swertia paniculate*. *Natural Product Communications, 5*, 907–910.

Ngouela, S., Ndjakou, B. L., Tchamo, D. N., Zelefack, F., Tsamo, E., & Connolly, J. D. (2005). A prenylated xanthone with antimicrobial activity from the seeds of *Symphonia globulifera*. *Natural Product Research, 19*(1), 23–27.

Nguemeving, J. R., Azebaze, A. G., Kuete, V., Eric Carly, N. N., Beng, V. P., Meyer, M., et al. (2006). Laurentixanthones A and B, antimicrobial xanthones from Vismia laurentii. *Phytochemistry, 67*(13), 1341–1346.

Nguyen, L. H. D., & Harrison, L. J. (2000). Xanthones and triterpenoids from the bark of *Garcinia vilersiana*. *Phytochemistry, 53*, 111–114.

Nielsen, H., & Arends, P. (1978). Structure of the xanthonolignoid kielcorin. *Phytochemistry, 17*, 2040–2041.

Nikolaeva, G. G., Glyzin, V. I., Mladentseva, M. S., Sheichenko, V. I., & Patudin, A. V. (1983). Xanthones of *Gentiana lutea*. *Chemistry of Natural Compounds, 19*, 106–107.

Nkengfack, A. E., Azebaze, G. A., Vardamides, J. C., Fomum, Z. T., & Heerden, F. R. (2002). A prenylated xanthone from *Allanblackia floribunda*. *Phytochemistry, 60*(4), 381–384.

Nkengfack, A. E., Mkounga, P., Meyer, M., Fomum, Z. T., & Bodo, B. (2002). Globulixanthones C, D and E: Three prenylated xanthones with antimicrobial properties from the root bark of *Symphonia globulifera*. *Phytochemistry, 61*(2), 181–187.

Nualkaew, N., Morita, H., Shimokawa, Y., Kinjo, K., Kushiro, T., De-Eknamkul, W., et al. (2012). Benzophenone synthase from *Garcinia mangostana* L. Pericarps. *Phytochemistry, 77*, 60–69.

Okorie, D. A. (1976). A new phthalide and xanthones from *Anthocleista djalonensis* and *Anthocleista vogelli*. *Phytochemistry, 15*(11), 1799–1800.

Oldenburg, T. B. P., Wilkes, H., Horsfield, B., Van Duin, A. C. T., Stoddart, D., & Wilhelms, A. (2002). Xanthones: Novel aromatic oxygen-containing compounds in crude oils. *Organic Geochemistry, 33*, 595–609.

Pana, E., Cao, S., Brodie, P. J., Miller, J. S., Rakotodrajaona, R., Ratovoson, F., et al. (2010). An antiproliferative xanthone of *Symphonia pauciflora* from the Madagascar rainforest. *Natural Product Communications, 5*(5), 751–754.

Pedro, M., Cerqueira, F., Sousa, M. E., Nascimento, M. S., & Pinto, M. (2002). Xanthones as inhibitors of growth of human cancer cell lines and their effects on the proliferation of human lymphocytes in vitro. *Bioorganic & Medicinal Chemistry, 10*(12), 3725–3730.

Peres, V., & Nagem, T. J. (1997). Naturally occurring pentaoxygenated, hexaoxygenated and dimeric xanthones: A literature survey. *Química Nova, 20*(4), 388–397.

Peres, V., Nagem, T. J., & De Oliveira, F. F. (2000). Tetraoxygenated naturally occurring xanthones. *Phytochemistry, 55*(7), 683–710.

Peters, S., Schmidt, W., & Beerhues, L. (1998). Regioselective oxidative phenol coupling of 2,304,6-tetrahydroxybenzophenone in cell cultures of *Centaurium erythraea* RAFN and *Hypericum androsaemum* L. *Planta, 204*(1), 64–69.

Pinheiro, L., Nakamura, C. V., Filho, B. P., Ferreira, A. G., Young, M. C. M., & Garcia Cortez, D. A. (2003). Antibacterial Xanthones from *Kielmeyera variabilis* Mart. (Clusiaceae). *Memorias do Instituto Oswaldo Cruz, 98*, 549–552.

Pinto, M. M., Sousa, M. E., & Nascimento, M. S. (2005). Xanthone derivatives: New insights in biological activities. *Current Medicinal Chemistry, 12*(21), 2517–2538.

Pontius, A., Krick, A., Kehraus Brun, R., & Konig, G. M. (2008). Antiprotozoal activities of heterocyclic substituted xanthones from the marine derived fungus *Chaetomium* sp. *Journal of Natural Products, 71*(9), 1579–1584.

Pouli, N., & Marakos, P. (2009). Fusedxanthone derivatives as antiproliferative agents. *Anti-cancer Agents in Medicinal Chemistry, 9*(1), 77–98.

Prox, A. (1968). Massenspektrometrische untersuchung einiger nat¨urlicher C-glukosyl-verbindungen. *Tetrahedron, 24*(9), 3697–3715.

Purev, O., Oyun, K., Odontuya, G., Tankhaeva, A. M., Nikolaeva, G. G., Khan, K. M., et al. (2002). Isolation and structure elucidation of two new xanthones from *Gentiana azurium* Bunge (Fam. Gentianaceae). *Zeitschrift fur Naturforschung B, 57*, 331–334.

Raistrick, H., Robinson, R., & White, D. (1936). Studies in the biochemistry of micro-organisms: Ravenelin (3-methyl-1:4:8-trihydroxyxanthone), a new metabolic product of *Helminthosporium ravenelii Curtis* and of *H. turcicum Passerini*. *Biochemical Journal, 30*(8), 1303–1314.

Rukachaisirikul, V., Kamkaew, M., Sukavisit, D., Phongpaichit, S., Sawangchote, P., & Taylor, W. C. (2003). Antibacterial xanthones from the leaves of *Garcinia nigrolineata*. *Journal of Natural Products, 66*, 1531–1535.

Schmidt, W., & Beerhues, L. (1997). Alternative pathway of xanthone biosynthesis in cell cultures of *Hypericum androsaemum* L. *FEBS Letters, 420*(2-3), 143–146.

Shan, W. G., Lin, T. S., Yu, N., Chen, Y., & Zhan, Z. J. (2012). Polyprenylated xanthones and benzophenones from the bark of *Garcinia oblongifolia*. *Helvetica Chimica Acta, 95*, 1442–1448.

Shan, Y., & Zhang, W. (2010). Preparative separation of major xanthones from mangosteen pericarp using highperformance centrifugal partition chromatography. *Journal of Separation Science, 33*, 1274–1278.

Shen, J., Tian, Z., & Yang, J. S. (2007). The constituents from the stems of *Garcinia cowa* Roxb. and their cytotoxic activities. *Die Pharmazie, 62*(7), 549–551.

Sob, S. V. T., Wabo, H. K., Tchinda, A. T., Tane, P., Ngadjui, B. T., & Ye, Y. (2010). Anthraquinones, sterols, triterpenoids and xanthones from Cassia obtusifolia. *Biochemical Systematics and Ecology, 38*(3), 342.

Song, Y., Hu, F., Lin, P., Lu, Y., & Shi, Z. (2004). Determination of xanthones in *Swertia mussotii* and *Swertia franchetiana* by high performance liquid chromatography. *Chinese Journal of Chromatography, 22*, 51–53.

Suksamrarn, S., Komutiban, O., Ratananukul, P., Chimnoi, N., Lartpornmatulee, N., & Suksamrarn, A. (2006). Cytotoxic prenylated xanthones from the young fruit of *Garcinia mangostana*. *Chemical and Pharmaceutical Bulletin, 54*(3), 301–305.

Sultanbawa, M. U. S. (1980). Xanthonoids of tropical plants. *Tetrahedron, 36*(1), 1465–1506.

Sun, R. R., Miao, F. P., Zhang, J., Wang, G., Yin, X. L., & Ji, N. Y. (2013). Three new xanthone derivatives from an algicolous isolate of *Aspergillus wentii*. *Magnetic Resonance in Chemistry, 51*, 65–68.

Suryawanshi, S., Mehrotra, N., Asthana, R. K., & Gupta, R. C. (2006). Liquid chromatography/tandemmass spectrometric study and analysis of xanthone and secoiridoid glycoside composition of *Swertia chirata*, a potent antidiabetic. *Rapid Communications in Mass Spectrometry, 20*, 3761–3768.

Suzuki, O., Katsumata, Y., Oya, M., Chari, V. M., Klapfenberger, R., Wagner, H., & Hostettman, K. (1980). Inhibition of type A and type B monoamine oxidase by isogentisin and its 3-o-glucoside. *Planta Medica, 39*, 19–23.

Suzuki, O., Katsumata, Y., Oya, M., Chari, V. M., Vermes, B., Wagner, H., & Hostettman, K. (1981). Inhibition of type A and type B monoamine oxidases by naturally occurring xanthones. *Planta Medica, 42*, 17–21.

Tala, F. M., Talontsi, M. F., Zeng, G.-Z., Wabo, K. H., Tan, N.-H., Spiteller, M., ... Tane, P. (2015). Antimicrobial and cytotoxic constituents from native Cameroonian medicinal plant *Hypericum riparium*. *Fitoterapia, 102*, 149–155.

Tala, M. F., Wabo, H. K., Zeng, G.-Z., Ji, C.-J., Tane, P., & Tan, N.-H. (2013). A prenylated xanthone and antiproliferative compounds from leaves of *Pentadesma butyracea*. *Phytochemistry Letters, 6*, 326–330.

Tanaka, N., Kashiwada, Y., Kim, S. Y., Sekiya, M., Ikeshiro, Y., & Takaishi, Y. (2009). Xanthones from *Hypericum chinense* and their cytotoxicity evaluation. *Phytochemistry, 70*(11-12), 1456–1461.

Terui, Y., Yiwen, C., Jun-Ying, L., Tsutomu, A., Yamamoto, H., Kawamura, Y., et al. (2003). Xantholipin, a novel inhibitor of HSP47 gene expression produced by *Streptomyces* sp. *Tetrahedron Letters, 44*, 5427–5430.

Valentao, P., Andrade, P. B., Silva, E., Vicente, A., Santos, H. L., Bastos, M., et al. (2002). Methoxylated xanthones in the quality control of small centaury (*Centaurium erythraea*) flowering tops. *Journal of Agricultural and Food Chemistry, 50*, 460–463.

Van, K. S., & Nessler, B. (1891). Ueber einige oxyxanthone. *Chemische Berichte, 24*, 3980–3984.

Velisek, J., Davidek, J., & Cejpek, K. (2008). Biosynthesis of food constituents: Natural pigments. *Czech Journal of Food Sciences, 26*(2), 73–98.

Verma, D. L., & Khetwal, K. S. (1985). Phenolics in the roots of *Swertia paniculata* Wall. *Science and Culture, 51*, 305–306.

Verney, A. M., & Debelmas, A. M. (1973). Xanthones of *Gentiana lutea, G. purpurea, G. punctata, G. pannonica*. *Annales Pharmaceutiques Francaises, 31*, 415–420.

Viera, L. M. M., & Kijjoa, A. (2005). Naturally occurring xanthone: Recent development. *Current Medicinal Chemistry, 12*(21), 2413–2446.

Vorobiof, D. A., & Abratt, R. (2007). The cancer burden in Africa. *South African Medical Journal, 97*(10), 937–939.

Wabo, H. K., Kikuchi, H., Katou, Y., Tane, P., & Oshima, Y. (2010). Xanthones and a benzophenone from the roots of Pentadesma butyracea and their antiproliferative activity. *Phytochemistry Letters, 3*(2), 104–107.

Wagenaar, M. M., & Clardy, J. (2001). Dicerandrols, new antibiotic and cytotoxic dimers produced by the fungus *Phomopsis longicolla* isolated from an endangered mint. *Journal of Natural Products, 64*(8), 1006–1009.

Wang, C.-Z., Maier, U. H., Keil, M., Zenk, M. H., Bacher, A., Rohdich, F., et al. (2003). Phenylalanine independent biosynthesis of 1,3,5,8-tetrahydroxyxanthone. A retrobiosynthetic NMR study with root cultures of *Swertia chirata*. *European Journal of Biochemistry, 270*(14), 2950–2958.

Wang, L. L., Liu, H. G., & Zhang, T. J. (2010). Advances in studies on xanthones. *Chinese Traditional and Herbal Drugs, 41*(7), 1196–1206.

Wang, S.-S., Zhao, W.-J., Han, X.-W., & Liang, X.-M. (2005). Two new iridoid glycosides from the Tibetan folk medicine *Swertia franchetiana*. *Chemical and Pharmaceutical Bulletin, 53*(6), 674–676.

Wedajo, F., Gure, A., Meshesha, M., Kedir, K., Frese, M., Sewald, N., & Abdissa, N. (2022). Cytotoxic compounds from the root bark of *Securidaca longipedunculata*. *Bulletin of the Chemical Society of Ethiopia, 36*(2), 417–422.

Whitman, W. E., & Wiles, L. A. (1956). The polarographic reduction of xanthone and methoxyxanthones. *Journal of Chemical Society*, 3016–3019.

Xue, Q.-C., Li, C.-J., Zuo, L., Yang, J.-Z., & Zhang, D.-M. (2009). Three new xanthones from the roots of *Polygala japonica* houtt. *Journal of Asian Natural Products Research, 11*(5), 465–469.

Yang, C.-H., Ma, L., Wie, Z.-P., Han, F., & Gao, J. (2012). Advances in isolation and synthesis of xanthones derivatives. *Chinese Herbal Medicines, 4*(2), 87–102.

Yang, H., Ding, C., Duan, Y., & Liu, J. (2005). Variation of active constituents of an important Tibet folk medicine *Swertia mussotii* Franch. (Gentianaceae) between artificially cultivated and naturally distributed. *Journal of Ethnopharmacology, 98*(1-2), 31–35.

Yang, Gao, H. Y. Y. H., Niu, D. Y., Yang, L.-Y., Gao, X.-M., Du, G., et al. (2013). Xanthone derivatives from the fermentation products of an endophytic fungus *Phomopsis* sp. *Fitoterapia, 13*, 239–246.

Yimdjo, M. C., Azebaze, A. G., Nkengfack, A. E., Meyer, A. M., Bodo, B., & Fomum, Z. T. (2004). Antimicrobial and cytotoxic agents from *Calophyllum inophyllum*. *Phytochemistry, 65*(20), 2789–2795.

Zelefack, F., Guilet, D., Fabre, N., Bayet, C., Chevalley, S., Ngouela, S., et al. (2009). Cytotoxic and antiplasmodial xanthones from *Pentadesma butyracea*. *Journal of Natural Products, 72*(5), 954–957.

Zhang, J.-S., Wang, X.-M., Dong, X.-H., Yang, H.-Y., & Li, G.-P. (2009). Studies on chemical constituents of *Swertia mussotii*. *Zhong Yao Cai = Zhongyaocai = Journal of Chinese Medicinal Materials, 32*(4), 511–514.

Zhou, T. T., Zhu, Z. Y., Wang, C., Fan, G., Peng, J., Chai, Y., et al. (2007). On-line puritymonitoring in high-speed counter-current chromatography: Application of HSCCC-HPLC-DAD for the preparation of 5-HMF, neomangiferin and mangiferin from *Anemarrhena asphodeloides* Bunge. *Journal of Pharmaceutical and Biomedical Analysis, 44*(1), 96–100.

Zou, Y. S., Hou, A. J., Zhu, G. F., Chen, Y. F., Sun, H. D., & Zhao, Q. S. (2004). Cytotoxic isoprenylated xanthones from *Cudrania tricuspidata*. *Bioorganic & Medicinal Chemistry, 12*(8), 1947–1953.

CHAPTER NINE

Alkaloids from African plants as pharmaceuticals to combat cancer drug resistance

Vaderament-A. Nchiozem-Ngnitedem[a,*,1], **Justus Mukavi**[b,1], **Leonidah K. Omosa**[c], **and Victor Kuete**[d]

[a]Institute of Chemistry, University of Potsdam, Potsdam-Golm, Germany
[b]Institute of Pharmaceutical Biology and Phytochemistry, University of Münster, Münster, Germany
[c]Department of Chemistry, Faculty of Science and Technology, University of Nairobi, Nairobi, Kenya
[d]Department of Biochemistry, Faculty of Science, University of Dschang, Dschang, Cameroon
*Corresponding author. e-mail address: n.vaderamentalexe@gmail.com

Contents

Abstract

Cancer continues to be one of the major causes of mortality globally despite impressive advancements in the development of new therapeutic medications. Over the years, there has been an increasing concern regarding multidrug-resistant (MDR) cancer cell lines. Alkaloids are among the many plants' secondary metabolites that have been identified from African medicinal plants as potential cancer preventive agents. In this chapter, we summarized the phytochemistry and cytotoxic potential of alkaloids derived from plants native to Africa towards both drug-sensitive and drug-resistant cancer cells. A systematic search of the literature (using Google Scholar, PubMed, Scopus, and Science Direct databases) revealed 152 different alkaloids from African flora with anticancer and/or cytotoxic properties. This comprises *naphthylisoquinoline* (42.76%), benzophenanthridine (9.87%),

[1] The two authors contributed equally.

Advances in Botanical Research, Volume 115
ISSN 0065-2296, https://doi.org/10.1016/bs.abr.2024.02.010

indolomonoterpenic alkaloids (9.87%), acridone (7.24%), furoquinoline (5.92%), bisbenzylisoquinoline (5.26%), steroidal (1.97%), β-carboline (1.32%) and other alkaloids (15.79%). Most studies on naphthyisoquinolines, which make up most anticancer alkaloids, have focused on drug-sensitive (CCRF-CEM) and multidrug-resistant (CEM/ADR5000) leukemia cell lines as well as Panc-1 human pancreatic cell lines. Based on the data presented here, alkaloids from African plants have shown highly promising anticancer potential. More in-depth studies should be undertaken to develop chemotherapeutic drugs to fight cancer including recalcitrant phenotypes.

Abbreviations

C_2H_5OH ethanol
CD circular dichroism
CH_2Cl_2 dichloromethane
CH_3CN acetonitrile
CH_3OH methanol
D.R. degree of resistance
GC-MS gas chromatography–mass spectrometry
H_2O water
HPLC-MS liquid chromatography–mass spectrometry
IC_{50} half inhibitory concentration
MDR multidrug-resistant
NH_3 ammonia
NIQ naphthalylisoquinoline
NMR nuclear magnetic spectroscopy
ODS octadecylsilyl
PC_{50} preferential cytotoxicity
PKS polyketide synthase
TFA trifluoroacetic acid
TLC thin-layer chromatography
μM micromolar

1. Introduction

Cancer is a broad category of disorders that can begin in practically any organ or tissue of the body and is defined by abnormal cell growth. A common cause of cancer-related death is metastasizing, which is the process by which tumors that form from malignant cells invade other body tissues and organs (World Health Organization, 2012). Roughly 10 million fatalities in 2020, or nearly one in every six, were due to cancer, making it one of the world's leading causes of death. Breast, lung, colon and rectum, and prostate cancers are the most prevalent types of cancer (Ferlay et al., 2021). According to empirical studies, the interplay between internal factors (such as hereditary mutations, hormones, and immunological disorders) and external/acquired factors (tobacco, chemicals, diet, radiation, and infectious organisms) is what causes the onset and progression of most malignancies (Anand et al., 2008). The choice of cancer treatment depends primarily on the type of cancer and how far along it is in the advancement process. Chemotherapy is the most extensively used form of cancer treatment, with surgery, radiation, and immunotherapy following closely (Abbas & Rehman, 2018). Medical plants have long been used to treat a wide range of disorders because of their therapeutic properties, which are regarded to be a source of a variety of bioactive chemicals (Roy, Ahuja, & Bharadvaja, 2017). Presently, a wide variety of herbal medicines and their active ingredients are attracting interest as potential pathways to the development of medications for the effective treatment of cancer (Sawadogo, Schumacher, Teiten, Dicato, & Diederich, 2012). Numerous secondary metabolites, including alkaloids, coumarins, flavonoids, phenolic acids, lignanolides, tannins, terpenoids, polyphenols, steroids, and quinones, among others, have been linked to these anticancer properties (Gezici & Şekeroğlu, 2019). Notably, more than 60% of the first-line anticancer treatments approved internationally were either natural products or their derivatives or were created using knowledge about tiny molecules or macromolecules found in nature (Cragg & Newman, 2005).

Alkaloids are primarily bitter substances that largely shield plants against pathogens and animal predation. Chemically speaking, they are a vast class of structurally varied nitrogen-containing compounds, comprising more than 20,000 bioactive substances that are produced by the biosynthetic processes of plants and microbes and are also found in mammals, insects, and marine organisms (Wansi, Devkota, Tshikalange, & Kuete, 2013). Due to the one or more nitrogen atoms present in the molecule, these compounds can form salts in the presence of acids; these salts are soluble in water but not organic solvents (Rainsford & Alamgir, 2017). It is impossible to classify all alkaloids according to a single, consistent taxonomic concept. Alkaloids are grouped whenever it is appropriate based on their chemical makeup, biochemical history, or natural origin (Hesse, 2002). Based on chemical structure and biogenesis, alkaloids are widely classified as true alkaloids, which include nitrogen in the heterocycle, and atypical alkaloids, which have nitrogen in the exocyclic position (Huang et al., 2022). True alkaloids, which originate from precursors of α amino acids, make up the majority of alkaloids. Due to their complex structural makeup, true alkaloids can be further divided into 14 subgroups based on their ring structures (Fig. 1), including pyrrolidines (**1**), pyrroles (**2**), imidazoles (**3**), piperidines (**4**), pyridines (**5**), indolizidines (**6**), pyrrolizidines (**7**), indoles (**8**), purines (**9**), quinolizidines (**10**), quinolines (**11**), isoquinolines (**12**), tropanes (**13**), and

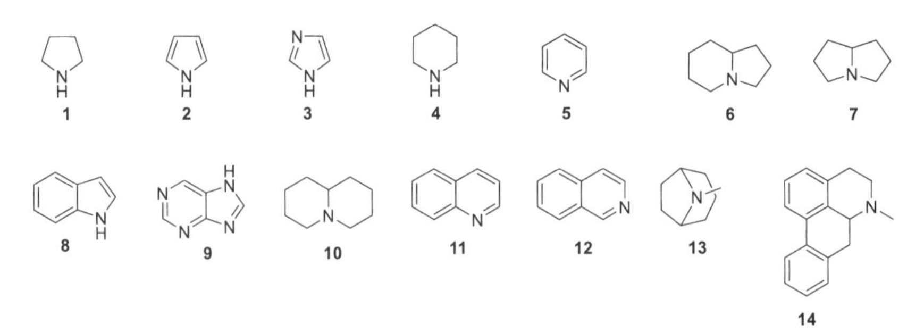

Fig. 1 The 14 divisions of alkaloids containing nitrogen in the heterocycle **1**: pyrrolidines; **2**: pyrroles; **3**: imidazoles; **4**: piperidines; **5**: pyridines; **6**: indolizidines; **7**: pyrrolizidines; **8**: indoles; **9**: purines; **10**: quinolizidines; **11**: quinolines; **12**: isoquinolines; **13**: tropanes; **14**: aporphine alkaloids.

aporphine alkaloids (**14**) (Othman, Sleiman, & Abdel-Massih, 2019). It is also feasible to classify alkaloids according to their natural origins because particular alkaloids are frequently restricted to particular sources (Evans, 2009).

Alkaloids easily form hydrogen bonds with proteins, enzymes, and receptors thanks to their proton-accepting nitrogen atom and one or more proton-donating amine hydrogen atoms. This explains the wide range of high bioactivity of the alkaloids together with the common occurrence of functional groups like phenolic hydroxyl and polycyclic moieties (Kittakoop, Mahidol, & Ruchirawat, 2014; Othman et al., 2019). Morphine, which is a strong opioid analgesic that is frequently used for the treatment of acute pain as well as chronic therapy of severe pain, was the first biologically active alkaloid to be isolated from nature, in 1805 (Pacifici, 2016). Following this, different alkaloids were extracted and utilized as medications to treat a variety of pathological conditions throughout the nineteenth and twentieth centuries. African medicinal plants offer a large number of alkaloids that have been reported to date with a variety of biological properties, including antimicrobial (Kaigongi et al., 2020; Karou et al., 2005), antimalarial (Amoa et al., 2013; Ancolio et al., 2002), antiprotozoal (Eze et al., 2020), anti-HIV (Bringmann et al., 2016; Mohammed et al., 2012), anticancer (Benamar, Melhaoui, Zyad, Bouabdallah, & Aziz, 2009; Nganou et al., 2019), enzyme inhibitory (Ka et al., 2020; Masi et al., 2021), anti-inflammatory (Elgorashi, Zschocke, & Van Staden Eloff, 2003), and antioxidant (Kiplimo, Islam, & Koorbanally, 2011).

Our earlier chapters on alkaloids derived from African medicinal plants mostly concentrated on their significant antimicrobial, cytotoxic, enzyme inhibitory, anti-inflammatory, and antioxidant activity (Wansi et al., 2013) as well as their poisonous and protective effects (Kuete, 2014). While a number of reviews have discussed alkaloids with anti-cancer characteristics (Graham, Quinn, Fabricant, & Farnsworth, 2000; Hashmi, Khan, Farooq, & Khan, 2018; Huang, Zhe-Ling, Yi-Tao, & Li-Gen, 2017; Nair & Van Staden, 2014; Nwodo, Ibezim, V. Simoben, & Ntie-Kang, 2015; Tao et al., 2020), none have provided a thorough overview of alkaloids from the African flora as therapeutics to counter cancer treatment resistance. In keeping with our ongoing efforts to provide knowledge about bioactive chemical compounds from African plants, the current chapter focuses on the cytotoxic effect of alkaloids from plants growing on the African continent toward drug-sensitive and multidrug-resistant cancer cells.

2. Chemistry and biosynthesis

The diverse classes of alkaloids have a distinct biosynthetic pathway, in contrast to most other kinds of secondary metabolites. They are biosynthesized from amino acids such phenylalanine (ephedrine, norpseudoephedrine and capsaicin) (Funayama & Cordell, 2014), tyrosine (benzylisoquinoline, tetrahydroisoquinoline, and amaryllidaceae alkaloids), tryptophan (monoterpene indole alkaloids) (Lichman, 2021), ornithine (pyrrolidine, tropane, and pyrrolizidine alkaloids) (Funayama & Cordell, 2014), and lysine (quinolizidine, lycopodium, piperidine, and indolizidine alkaloids) (Bunsupa, Yamazaki, & Saito, 2017). In addition, alkaloid structures also typically contain components from the acetate, shikimate, or deoxyxylulose phosphate pathways. In our last chapter, we highlighted these amino acid precursors and the related alkaloids (Wansi et al., 2013). We describe herein a brief overview of the chemistry and biosynthesis of a number of cytotoxic alkaloids found in the flora of Africa, including naphthylisoquinoline, isoquinoline, acridone, and benzophenanthridine alkaloids.

2.1 Naphthylisoquinoline alkaloids

Only the plant families Ancistrocladaceae and Dioncophylaceae from the tropics have produced naphthalylisoquinoline (NIQ) alkaloids. This group of natural compounds is composed chemically of a naphthalene moiety and an isoquinoline subunit that are connected by a biaryl axis. Within this family of compounds, there is a lot of structural diversity due to the various options for the location of the biaryl axis between the two moieties, the pattern of oxygen substitution, the stereocenters in the isoquinoline part, and the site of dimerization (Moyo et al., 2020). The biosynthesis of naphthylisoquinoline alkaloids begins with a pentaketide intermediate that is created by incomplete reduction and dehydration processes carried out by polyketide synthase (PKS) from acetyl-CoA and malonyl-CoA. Following this, the biosynthesis of the naphthalene and isoquinoline moieties, which are both produced from six acetate units *via* the identical β-pentaketo precursor, proceeds *via* an acetate-malonate pathway (Bringmann et al., 2007; Ibrahim & Mohamed, 2015). Consequently, they are both generated through cyclization, though in different ways, in the case of the naphthalene, by a second aldol condensation, and in the case of the isoquinoline moiety, by nitrogen integration. To produce the full NIQ molecule, they are then convergently coupled together in accordance with the phenoloxidative biaryl coupling principle (Ibrahim & Mohamed, 2015). Different NIQs with various structural properties, such as dioncophylline (**15**), ancistrocladine (**16**), and ancisctrocladidne (**17**), are produced by a variety of metabolic mechanisms (Scheme 1).

2.2 Isoquinoline alkaloids

Isoquinoline alkaloids, one of the largest classes of biologically active alkaloids, are most frequently found in the Ranunculales, Menispermaceae, Berberidaceae, Papaveraceae, Hernandiaceae, and Monimiaceae plant families (Chou, Lin, Chen, & Chen, 1994; Krane & Shamma, 1982). The fundamental structural unit of naturally occurring isoquinoline alkaloids is 1-benzylisoquinoline. Due to their diverse structural makeup, isoquinoline alkaloids have been divided into two main groups: simple isoquinolines, which are made up of a fused benzene ring and pyridine ring, and benzylisoquinolines, which feature an extra aromatic ring (Singh, Pathak, Fatima, & Negi, 2021). The biosynthesis of the isoquinoline alkaloids basically starts with the conversion of the vital amino acid tylosine to dopamine and *p*-hydroxyphenylacetaldehyde through decarboxylation, orthohydroxylation, and deamination (Dey et al., 2020; Singh et al., 2021). Dopamine and *p*-hydroxyphenylacetaldehyde are combined by the enzyme norcoclaurine synthase (NCS) to create (*S*)-norcoclaurine, which serves as the main building block for all isoquinoline alkaloids. (*S*)-Norcoclaurine is successively transformed into (*S*)-coclaurine by the enzyme *O*-methyltransferase (OMT), followed by (*S*)-*N*-methylcoclaurine by the enzyme coclaurine *N*-methyltransferase (CNMT), (*S*)− 3′-hydroxy-*N*-methyl coclaurine by the enzyme P-450 hydroxylase, and finally (*S*)-reticuline by the enzyme 3′-hydroxy *N*-methylcoclaurine 4′-*O*-methyltransferase (4′-OMT) (Choi, Morishige, Shitan, Yazaki, & Sato, 2002; Morishige, Tsujita, Yamada, & Sato, 2000; Pauli & Kutchan, 1998). The usual biosynthetic intermediate for this pathway is (*S*)-reticuline, from which further transformational processes

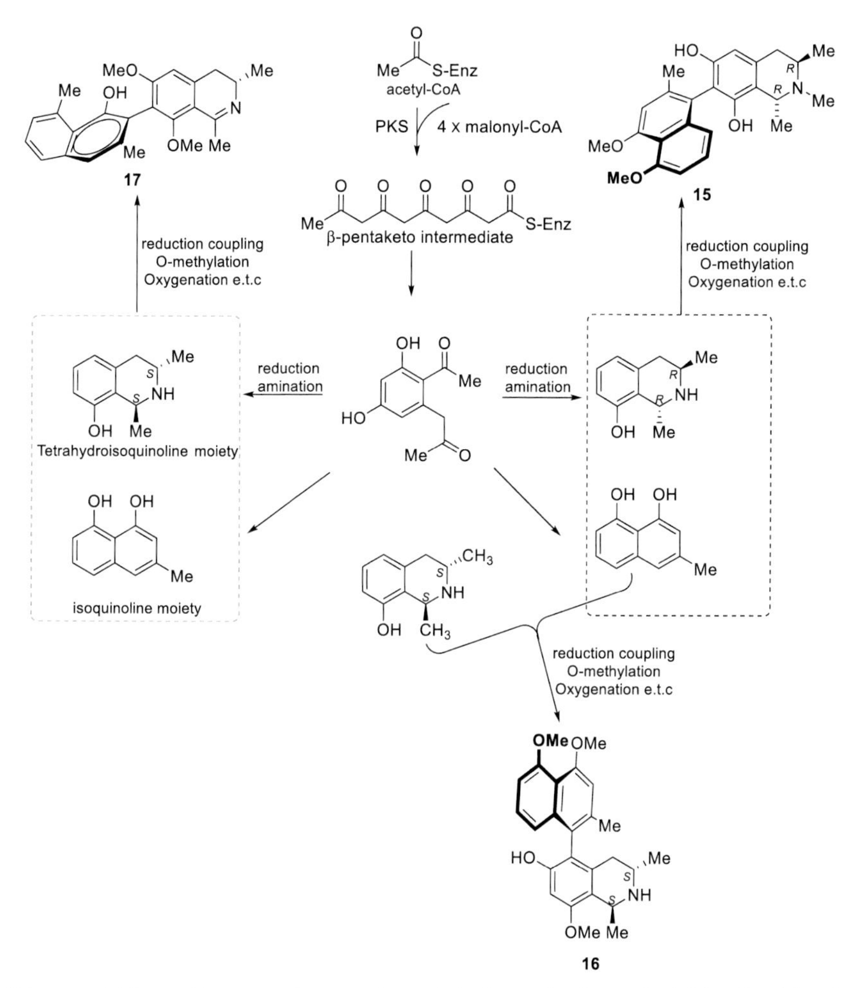

Scheme 1 Biosynthetic pathway of naphthylisoquinoline alkaloids (Ibrahim & Mohamed, 2015).

result in the production of diverse structural classes of isoquinoline alkaloids, including berberine (**18**) (Scheme 2)

2.3 Benzophenanthridine alkaloids

Benzophenanthridine alkaloids, which are members of the benzylisoquinoline alkaloid family, are mostly found in plants from the Rutaceae, Papaveraceae, and Fumariaceae families. This group of alkaloids are tetracyclic molecules that contain a heterocyclic framework that is not aromatic (Laines-Hidalgo, Muñoz-Sánchez, Loza-Müller, & Vázquez-Flota, 2022).

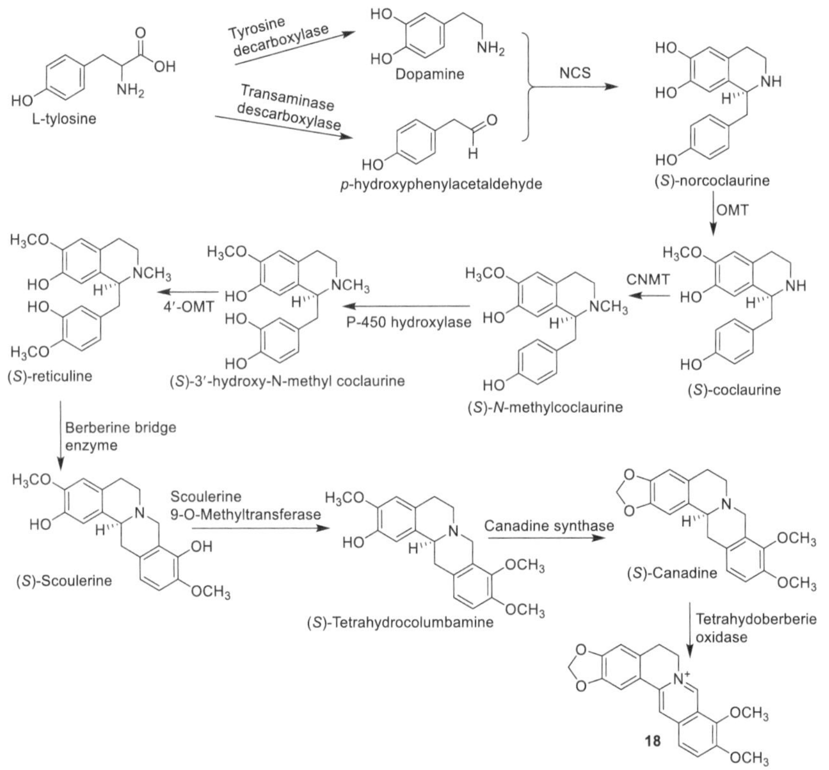

Scheme 2 Biosynthetic pathway of isoquinoline alkaloids.

Based on the arrangement of their rings, benzophenanthridine alkaloids are classified into three classes (Fig. 2): type I (hexahydrobenzo[*c*]phenanthridines), type II (dihydrobenzo[*c*]phenanthridines), and type III (quaternary benzo[*c*]phenathridine). Two aromatic and aliphatic rings make up type I, and the nitrogen atom at position B is typically methylated. The structural framework of type II consists of three aromatic systems, one alkane ring, and a methylated nitrogen atom. In type II alkaloids, the C-8 position is typically modified with an acetonyl or methoxyl group. Since the C7-C8 bond may be open (type II-1), or the C-8 may be polymerized (type II-2), they are further divided into two groups. Type III is generated when type I or type II benzophenanthridine alkaloids are N protonated, producing the appropriate ammonium quaternary salts (Bisai, Saina Shaheeda, Gupta, & Bisai, 2019; Han, Yang, Liu, Liu, & Yin, 2016). Like benzylisoquinoline alkaloids, early steps in the biosynthesis of benzophenanthridine alkaloids involve the intermediate

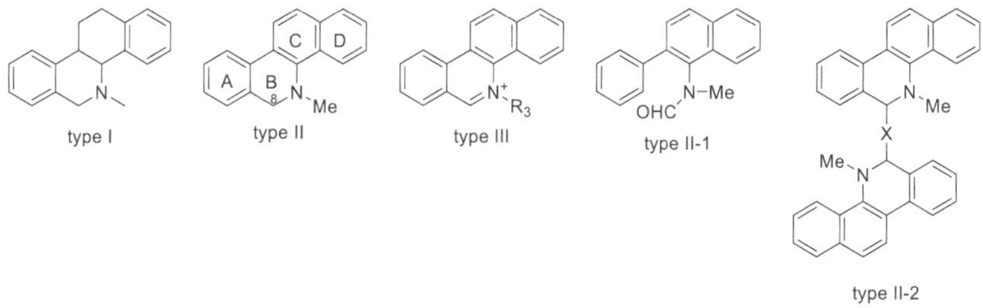

Fig. 2 Structural classification of benzophenanthridine alkaloids.

(*S*)-norcoclaurine, which is created *via* the condensation of dopamine and *p*-hydroxyphenylacetaldehyde (Scheme 3). Up until the synthesis of (*S*)-scoulerine, which is then oxidized by stylopine synthase (StySyn) to yield (*S*)-stylopine, the pathway proceeds similarly to that shown in Scheme 2. Protopine is formed when stylopine is *N*-methylated by s-tetrahydroprotoberberine *N*-methyltransferase (TNMT) to (*S*)-*N*-methylstylopine and then subsequently oxidized by methyltetrahydroprotoberberine 14-monooxygenase (MSH). A spontaneous ring opening, rearrangement, and dehydration arise from the hydroxylation of protopine, forming a benzophenanthridine alkaloid, dihydrosanguinarine (**19**) (Hagel & Facchini, 2013).

2.4 Acridone alkaloids

The acridone alkaloids are exclusive to plants of the Rutaceae family. They are heterocyclic alkaloids made up of the parent scaffold acridin-9(10*H*)-one (**20**), which has a carbonyl group at C-9 position and nitrogen atom at position 10 of ring B (Alipour et al., 2014). The majority of acridone alkaloids found in nature possess an oxygen functionality in the C-1 and C-3 positions and are nitrogen-methylated. These OH groups might be free, alkylated, or joined to furan or pyran rings (Michael, 2017). Acridones can be divided into two major categories. The first class consists of acridine alkaloids that have rings A or B oxygenated. Acyclic or cyclic C-5 side chains can be found in ring A of the second group (Schmidt & Liu, 2015).

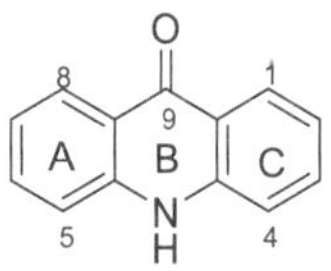

20

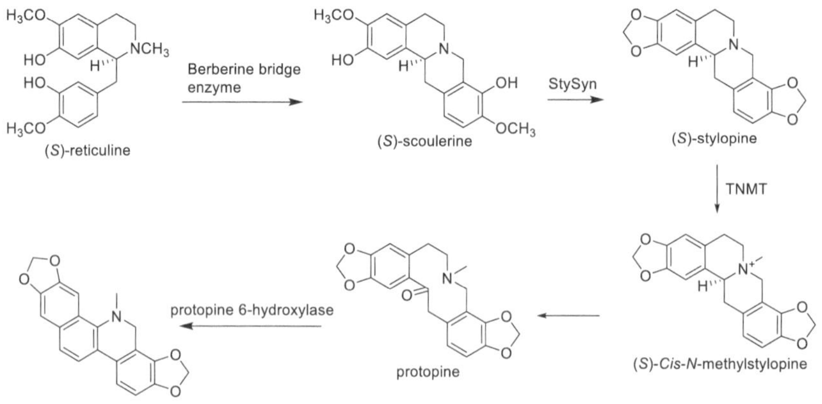

Scheme 3 Biosynthetic pathway of benzophenanthridine alkaloids.

From a biogenetic approach, the tricyclic nucleus, which is typically oxygenated at C-1 and C-3, is produced by the condensation of an anthranilic acid unit with three acetate units, which subsequently forms the 9(10*H*)-acridinone skeleton. Choi et al. have well described the biosynthesis process for this class of alkaloids with the use of a plant type III polyketide synthase that is heterologously expressed in *Escherichia coli* (Scheme 4) (Choi, Choo, Kim, & Ahn, 2020). Briefly, anthranilate *N*-methyltransferase (ANMT) converts anthranilate to *N*-methylanthranilate, which is the first significant step in the biosynthesis of acridone alkaloids. The subsequent step involves utilizing coenzyme A (CoA) to create *N*-methylanthraniloyl-CoA. The condensation of *N*-methylanthraniloyl-CoA and malonyl-CoA is then catalyzed by acridone synthase (ACS), one of the plant polyketide synthases (PKSs) (Choi et al., 2020; Schmidt & Liu, 2015).

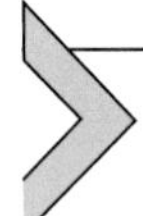

3. Phytochemical characterization of alkaloids from a crude plant extract

Alkaloid identification frequently uses conventional separation and analysis techniques, with hyphenated techniques like liquid chromatography–mass spectrometry (HPLC-MS) and gas chromatography–mass spectrometry (GC-MS) having special value. Since an alkaloid-bearing plant normally contains a complex mixture of many alkaloids with other bioactive compounds, alkaloid isolation, and structural elucidation are typically laborious (Talaty, Takáts, & Cooks, 2005). The qualitative and quantitative analysis of

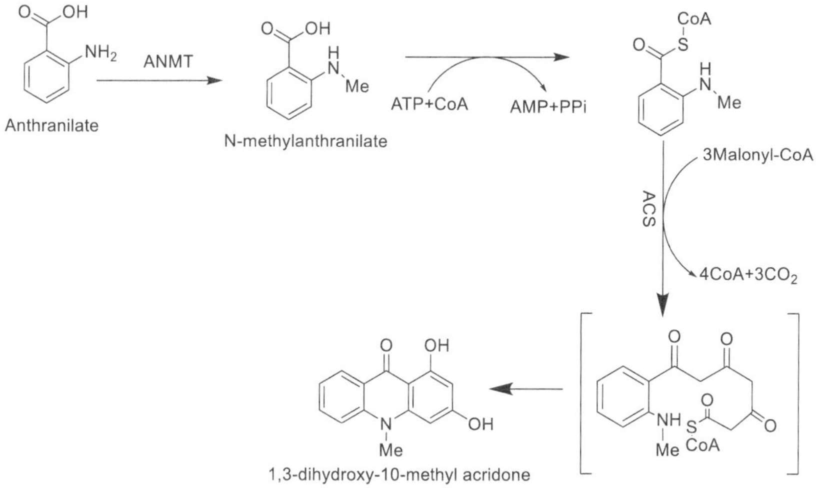

Scheme 4 Biosynthesis pathway for generating acridone alkaloids (Choi et al., 2020).

these chemicals in a range of sample matrices has been accomplished using a range of different chromatographic techniques. For instance, thin-layer chromatography (TLC) offers a low-cost analytical technique that is particularly helpful for qualitative analysis. The main benefit of TLC is the availability of specialized spray reagents that produce distinct colors with various alkaloids and can serve as confirmation for qualitative analyzes based on retardation factors (McCalley, 2002). Dragendorff's, Mayer's and Marquis reagents are some of the most frequently used detection reagents for alkaloids (Wansi et al., 2013). Other analytical techniques, including capillary electrophoresis, UV, fluorescence, and mass spectrometry detection, as well as reversed phase HPLC using octadecylsilyl (ODS) columns in combination with acidic mobile phases, have also been utilized extensively. Alkaloids' structural elucidation frequently involves the use of spectroscopic methods like nuclear magnetic spectroscopy (NMR) and circular dichroism (CD) (Talaty et al., 2005; Unger & Stöckigt, 1997; Wang et al., 2014).

One of the most significant cytotoxic alkaloids reported from African medicinal plants are naphthylisoquinoline alkaloids (Awale et al., 2018; Fayez et al., 2018a, 2018b, 2021, Kavatsurwa et al., 2018; Li et al., 2017a, 2017b; Lombe, Feineis, & Bringmann, 2019; Tshitenge et al., 2018, 2019). Typically, they are mostly extracted from plant crude extracts with CH_2Cl_2, CH_3OH, 95% C_2H_5OH, CH_2Cl_2/CH_3OH (1:1 or 6:4), CH_2Cl_2/NH_3, or acidified water (2% HCl) following defatting the extract with petroleum ether. Naphthylisoquinolines are primarily purified using preparative HPLC with (A) CH_3CN/H_2O (9:1) + 0.05% TFA and (B) H_2O/CH_3CN (9:1) + 0.05% TFA, or (A) CH_3OH/H_2O (9:1) + 0.05% TFA and (B) H_2O/CH_3OH (9:1) + 0.05% TFA, based on the resolution of the separation on a reverse-phase column. Other techniques for separating naphthylisoquinolines include recrystallization, multilayer counter current chromatography, preparative TLC, centrifugal partition chromatography, cation-exchange chromatography (Amberlyst-15), high-speed counter-current chromatography, and column chromatography on silica gel (Bringmann et al., 1998; Xu et al., 2010).

4. Cancer cell lines used to assess the cytotoxicity of alkaloids isolated from African medicinal plants

The following cell lines were investigated: human malignant melanoma (SK-MEL, MV4–11, UACC-257, INA-6), epidermoid (KB), ductal (BT-549), kidney epithelial (LLC-PK11), lung (MCF-7, BT549, HCC 1395, H460), breast (HCC, HepG2, HCC 1395), liver (A549, HuH-7), prostate (PC-3, DU 145), ovary (OVCAR-3, A2780, Ovcar-8, A2780, Ovcar-4, Igrov-1, SK-OV-3), colon (HT-29, HCT116, SW620), neuroblastoma (SH-SY5Y, HeLa), noncancerous fibroblast (KMST-6), drug sensitive leukemia (CCRF-CEM) and its (CEM/ADR5000) MDR P-glycoprotein-over-expressing subline, breast (MDAMB-231-*pcDNA3*) and its resistant subline (MDA-MB-231-*BCRP* clone 23), colon (HCT116 $p53^{+/+}$) and its knockout clone (HCT116 $p^{53-/-}$), glioblastoma (U87MG) and its resistant subline (U87MG.Δ*EGFR*), mouse lymphoma cell lines (PAR-L5178) and its MDR-L5178 cell line, mastocytoma (P815), and laryngeal (Hep).

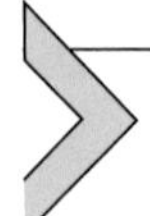

5. Cytotoxic alkaloids isolated from the African medicinal plants towards drug sensitive and MDR cancer cells

Over the past decades, substantial efforts have been made by diverse research groups in Africa as part of their investigation campaign in the fight against cancers. In this regard, numerous plant-based secondary metabolites *inter alia*, alkaloids, have been isolated and characterized. Their cytotoxicity was further investigated against a panel of cancer cell lines, including drug sensitive and their multidrug-resistant (MDR) sublime. All alkaloids reported in this section were obtained from African medicinal plants; their origin is summarized in Table 1, while the chemical structures are found in Fig. 3. Based on our literature survey, 65 structurally diverse naphthyldihydroisoquinolines were purified from the various tissues of *Ancistracladus* species. For instance, ancistrolikokine E_3 (**21**), a 5,8′-coupled naphthylisoquinoline isolated from the Congolese liana *A. likoko*, demonstrated great preferential cytotoxicity against Panc-1 (human pancreatic) cell line with a PC_{50} value of 2.5 μM without significant toxicity on normal cells. Mechanistically, the mitochondrial damage pathway was also suggested to be involved in ancistrolikokine E_3-induced apoptosis. Moreover, it inhibited significantly PANC-1 cell migration and colony formation in a concentration-dependent manner (Awale et al., 2018). Furthermore, within, this subgroup of alkaloids, another set of 5,8′-coupled naphthylisoquinolines from the roots of the West African liana *A. abbreviatus* was found to be cytotoxic against human PANC-1 pancreatic cancer cells under nutrient-derived conditions with PC_{50} ranging from 7.50 to 45.5 μM (Fayez et al., 2018b). The most cytotoxic alkaloids of *A. likoko* displayed 5,1′-linkage and were identified as 6-*O*-methyl- 4′-*O*-demethylhamatine (**22**, PC_{50} = 7.50 μM) and 5-*epi*-ancistectorine A_2 (**23**, PC_{50} = 11.5 μM). In the same line, other small library of 5,1′-alkaloids and their atropodiastereomers (**24–33**) exhibited moderate to weak antiausterity activities (Fig. 3). Few years later, ancistrobrevidines A-C (**34–36**) and ancistrobreviquinone A (**37**), bearing a 3, 4-naphthoquinone core coupled to a tetrahydroisoquinoline fragment were purified from the leaves and roots of *A. abbreviatus*, respectively (Fayez et al., 2021). From a biological standpoint, ancistrobrevidine C (**36**), showed cytotoxicity towards HeLa cells in a concentration-dependent manner and induced dramatic alteration in cell morphology, leading to cell death with IC_{50} of 7.2 μM against HeLa and PC_{50} of 13.9 μM towards PANC-1. Structurally related analogs (**34–35**), and hybrid-type (**37**) exerted disappointing results against both cell lines, therefore were considered inactive (Fayez et al., 2021).

Benzophenanthridine which represents one of 11 classes of benzylisoquinoline alkaloids represents the second class of alkaloids, largely found in plants growing in Africa. Plants from the family Rutaceae in general and from the genus *Zanthoxylum* have widely been reported as promising sources of bioactive compounds against recalcitrant cell lines. The antineoplastic potency of *Z. paracanthum*, a plant endemic in Kenya was reported on various drug-resistant cancer cell lines. The active ingredients were identified as dihydrochelerythrine (**86**), 6-hydroxymethyldihydronitidine (**87**), and bis-[6-(5,6-dihydrochelerythrinyl)] ether (**88**) displaying potent

Table 1 Cytotoxic alkaloids from African medicinal plants and their effects on sensitive and drug resistant cancer cell lines.

Compounds names	Source (Plant part)	Country	Cancer Cell Lines and IC_{50} values (μM) and degree of resistance (D.R.) in bracket	References
Naphthylisoquinoline				
4′-*O*-demethylancistrocladine (**78**)	*Ancistrocladus sp* (Stem bark)	DR Congo	11.2 (PANC-1)	Kavatsurwa et al. (2018)
4′-*O*-demethyl-7-epi-dioncophylline A (**63**)	*Ancistrocladus ileboensis* (Root bark, leaves)	DR Congo	16.0 (INA-6)	Li et al. (2017a)
4′-*O*-demethyldioncophylline A (**62**)	*Ancistrocladus ileboensis* (Root bark, leaves)	DR Congo	2.7 (INA-6)	Li et al. (2017a)
5′-*O*-methyldioncophylline D (**59**)	*Ancistrocladus ileboensis* (Root bark, leaves)	DR Congo	2.6 (INA-6); 1.85 (CCRF-CEM) *vs* 5.31 (CEM/ADR5000) (D.R.: 2.9)	Li et al. (2017a)
6,5′-*O*,*O*-Didemethyl ancistroealaine A (**83**)	*Ancistrocladus sp* (Stem bark)	DR Congo	9.8 (PANC-1)	Kavatsurwa et al. (2018)
6-*O*-Demethylancistroealaine A (**84**)	*Ancistrocladus sp* (Stem bark)	DR Congo	14.0 (PANC-1)	Kavatsurwa et al. (2018)
6-*O*-methylhamateine (**51**)	*Ancistrocladus abbreviatus* (Roots)	Ivory Coast	3.94 (CCRF-CEM) *vs* 5.52 (CEM/ADR5000) (D.R.: 1.40)	Fayez et al. (2019)
6-*O*-Methylhamatine (**77**)	*Ancistrocladus sp* (Stem bark)	DR Congo	31.9 (PANC-1)	Kavatsurwa et al. (2018)

7-*Epi*-ancistrobrevine D (**85**)	*Ancistrocladus sp* (Stem bark)	DR Congo	29.9 (PANC-1)	Kavatsurwa et al. (2018)
Ancistrobertsonine A (**80**)	*Ancistrocladus sp* (Stem bark)	DR Congo	11.8 (PANC-1)	Kavatsurwa et al. (2018)
Ancistrobreveine A (**54**)	*Ancistrocladus abbreviatus* (Roots)	Ivory Coast	24.78 (CCRF-CEM) *vs* >100 (CEM/ADR5000)	Fayez et al. (2019)
Ancistrobreveine B (**55**)	*Ancistrocladus abbreviatus* (Roots)	Ivory Coast	39.94 (CCRF-CEM) *vs* 50.64 (CEM/ADR5000) (D.R.: 1.27)	Fayez et al. (2019)
Ancistrobreveine C (**56**)	*Ancistrocladus abbreviatus* (Roots)	Ivory Coast	12.44 (CCRF-CEM) *vs* 12.64 (CEM/ADR5000) (D.R.: 1.02)	Fayez et al. (2019)
Ancistrobreveine D (**52**)	*Ancistrocladus abbreviatus* (Roots)	Ivory Coast	24.27 (CCRF-CEM) *vs* 29.51 (CEM/ADR5000) (D.R.: 1.21)	Fayez et al. (2019)
Ancistrobrevine B (**81**)	*Ancistrocladus sp* (Stem bark)	DR Congo	20.2 (PANC-1)	Kavatsurwa et al. (2018)
Ancistrocladisine A (**60**)	*Ancistrocladus ileboensis* (Root bark, Leaves)	DR Congo	4.8 (INA-6)	Li et al. (2017a)
Ancistroealaine F (**38**)	*Ancistrocladus likoko* (Twigs)	DR Congo	11.69 (CCRF-CEM) *vs* 19.94 (CEM/ADR5000) (D.R.: 1.7)	Fayez et al. (2018a)
Ancistroguineine A (**79**)	*Ancistrocladus sp* (Stem bark)	DR Congo	15.8 (PANC-1)	Kavatsurwa et al. (2018)

(*continued*)

Table 1 Cytotoxic alkaloids from African medicinal plants and their effects on sensitive and drug resistant cancer cell lines. (*cont'd*)

Compounds names	Source (Plant part)	Country	Cancer Cell Lines and IC_{50} values (μM) and degree of resistance (D.R.) in bracket	References
Ancistrolikokine C (**42**)	*Ancistrocladus likoko* (Twigs)	DR Congo	21.62 (CCRF-CEM) *vs* 25.47 (CEM/ADR5000) (D.R.: 1.18)	Fayez et al. (2018a)
Ancistrolikokine C_2 (**43**)	*Ancistrocladus likoko* (Twigs)	DR Congo	23.63 (CCRF-CEM) *vs* 49.95 (CEM/ADR5000) (D.R.: 2.11)	Fayez et al. (2018a)
Ancistrolikokine C_3 (**45**)	*Ancistrocladus likoko* (Twigs)	DR Congo	52.46 (CCRF-CEM) *vs* >100 (CEM/ADR5000)	Fayez et al. (2018a)
Ancistrolikokine E (**39**)	*Ancistrocladus likoko* (Twigs)	DR Congo	>100 (CCRF-CEM) *vs* >100 (CEM/ADR5000)	Fayez et al. (2018a)
Ancistrolikokine E_2 (**40**)	*Ancistrocladus likoko* (Twigs)	DR Congo	24.13 (CCRF-CEM) *vs* 82.78 (CEM/ADR5000) (D.R.: 3.43)	Fayez et al. (2018a)
Ancistrolikokine E_3 (**21**)	*Ancistrocladus likoko* (Twigs)	DR Congo	4.36 (CCRF-CEM) *vs* 18.34 (CEM/ADR5000) (D.R.: 4.22)	Fayez et al. (2018a)
Ancistrolikokine G (**41**)	*Ancistrocladus likoko* (Twigs)	DR Congo	4.73 (CCRF-CEM) *vs* 7.73 (CEM/ADR5000) (D.R.: 1.63)	Fayez et al. (2018a)
Ancistrolikokine I (**44**)	*Ancistrocladus likoko* (Twigs)	DR Congo	4.48 (CCRF-CEM) *vs* 26.48 (CEM/ADR5000) (D.R.: 5.90)	Fayez et al. (2018a)

Ancistrolikokine J (**46**)	*Ancistrocladus likoko* (Twigs)	DR Congo	24.89 (CCRF-CEM) *vs* 30.16 (CEM/ADR5000) (D.R.: 1.21)	Fayez et al. (2018a)
Ancistrolikokine J_2 (**47**)	*Ancistrocladus likoko* (Twigs)	DR Congo	23.34 (CCRF-CEM) *vs* 20.06 (CEM/ADR5000) (D.R.: 0.86)	Fayez et al. (2018a)
Ancistrotectoriline A (**82**)	*Ancistrocladus sp* (Stem bark)	DR Congo	67.8 (PANC-1)	Kavatsurwa et al. (2018)
Ancistroyafungine A (**73**)	*Ancistrocladus sp* (Stem bark)	DR Congo	22.7 (PANC-1)	Kavatsurwa et al. (2018)
Ancistroyafungine B (**74**)	*Ancistrocladus sp* (Stem bark)	DR Congo	7.6 (PANC-1)	Kavatsurwa et al. (2018)
Ancistroyafungine C (**75**)	*Ancistrocladus sp* (Stem bark)	DR Congo	15.0 (PANC-1)	Kavatsurwa et al. (2018)
Ancistroyafungine D (**76**)	*Ancistrocladus sp* (Stem bark)	DR Congo	9.7 (PANC-1)	Kavatsurwa et al. (2018)
Dioncophylline A (**61**)	*Ancistrocladus ileboensis* (Root bark, Leaves)	DR Congo	0.22 (INA-6); 0.24 (CCRF-CEM) *vs* 0.52 (CEM/ADR5000) (D.R.: 2.1)	Li et al. (2017a)
Dioncophylline D_2 (**58**)	*Ancistrocladus ileboensis* (Root bark, Leaves)	DR Congo	32.0 (INA-6)	Li et al. (2017a)
Dioncophylline F (**57**)	*Ancistrocladus ileboensis* (Root bark, Leaves)	DR Congo	21.0 (INA-6)	Li et al. (2017a)

(continued)

Table 1 Cytotoxic alkaloids from African medicinal plants and their effects on sensitive and drug resistant cancer cell lines. (*cont'd*)

Compounds names	Source (Plant part)	Country	Cancer Cell Lines and IC_{50} values (μM) and degree of resistance (D.R.) in bracket	References
Ealamine A (**67**)	*Ancistrocladus ealaensis* (Twigs, Leaves)	DR Congo	12.5 (PANC-1)	Tshitenge et al. (2019)
Ealamine B (**68**)	*Ancistrocladus ealaensis* (Twigs, Leaves)	DR Congo	30.6 (PANC-1)	Tshitenge et al. (2019)
Ealamine C (**69**)	*Ancistrocladus ealaensis* (Twigs, Leaves)	DR Congo	9.9 (PANC-1)	Tshitenge et al. (2019)
Ealamine D (**70**)	*Ancistrocladus ealaensis* (Twigs, Leaves)	DR Congo	60.4 (PANC-1)	Tshitenge et al. (2019)
Ealamine G (**71**)	*Ancistrocladus ealaensis* (Twigs, Leaves)	DR Congo	33.4 (PANC-1)	Tshitenge et al. (2019)
Ent-dioncophylleine A (**53**)	*Ancistrocladus abbreviatus* (Roots)	Ivory Coast	26.88 (CCRF-CEM) *vs* 31.41 (CEM/ADR5000) (D.R.: 1.17)	Fayez et al. (2019)
Jozilebomine A (**64**)	*Ancistrocladus ileboensis* (Root bark)	DR Congo	1.08 (HeLa); 2.24 (PANC-1)	Li et al. (2017b)
Jozilebomine B (**65**)	*Ancistrocladus ileboensis* (Root bark)	DR Congo	0.68 (HeLa); 0.87 (PANC-1)	Li et al. (2017b)

Jozimine A_2 (**66**)	*Ancistrocladus ileboensis* (Root bark)	DR Congo	0.22 (HeLa); 0.10 (PANC-1)	Li et al. (2017b)
Mbandakamine A (**43**)	*Ancistrocladus ealaensis* (Leaves)	DR Congo	7.40 (CCRF-CEM) *vs* 23.88 (CEM/ADR5000) (D.R.: 3.2)	Tshitenge et al. (2018)
Mbandakamine C (**49**)	*Ancistrocladus ealaensis* (Leaves)	DR Congo	1.49 (CCRF-CEM) *vs* 27.71 (CEM/ADR5000) (D.R.: 18.5)	Tshitenge et al. (2018)
Mbandakamine D (**50**)	*Ancistrocladus ealaensis* (Leaves)	DR Congo	2.95 (CCRF-CEM) *vs* 19.03 (CEM/ADR5000) (D.R.: 6.4)	Tshitenge et al. (2018)
Yaoundamine A (**72**)	*Ancistrocladus ealaensis* (Twigs, Leaves)	DR Congo	37.8 (PANC-1)	Tshitenge et al. (2019)
Benzophenanthridine				
8-Acetonyldihydrochelerythrine (**93**)	*Zanthoxylum paracanthum* (Root bark)	Kenya	25 (HCC 1395); 165 (DU 145)	Kaigongi et al. (2020)
8-Oxochelerythrine (**95**)	*Zanthoxylum paracanthum* (Root bark)	Kenya	39 (HCC 1395); 175 (DU 145)	Kaigongi et al. (2020)
Arnottianamide (**94**)	*Zanthoxylum paracanthum* (Root bark)	Kenya	94 (HCC 1395); 207 (DU 145)	Kaigongi et al. (2020)

(continued)

Table 1 Cytotoxic alkaloids from African medicinal plants and their effects on sensitive and drug resistant cancer cell lines. (*cont'd*)

Compounds names	Source (Plant part)	Country	Cancer Cell Lines and IC_{50} values (μM) and degree of resistance (D.R.) in bracket	References
Buesgenine (**96**)	*Zanthoxylum buesgenii* (Aerial)	Cameroon	0.24 (CCRF-CEM) *vs* 31.58 (CEM/ADR5000) (D.R.: 131.5); 30.14 (MDA-MB231-*pcDNA*) *vs* 65.01 (MDA-MB231/*BCRP*) (D.R.: 2.16); 42.46 (HCT116 $p53^{+/+}$) *vs* 62.34 (HCT116 $p53^{-/-}$) (D.R.: 1.47); 60.55 (U87MG) *vs* 61.84 (U87MG.Δ*EGFR*) (D.R.: 1.02)	Sandjo et al. (2014)
Fagaridine chloride (**100**)	*Fagara tessmannii* (Bark)	Cameroon	1.69 (CCRF-CEM) *vs* 9.77 (CEM/ADR5000) (D.R.: 5.78); 13.32 (MDA-MB231-*pcDNA*) *vs* 16.42 (MDA-MB231/*BCRP*) (D.R.: 1.23); 18.60 (HCT116 $p53^{+/+}$) *vs* 14.45 (HCT116 $p53^{-/-}$) (D.R.: 0.78); 9.27 (U87MG) *vs* 13.13 (U87MG.Δ*EGFR*) (D.R.: 1.42)	Mbaveng et al. (2019a)
Isofagaridine (**97**)	*Zanthoxylum buesgenii* (Aerial)	Cameroon	0.30 (CCRF-CEM) *vs* 20.37 (CEM/ADR5000) (D.R.: 67.9); 41.38 (MDA-MB231-*pcDNA*) *vs* 113.98 (MDA-MB231/*BCRP*) (D.R.: 2.75); 87.08 (HCT116 $p53^{+/+}$) *vs* >119.76 HCT116 ($p53^{-/-}$); 105.19 (U87MG) *vs* 115.30 (U87MG.Δ*EGFR*) (D.R.: 1.1)	Sandjo et al. (2014)

Nitidine chloride (**99**)	*Fagara tessmannii* (Bark)	Cameroon	1.75 (CCRF-CEM) *vs* 6.62 (CEM/ADR5000) (D.R.: 3.78); 6.28 (MDA-MB231-*pcDNA*) *vs* 3.87 (MDA-MB231/*BCRP*) (D.R.: 0.62); 14.38 (HCT116 $p53^{+/+}$) *vs* 17.87 (HCT116 $p53^{-/-}$) (D.R.: 1.24); 4.20 (U87MG) *vs* 23.52 (U87MG.Δ*EGFR*) (D.R.: 5.60)	Mbaveng et al. (2019a)
Zanthoxyline (**98**)	*Fagara tessmannii* (Bark)	Cameroon	97.74 (CCRF-CEM) *vs* 78.89 (CEM/ADR5000) (D.R.: 0.81); 91.20 (MDA-MB231-*pcDNA*) *vs* 95.83 (MDA-MB231/*BCRP*) (D.R.: 1.05); >100 (HCT116 $p53^{+/+}$) *vs* 72.56 (HCT116 $p53^{-/-}$) (D.R.: <0.73); 59.19 (U87MG) *vs* >100 (U87MG.Δ*EGFR*) (D.R.: >1.69)	Mbaveng et al. (2019a)
Furoquinoline				
Kokusaginine B (**108**)	*Araliopsis soyauxii* (Stem bark)	Cameroon	89.54 (CCRF-CEM) *vs* >154.44 (CEM/ADR5000) (D.R.: >1.72)	Nganou et al. (2019)

(*continued*)

Table 1 Cytotoxic alkaloids from African medicinal plants and their effects on sensitive and drug resistant cancer cell lines. (*cont'd*)

Compounds names	Source (Plant part)	Country	Cancer Cell Lines and IC_{50} values (μM) and degree of resistance (D.R.) in bracket	References
Montrifoline (**109**)	*Oricia suaveolens* (Roots)	Cameroon	46.83 (CCRF-CEM) *vs* 74.06 (CEM/ADR5000) (D.R.: 1.58); 53.34 (MDA-MB231-*pcDNA*) *vs* 76.71 (MDA-MB231/*BCRP*) (D.R.: 1.44); 59.22 (HCT116 $p53^{+/+}$) *vs* 90.63 (HCT116 $p53^{-/-}$) (D.R.: 1.53); 90.36 (U87MG) *vs* 66.69 (U87MG.Δ*EGFR*) (D.R.: 0.74)	Kuete et al. (2015a)
Acridone				
1,3-Dimethoxy-10-methylacridone (**113**)	*Oricia suaveolens* (Roots)	Cameroon	12.08 (CCRF-CEM) *vs* 58.10 (CEM/ADR5000) (D.R.: 4.81); 10.30 (MDA-MB231-*pcDNA*) *vs* 3.38 (MDA-MB231/*BCRP*) (D.R.: 0.33); 9.07 (HCT116 $p53^{+/+}$) *vs* 8.81 (HCT116 $p53^{-/-}$) (D.R.: 0.97); 17.55 (U87MG) *vs* 8.88 (U87MG.Δ*EGFR*) (D.R.: 0.51)	Kuete et al. (2015a)

1-Hydroxy-4-methoxy-10-methylacridone (**111**)	*Oricia suaveolens* (Roots)	Cameroon	17.06 (CCRF-CEM) *vs* 43.80 (CEM/ADR5000) (D.R.: 5.57); 106.47 (MDA-MB231-*pcDNA*) *vs* 18.82 (MDA-MB231/*BCRP*) (D.R.: 0.18); 17.33 (HCT116 $p53^{+/+}$) *vs* 6.78 (HCT116 $p53^{-/-}$) (D.R.: 0.39); 53.25 (U87MG) *vs* 11.33 (U87MG.Δ*EGFR*) (D.R.: 0.21)	Kuete et al. (2015a)
5-Hydroxynoracronycine (**117**)	*Citrus aurantium* (Stem bark)	Nigeria	24.03 (MCF7); 25.94 (A549); 28.10 (PC3); 30.09 (HepG2); 190.3 (PNT2)	Segun et al. (2018)
Arborinine (**114**)	*Uapaca togoensis* (Woods)	Cameroon	31.77 (CCRF-CEM) *vs* 3.55 (CEM/ADR5000) (D.R.: 0.11); 8.88 (MDA-MB231-*pcDNA*) *vs* 7.76 (MDA-MB231/*BCRP*) (D.R.: 0.87); 6.01 (HCT116 $p53^{+/+}$) *vs* 8.67 (HCT116 $p53^{-/-}$) (D.R.: 1.44); 20.41 (U87MG) *vs* 6.89 (U87MG.Δ*EGFR*) (D.R.: 0.34)	Kuete et al. (2015c)
Citracridone-I (**116**)	*Citrus aurantium* (Stem bark)	Nigeria	12.65 (MCF7); 14.02 (A549); 14.88 (PC3); 22.42 (HepG2); 151.5 (PNT2)	Segun et al. (2018)
Citracridone-III (**120**)	*Citrus aurantium* (Stem bark)	Nigeria	20.91 (MCF7); 20.36 (A549); 18.23 (PC3); 32.62 (HepG2); 175.4 (PNT2)	Segun et al. (2018)

(*continued*)

Table 1 Cytotoxic alkaloids from African medicinal plants and their effects on sensitive and drug resistant cancer cell lines. (*cont'd*)

Compounds names	Source (Plant part)	Country	Cancer Cell Lines and IC_{50} values (μM) and degree of resistance (D.R.) in bracket	References
Citrusinine-I (**115**)	*Citrus aurantium* (Stem bark)	Nigeria	44.53 (MCF7); 35.76 (A549); 38.23 (PC3); 50.74 (HepG2); 179 (PNT2)	Segun et al. (2018)
Evoxanthine (**110**)	*Oricia suaveolens* (Roots)	Cameroon	24.59 (CCRF-CEM) *vs* 33.53 (CEM/ADR5000) (D.R.: 1.36); 14.45 (MDA-MB231-*pcDNA*) *vs* 30.95 (MDA-MB231/*BCRP*) (D.R.: 2.14); 6.11 (HCT116 $p53^{+/+}$) *vs* 11.63 (HCT116 $p53^{-/-}$) (D.R.: 1.90); 37.63 (U87MG) *vs* 16.89 (U87MG.Δ*EGFR*) (D.R.: 0.45)	Kuete et al. (2015a)
Glycofolinine (**119**)	*Citrus aurantium* (Stem bark)	Nigeria	26.87 (MCF7); 25.02 (A549); 24.29 (PC3); 26.38 (HepG2); 222.7 (PNT2)	Segun et al. (2018)
Natsucitrine-I (**118**)	*Citrus aurantium* (Stem bark)	Nigeria	31.11 (MCF7); 29.69 (A549); 27.44 (PC3); 37.45 (HepG2); 205.1 (PNT2)	Segun et al. (2018)
Norevoxanthine (**112**)	*Oricia suaveolens* (Roots)	Cameroon	137.62 (CCRF-CEM) *vs* 103.35 (CEM/ADR5000) (D.R.: 0.75); 73.27 (MDA-MB231-*pcDNA*) *vs* 86.02 (MDA-MB231/*BCRP*) (D.R.: 1.17); 20.86 (HCT116 $p53^{+/+}$) *vs* 60.59 (HCT116 $p53^{-/-}$) (D.R.: 2.90); 89.29 (U87MG) *vs* 5.72 (U87MG.Δ*EGFR*) (D.R.: 0.06)	Kuete et al. (2015a)

Indole

(19′*S*)-Hydroxytabernaelegantine A (**125**)	*Tabernaemontana elegans* (Roots)	Mozambique	8.4 (HCT116)	Paterna et al. (2016a)
3′-Oxotabernaelegantine C (**126**)	*Tabernaemontana elegans* (Roots)	Mozambique	>10 (HCT116); >10 (HepG2)	Paterna et al. (2016a)
3′-Oxotabernaelegantine D (**127**)	*Tabernaemontana elegans* (Roots)	Mozambique	8.8 (HCT116); >10 (HepG2)	Paterna et al. (2016a)
Soyauxinium chloride (**132**)	*Araliopsis soyauxii* (Stem bark)	Cameroon	3.64 (CCRF-CEM) *vs* 4.64 (CEM/ADR5000) (D.R.: 1.17); 4.83 (MDA-MB231-*pcDNA*) *vs* 4.93 (MDA-MB231/*BCRP*) (D.R.: 1.02); 6.01 (HCT116 $p53^{+/+}$) *vs* 5.22 (HCT116 $p53^{-/-}$) (D.R.: 0.87); 9.87 (U87MG) *vs* 9.54 (U87MG.Δ*EGFR*) (D.R.: 0.97)	Mbaveng et al. (2021b)
Tabernaelegantine A (**128**)	*Tabernaemontana elegans* (Roots)	Mozambique	>10 (HCT116); >10 (HepG2)	Paterna et al. (2016a)
Tabernaelegantine D (**129**)	*Tabernaemontana elegans* (Roots)	Mozambique	8.5 (HCT116); >10 (HepG2)	Paterna et al. (2016a)
Tabernine A (**133**)	*Tabernaemontana elegans* (Leaves)	Mozambique	45.9 (PAR-L5178); 37.5 (MDR-L5178)	Mansoor et al. (2009a)

(continued)

Table 1 Cytotoxic alkaloids from African medicinal plants and their effects on sensitive and drug resistant cancer cell lines. (*cont'd*)

Compounds names	Source (Plant part)	Country	Cancer Cell Lines and IC_{50} values (μM) and degree of resistance (D.R.) in bracket	References
Tabernine B (**134**)	*Tabernaemontana elegans* (Leaves)	Mozambique	46.6 (PAR-L5178); 39.7 (MDR-L5178)	Mansoor et al. (2009a)
Tabernine C (**135**)	*Tabernaemontana elegans* (Leaves)	Mozambique	70.6 (PAR-L5178); 51.5 (MDR-L5178)	Mansoor et al. (2009a)
Bisbenzylisoquinoline				
2′-Norcocsuline (**143**)	*Triclisia subcordata* (Roots)	Nigeria	1.9 (Ovcar-8); 2.9 (A2780); 6.2 (Ovcar-4); 0.8 (Igrov-1)	Uche et al. (2017)
Cycleanine (**140**)	*Triclisia subcordata* (Root bark)	Nigeria	10 (Ovcar-8); 7.6 (A2780); 14 (Ovcar-4); 7.2 (Igrov-1)	Uche et al. (2016)
Isochondodendrine (**142**)	*Triclisia subcordata* (Roots)	Nigeria	8.0 (Ovcar-8); 3.5 (A2780); 17 (Ovcar-4); 9.0 (Igrov-1)	Uche et al. (2017)
Tetrandrine (**141**)	*Triclisia subcordata* (Root bark)	Nigeria	12 (Ovcar-8); 7.7 (A2780); 6.9 (Ovcar-4); 9.3 (Igrov-1)	Uche et al. (2016)
Steroidal				
Funtumine (**146**)	*Holarrhena floribunda* (Leaves)	Nigeria	22.36 (HT-29); 46.17 (HeLa); 52.69 (MCF-7); 85.45 (KMST-6)	Badmus et al. (2019)

Holamine (**145**)	*Holarrhena floribunda* (Leaves)	Nigeria	31.06 (HT-29); 51.42 (HeLa); 42.88 (MCF-7); 102.95 (KMST-6)	Badmus et al. (2019)
β-Carboline				
10-Methoxycanthin-6-one (**147**)	*Zanthoxylum paracanthum* (Root bark)	Kenya	59 (HCC1395); 6 (DU145); 216 (Vero E6)	Kaigongi et al. (2020)
Canthin-6-one (**148**)	*Zanthoxylum paracanthum* (Root bark)	Kenya	37 (HCC1395); 43 (DU145); 190 (Vero E6)	Kaigongi et al. (2020)
Canthin-6-one (**148**)	*Zanthoxylum paracanthum* (Stem bark)	Kenya	15.82 (CCRF-CEM) *vs* 10.52 (CEM/ADR5000) (D.R.: 0.66)	Omosa et al. (2021)
Miscellaneous				
11-Methoxyerysodine (**164**)	*Erythrina abyssinica* (Seeds)	Sudan	35 (Hep-G2); 35 (HEP-2)	Mohammed et al. (2012)
2*R*-Bgugaine (**153**)	*Arisarum vulgare* (Tubers)	Morocco	36 (P815); 18 (Hep)	Benamar et al. (2009)
2*S*-Bgugaine (**154**)	*Arisarum vulgare* (Tubers)	Morocco	18 (P815); 355 (Hep)	Benamar et al. (2009)
4-(Isoprenyloxy)-3-methoxy-3,4-deoxymethylenedioxyfagaramide (**152**)	*Zanthoxylum paracanthum* (Stem bark)	Kenya	29.13 (CCRF-CEM) *vs* 31.00 (CEM/ADR5000) (D.R.: 1.06)	Omosa et al. (2021)
4-Methoxy-1-methyl-2(1*H*) quinolinone (**150**)	*Araliopsis soyauxii* (Bark)	Cameroon	51.53 (CCRF-CEM) *vs* >100 (CEM/ADR5000) (D.R.: >1.94); 88.36 (MDA-MB231-*pcDNA*) *vs* >100 (MDA-MB231/*BCRP*) (D.R.: >1)	Mbaveng et al. (2021a)

(continued)

Table 1 Cytotoxic alkaloids from African medicinal plants and their effects on sensitive and drug resistant cancer cell lines. (*cont'd*)

Compounds names	Source (Plant part)	Country	Cancer Cell Lines and IC_{50} values (μM) and degree of resistance (D.R.) in bracket	References
5,14-Dimethylbudmunchiamine L1 (**165**)	*Albizia schimperiana* (Stem bark)	Kenya	4 (SK-MEL); 4 (KB); 4 (BT-549); 4 (SK-OV-3); 3 (LLC-PK11)	Samoylenko, et al. (2009)
5-Normethylbudmunchiamine K (**167**)	*Albizia schimperiana* (Stem bark)	Kenya	4 (SK-MEL); 5 (KB); 4 (BT-549); 10 (SK-OV-3); 5 (LLC-PK11)	Samoylenko, et al. (2009)
6-Hydroxy-5-normethyl budmunchiamine K (**168**)	*Albizia schimperiana* (Stem bark)	Kenya	11 (SK-MEL); 11 (KB); 12 (BT-549); 12 (SK-OV-3); 12 (LLC-PK11)	Samoylenko, et al. (2009)
6-Hydroxybudmunchiamine K (**166**)	*Albizia schimperiana* (Stem bark)	Kenya	11 (SK-MEL); 11 (KB); 11 (BT-549); 12 (SK-OV-3); 10 (LLC-PK11)	Samoylenko, et al. (2009)
6-Hydroxycrinamine (**169**)	*Crinum abyscinicum* (Bulbs)	Ethiopia	9 (A2780); 17 (MV4-11)	Abebe et al. (2020)
8-Oxoerythraline (**163**)	*Erythrina abyssinica* (Seeds)	Sudan	12 (Hep-G2); 59 (HEP-2)	Mohammed et al. (2012)
Albomaculine (**156**)	*Haemanthus humilis* (Bulbs)	South Africa	>50 (A549); >50 (HCT-15); >50 (SK-MEL-28); >50 (MCF7); >50 (MDA-MB-231); >50 (Hs578T)	Masi et al. (2019)
Coccinine (**157**)	*Haemanthus humilis* (Bulbs)	South Africa	5.9 (A549); 16.8 (HCT-15); >50 (SK-MEL-28); 7.9 (MCF7); 13.8 (MDA-MB-231); 5.3 (Hs578T)	Masi et al. (2019)

Erysodine (**161**)	*Erythrina abyssinica* (Seeds)	Sudan	39 (Hep-G2); 66 (HEP-2)	Mohammed et al. (2012)
Erysotrine (**162**)	*Erythrina abyssinica* (Seeds)	Sudan	50 (Hep-G2); 69 (HEP-2)	Mohammed et al. (2012)
Erythraline (**160**)	*Erythrina abyssinica* (Seeds)	Sudan	59 (Hep-G2); 54 (HEP-2)	Mohammed et al. (2012)
Incartine (**155**)	*Haemanthus humilis* (Bulbs)	South Africa	>50 (A549); >50 (HCT-15); >50 (SK-MEL-28); >50 (MCF7); >50 (MDA-MB-231); >50 (Hs578T)	Masi et al. (2019)
Isotetrandrine (**171**	*Xylopia aethiopica* (Seeds)	Cameroon	1.53 (CCRF-CEM) *vs* 2.36 (CEM/ADR5000) (D.R.: 1.54); 7.28 (MDA-MB231-*pcDNA*) *vs* 6.70 (MDA-MB231/*BCRP*) (D.R.: 0.92); 2.39 (HCT116 $p53^{+/+}$) *vs* 4.55 (HCT116 $p53^{-/-}$) (D.R.: 1.90); 3.84 (U87MG) *vs* 1.45 (U87MG.Δ*EGFR*) (D.R.: 0.38)	Kuete et al. (2015b)
Lycorine (**170**)	*Crinum abyscinicum* (Bulbs)	Ethiopia	9 (A2780); 11(MV4-11)	Abebe et al. (2020)
Montanine (**158**)	*Haemanthus humilis* (Bulbs)	South Africa	1.9 (A549); 6.8 (HCT-15); 23.2 (SK-MEL-28); 4.4 (MCF7); 3.4 (MDA-MB-231); 3.6 (Hs578T)	Masi et al. (2019)

(*continued*)

Table 1 Cytotoxic alkaloids from African medicinal plants and their effects on sensitive and drug resistant cancer cell lines. (*cont'd*)

Compounds names	Source (Plant part)	Country	Cancer Cell Lines and IC_{50} values (μM) and degree of resistance (D.R.) in bracket	References
N-p-coumaroyltyramine (**159**)	*Crinum biflorum* (Bulbs)	Senegal	>10 (A431); 7.5 (HeLa)	Masi et al. (2021)
Roeharmine (**151**)	*Oricia suaveolens* (Leaves)	Cameroon	11.4 (HepG2); 13.6 (HT-29); 11.9 (PC-3); 14.7 (Ovcar-3); 12.7 (MCF-7); 7.6 (MRC-5)	Nouga et al. (2016)
Ungeremine (**172**)	*Crinum zeylanicum* (Leaves)	Cameroon	4.89 (CCRF-CEM) *vs* 75.24 (CEM/ADR5000) (D.R.: 15.40); 5.47 (MDA-MB231-*pcDNA*) *vs* 3.67 (MDA-MB231/*BCRP*) (D.R.: 0.67); 6.45 (HCT116 $p53^{+/+}$) *vs* 7.06 (HCT116 $p53^{-/-}$) (D.R.: 1.09); 5.38 (U87MG) *vs* 22.48 (U87MG.Δ*EGFR*) (D.R.: 4.18)	Mbaveng et al. (2019b)
Veprisazole (**149**)	*Araliopsis soyauxii* (Stem bark)	Cameroon	50.24 (CCRF-CEM) *vs* 64.49 (CEM/ADR5000) (D.R.: 1.28); >97.53 (MDA-MB231-*pcDNA*) *vs* 71.07 (MDA-MB231/*BCRP*) (D.R.: <0.73); >97.53 (HCT116 $p53^{+/+}$) *vs* 30.34 (HCT116 $p53^{-/-}$) (D.R.: <0.31); >97.53 (U87MG) *vs* 53.22 (U87MG.Δ*EGFR*) (D.R.: <0.55)	Nganou et al. (2019)

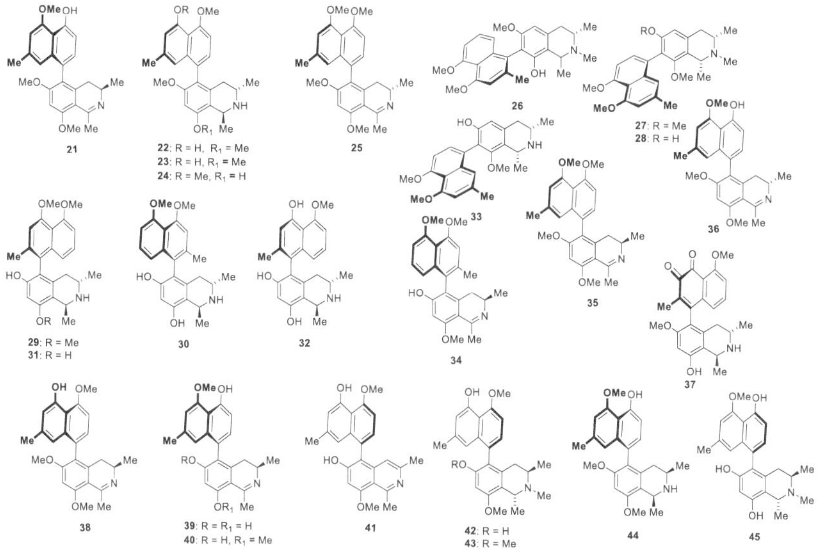

Fig. 3 Chemical structures of cytotoxic alkaloids from African medicinal plants. Naphthylisoquinoline (**21–85**); benzophenanthridine (**86–100**); furoquinoline (**101–109**); acridone (**110–120**); indole (**121–135**); bisbenzylisoquinoline (**136–143**); steroidal (**144–146**); β-carboline (**147– 148**); miscellaneous (**149–172**): **21**: ancistrolikokine E_3; **22**: 6-*O*-methyl-4′-*O*-demethylhamatine; **23**: 5-*epi*-ancistectorine A_2; **24**: 6-*O*-methylhamatine; **25**: 6-*O*-methylhamatinine; **26**: ancistrobrevine D; **27**: ancistrobrevine A; **28**: 6-*O*-demethylancistrobrevine A; **29**: hamatine; **30**: ancistrobrevine E; **31**: 5-*epi*-ancistrobrevine E; **32**: 5-*epi*-ancistrobrevine F; **33**: 6-*O*-demethylancistrobrevine H; **34–36**: ancistrobrevidines A-C; **37**: ancistrobreviquinone A; **38**: ancistroealaine F; **39**: ancistrolikokine E; **40**: ancistrolikokine E_2; **41**: ancistrolikokine G; **42**: ancistrolikokine C; **43**: ancistrolikokine C_2; **44**: ancistrolikokine I; **45**: ancistrolikokine C_3; **46**: ancistrolikokine J; **47**: ancistrolikokine J_2; **48**: mbandakamine A; **49**: mbandakamine C; **50**: mbandakamine D; **51:** 6-*O*-methylhamateine; **52**: ancistrobreveine D; **53**: *ent*-dioncophylleine A; **54**: ancistrobreveine A; **55**: ancistrobreveine B; **56**: ancistrobreveine C; **57**: dioncophylline F; **58**: dioncophylline D_2; **59:** 5′-*O*-methyldioncophylline D; **60**: ancistrocladisine A; **61**: dioncophylline A; **62**: 4′-*O*-demethyldioncophylline A; **63**: 4′-*O*-demethyl-7-*epi*-dioncophylline A; **64**: jozilebomine A; **65**: jozilebomine B; **66**: jozimine A_2; **67**: ealamine A; **68**: ealamine B; **69**: ealamine C; **70**: ealamine D; **71**: ealamine G; **72**: yaoundamine A; **73**: ancistroyafungine A; **74**: ancistroyafungine B; **75**: ancistroyafungine C; **76**: ancistroyafungine D; **77**: 6-*O*-methylhamatine; **78**: 4′-*O*-demethylancistrocladine; **79**: ancistroguineine A; **80**: ancistrobertsonine A; **81**: ancistrobrevine B; **82**: ancistrotectoriline A; **83**: 6,5′-*O*,*O*-didemethyl ancistroealaine A; **84**: 6-*O*-demethylancistroealaine A; **85**: 7-*epi*-ancistrobrevine D; **86**: dihydrochelerythrine; **87**: 6-hydroxymethyldihydronitidine; **88**: bis-[6-(5,6-dihydrochelerythrinyl)] ether; **89**: 6-acetonyldihydrochelerythrine; **90**: decarine; **91**: zanthocapensine; **92**: chelerythrine; **93**: 8-acetonyldihydrochelerythrine; **94**: arnottianamide; **95**: 8-oxochelerythrine; **96**: buesgenine; **97**: isofagaridine; **98**: zanthoxyline; **99**: nitidine chloride; **100**: fagaridine chloride; **101**: maculine B; **102**: maculine; **103**: kokusaginine; **104**: flindersiamine; **105**: dictamine; **106**: 6,7-methylenedioxy-5-hydroxy-8-methoxy-dictamnine;

(Continued)

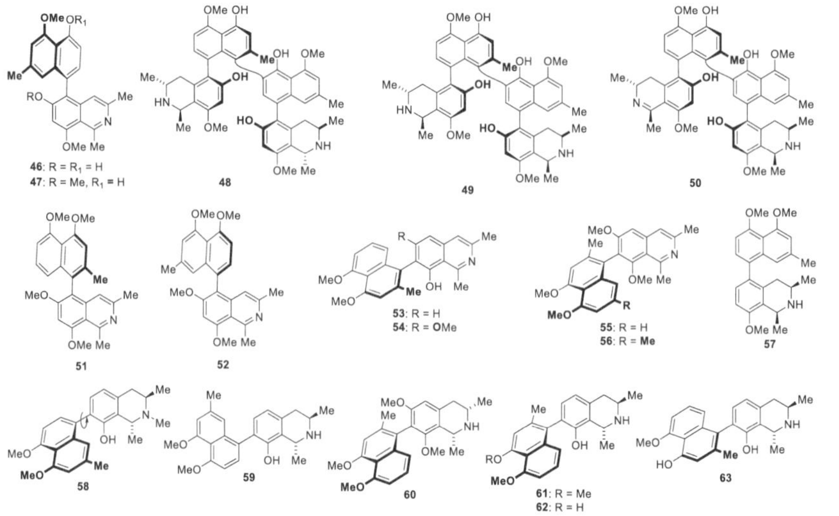

Fig. 3 (*Continued*)

cytotoxicity against CCRF-CEM (IC_{50} of 2.0, 2.31, 0.11 μM, respectively) and moderate to potent activity towards CEM/ADR5000 (IC_{50} of 55.6, 35.0, 2.3 μM, respectively) (Omosa et al., 2022). In an earlier study, dihydrochelerythrine (**86**) inhibited the proliferation of breast cancer cells BT549

Fig. 3—Cont'd **107**: γ-fagarine; **108**: kokusaginine B; **109**: montrifoline; **110**: evoxanthine; **111**: 1-hydroxy-4-methoxy-10-methylacridone; **112**: norevoxanthine; **113**: 1,3-dimethoxy-10-methylacridone; **114**: arborinin; **115:** citrusinine-I; **116**: citracridone-I; **117**: 5-hydroxynoracronycine; **118**: natsucitrine-I; **119**: glycofolinine; **120**: citracridone-III; **121**: isostrychnopentamine; **122**: tabernaelegantine C; **123**: tabernaelegantinine B; **124**: (3′*R*)-hydroxytaberanelegantine C; **125**: (19′*S*)-hydroxytabernaelegantine A; **126**: 3′-oxotabernaelegantine C; **127**: 3′-oxotabernaelegantine D; **128**: tabernaelegantine A; **129**: tabernaelegantine D; **130**: tabernaemontanine; **131**: vobasine; **132**: soyauxinium chloride; **133**: tabernine A; **134**: tabernine B; **135**: tabernine C; **136**: (+)– 1,2-dehydrotelobine; **137:** (+)– 2′-norcocsuline and **138**: stebisimine; **139**: puetogaline B; **140**: cycleanine; **141**: tetrandrine; **142**: isochondodendrine; **143**: 2′-norcocsuline; **144**: solamargine; **145**: holamine; **146**: funtumine; **147**: 10-methoxycanthin-6-one; **148**: canthin-6-one; **149**: veprisazole; **150**: 4-methoxy-1-methyl-2(1*H*) quinolinone; **151**: roeharmine; **152**: 4-(isoprenyloxy)– 3-methoxy-3,4-deoxymethylenedioxyfagaramide; **153**: 2*R*-bgugaine; **154**: 2*S*-bgugaine; **155**: incartine; **156**: albomaculine; **157**: coccinine; **158**: montanine; **159**: *N*-*p*-coumaroyltyramine; **160**: erythraline; **161**: erysodine; **162**: erysotrine; **163**: 8-oxoerythraline; **164**: 11-methoxyerysodine; **165**: 5,14-dimethylbudmunchiamine L1; **166**: 6-hydroxybudmunchiamine K; **167**: 5-normethylbudmunchiamine K; **168**: 6-hydroxy-5-normethylbudmunchiamine K; **169**: 6-hydroxycrinamine; **170**: lycorine; **171**: isotetrandrine: **172**: ungeremine.

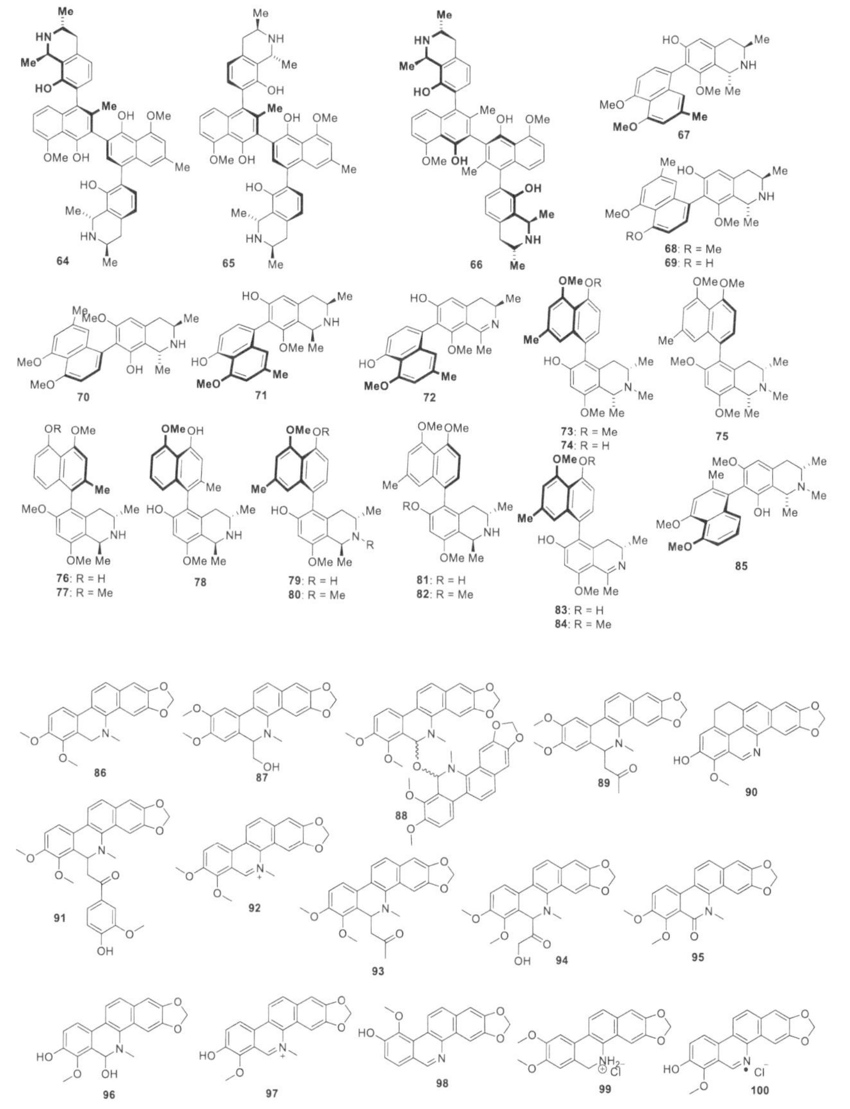

Fig. 3 (Continued)

(IC_{50} of 21.2 μM), hepatocellular carcinoma (HCC) (IC_{50} of 8.9 μM) and showed low cytotoxicity against HEp_2 cells (IC_{50} of 64.0 μM) (Andima et al., 2019). 6-Acetonyldihydrochelerythrine (**89**) from *Z. capense*, at 20 μM, induced 11-fold increases in apoptotic HCT116 cells as compared to the

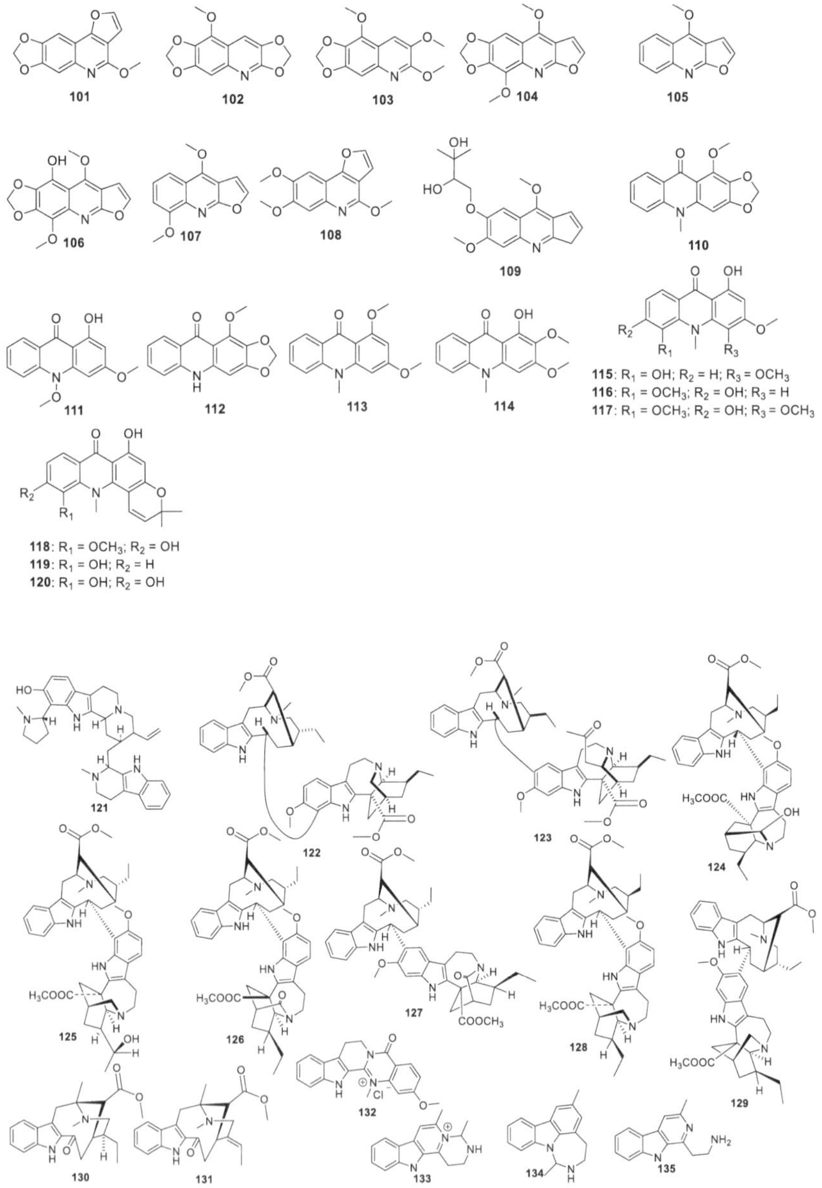

101 102 103 104 105

106 107 108 109 110

111 112 113 114

115: R_1 = OH; R_2 = H; R_3 = OCH_3
116: R_1 = OCH_3; R_2 = OH; R_3 = H
117: R_1 = OCH_3; R_2 = OH; R_3 = OCH_3

118: R_1 = OCH_3; R_2 = OH
119: R_1 = OH; R_2 = H
120: R_1 = OH; R_2 = OH

121 122 123 124 125 126 127 128 129 130 131 132 133 134 135

Fig. 3 (*Continued*)

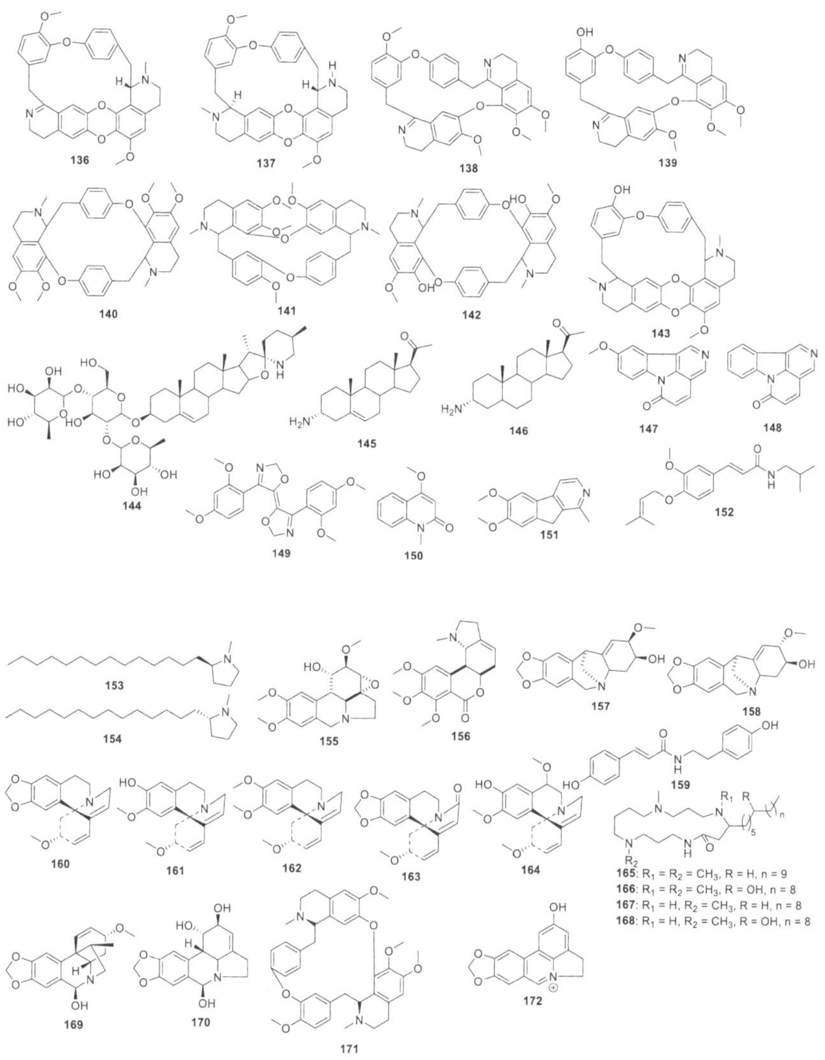

165: $R_1 = R_2 = CH_3$, R = H, n = 9
166: $R_1 = R_2 = CH_3$, R = OH, n = 8
167: R_1 = H, $R_2 = CH_3$, R = H, n = 8
168: R_1 = H, $R_2 = CH_3$, R = OH, n = 8

Fig. 3 (*Continued*)

control. Apoptosis induction of **89** was further confirmed by caspase-3-like activity assays, which showed 2-fold increases in caspase-3-like activity compared to vehicle control. In addition, decarine (**90**), zanthocapensine (**91**), from the roots of the *Zanthoxylum capense*, exhibited moderate inhibition at 20 μM against HCT116 colon carcinoma cells (Mansoor et al., 2013a). Structure-activity relationships reveal that the activities of the benzophenanthridine alkaloids are mostly influenced by various substituents on the isoquinoline motif. The quaternary form has also been found to be vital against cell lines. Hence, chelerythrine (**92**) from *Z. capense* induced cell death up to at least 66% in a concentration-dependent manner. IC_{50} values were recorded at 95 (MCF-7) and 154 (Caco-2) μM (Bodede, Shaik, Singh, & Moodley, 2017).

Furoquinoline and acridone (Table 1 and Fig. 3) alkaloids are widely distributed in plants of the family Rutaceae (including the genera *Zanthoxylum, Oricia,* and *Citrus*) and Euphorbiaceae (genus *Uapaca*). Furoquinoline such as maculine B (**101**), displayed cytotoxic effects towards CCRF-CEM cells (IC_{50} of 28.40 μM), CEM-ADR5000 cells (IC_{50} of 110.89 μM), MDA-MB231 (IC_{50} of 56.50 μM), MDA-MB231/*BCRP* (IC_{50} of 34.32 μM), HCT116 $p53^{+/+}$ cells (IC_{50} of 56.34 μM), HCT116 $p53^{-/-}$cells (IC_{50} of 33.42 μM), U87MG cells (IC_{50} of 24.86 μM) and U87MG.Δ*EGFR* cells (IC_{50} of 50.99 μM) with significantly low cytotoxicity in normal AML12 hepatocytes ($IC_{50} > 164.51$ μM) (Nganou et al., 2019). From *Zanthoxylum buesgenii*, maculine (**102**) and kokusaginine (**103**) showed selective cytotoxicity against CCRF-CEM cells (IC_{50} of 89.09 and 49.81 μM, respectively) and CEM-ADR5000 cells (IC_{50} of 63.09 and 44.56 μM, respectively) (Sandjo, Kuete, Tchangna, Efferth, & Ngadjui, 2014). While tested in a panel of eight tumor cell lines, flindersiamine (**104**) and dictamine (**105**) obtained from *Araliopsis soyauxii* Engl exhibited selective IC_{50} values of 96.34 and 97.57 μM, respectively against CCRF-CEM cells (Mbaveng et al., 2021a). The anticancer activity of phytochemicals from *Oricia* species was also assessed using MTT assay against different human kinds of cancer cells. As a result, 6,7-methylenedioxy-5-hydroxy-8-methoxy-dictamnine (**106**) displayed potent activity against 5/5 cells lines tested notably towards epatocellular carcinoma (Hep- G2), prostate cancer (PC-3), human ovary carcinoma (OVCAR-3), human breast adenocarcinoma (MCF-7), human colon carcinoma (HT-29) and normal human fetal lung cell line (MRC-5) with IC_{50} values of 8.7, 11.2, 9.6, 10.4 10.8 and 0.67 μM, respectively (Nouga et al., 2016). A similar level of activity was also recorded for γ-fagarine (**107**) with IC_{50} values of 10.4, 11.8, 9.7, 11.7, 11.5, and 10.4 μM, respectively (Nouga et al., 2016). Important activity was also noted for acridone, evoxanthine (**110**) with IC_{50} values of 11.1 (Hep- G2), 12.4 (PC-3), 10.6 (OVCAR-3), 11.7 (MCF-7), 11.9 (HT-29) and 10.9 μM (MRC-5) (Nouga et al., 2016). In addition, more related analogs sharing the acridone skeleton were further isolated and tested as cytotoxic agents. In their investigation endeavor, Kuete and co-workers reported an $IC_{50} <$ 138 μM for evoxanthine (**110**),

1-hydroxy-4-methoxy-10-methylacridone (**111**), norevoxanthine (**112**), and 1,3-dimethoxy-10-methylacridone (**113**) against 9 tested cancer cell lines (Table 1 and Fig. 3). 1,3-Dimethoxy-10-methylacridone (**113**) showed the best activity, with IC_{50} values ranging from 3.38 μM (towards MDA-MB-231-BCRP breast adenocarcinoma cells) to 58.10 μM (towards leukemia CEM/ADR5000 cells) (Kuete et al., 2015a). This compound induced apoptosis in CCRF-CEM leukemia cells, mediated by increased reactive oxygen species product (Kuete et al., 2015a). However, neither the activation of caspases nor the disruption of MMP were observed with this compound. Comparable observations were also made with arborinine (**114**) from *Uapaca togoensis* despite having $IC_{50} < 10$ μM on 7/9 recalcitrant cell lines. Its IC_{50} values ranged from 3.55 μM (against CEM/ADR5000 cells) to 31.77 μM (against CCRF-CEM cells) compared to 0.20 μM (against CCRF-CEM cells) to 195.12 μM (against CEM/ADR5000 cells) for doxorubicin (Kuete et al., 2015c). The mechanistic study further demonstrated that arborinine (**114**) also strongly induced apoptosis in CCRF-CEM cells and cell cycle arrest in the G0/G1 and S phases.

Indolomonoterpenic alkaloids are well known for their extrinsic and intrinsic mechanism of action against cancer cell lines. Their chemotherapeutic effect is mediated by different biochemical modes of action, such as inhibition of topoisomerases I and II and inhibition of microtubules formation. Isostrychnopentamine (**121**) is an indolomonoterpenic alkaloid purified from the leaves of *Strychnos usambarensis,* which induces cell cycle arrest and apoptosis in human HCT-116 colon cancer cells. Besides being an interesting lead for cancer chemotherapy, **121** displayed no inhibition effect on DNA topoisomerases I or II (Frédérich et al., 2003). In their continued search for apoptosis-inducing compounds from Mozambican medicinal plants, Mansoor et collaborators reported tabernaelegantine C (**122**) and tabernaelegantinine B (**123**), two bisindole alkaloids from the roots extract of *Tabernaemontana elegans* (Mansoor et al., 2013b). Compounds **122** and **123** induced characteristic patterns of apoptosis in HCT116 cancer cells including, cells linkage, condensation, fragmentation of the nucleus, blebbing of the plasma membrane, and chromatin condensation which provides new lead compounds for further investigation against HCT116 colon cancer cells. In their subsequent study on *T. elegans,* (3′*R*)-hydroxytaberanelegantine C (**124**) displayed strong apoptosis induction activity comparable to 5-fluorouracil (5-FU). After 72 h of cell treatment, IC_{50} values of **124** with some degree of selectivity > 2.8 were observed at IC_{50} of 3.20 (HCT116), 3.49 (SW620), 2.91 (HepG2) and

10.17 μM (normal cell: CCD18co) compared with 5-fluorouracil (IC_{50} values of 2.49, 5.39, 12.2 and >100 μM, respectively) (Paterna et al., 2016b). Subsequent acid–base extraction of the methanolic extract from the roots *T. elegans* afforded vobasinyl–Iboga type alkaloids identified as (19′*S*)-hydroxytabernaelegantine A (**125**), 3′-oxotabernaelegantine C (**126**), 3′-oxotabernaelegantine D (**127**), tabernaelegantine A (**128**) and tabernaelegantine D (**129**). All alkaloids showed cytotoxic activity against HCT116 colon and HepG2 liver carcinoma cells. Mechanistic investigation demonstrates that vobasinyl–iboga alkaloids **125–127** and **129** are powerful inducers of apoptosis and cell cycle arrest in HCT116 colon cancer cells (Paterna et al., 2016a). Further, monoterpene indole alkaloids, tabernaemontanine (**130**), and vobasine (**131**) displayed promising apoptosis induction profiles in human hepatoma HuH-7 cells which was further correlated with a significant increase of caspase-3 activity (Mansoor, Ramalho, Mulhovo, Rodrigues, & Ferreira, 2009b). Soyauxinium chloride (**132**), exhibited cytotoxicity against a large panel of animal and human cancer cell lines, including various sensitive and MDR sublime. Soyauxinium chloride (**132**) displayed cytotoxicity with IC_{50} values recorded from 3.64 μM (CCRF-CEM cells) to 16.86 μM (against the BRAF-wildtype SKMel-505 melanoma cells) and further induced apoptosis in CCRF-CEM cells *via* caspases 3/7-, 8- and 9-activation. The mitochondrial damage pathway and increased ROS production, and otherwise ferroptosis and necroptosis were also involved in 15-induced apoptosis (Mbaveng et al., 2021b).

Bisbenzylisoquinoline alkaloids mostly occur in the Menispermaceae family and have been reported as a promising class of alkaloids to overcome drug resistance in cancer therapy. Through activity-guided isolation, (+)− 1,2-dehydrotelobine (**136**), (+)− 2′-norcocsuline (**137**) and stebisimine (**138**), puetogaline B (**139**), were screened for their antiproliferative activity against four human cancer cells lines, A2780, H460, MCF-7 and UACC-257 (Liu et al., 2013). Strong activities were observed for **136** with IC_{50} values of 4.10, 2.18, 4.23 and 3.37 μM, respectively, followed by **137** (IC_{50} values of 2.70, >3.33, >1.11, 6.37 μM, respectively). On the contrary, **138** and **139** exhibited moderate activities against the tested cells lines with IC_{50} > 10 μM (Liu et al., 2013). Other medicinal spices harvested in Nigeria, such as *Triclisia subcordata* are well known as a source of bisbenzylisoquinoline alkaloids as showcased by cycleanine (**140**) and tetrandrine (**141**) which showed selective activity against Ovcar-8, A2780, Ovcar-4, and Igrov-1 cancer cell lines compared to human normal ovarian surface epithelial cells with IC_{50} value of 35 μM for **140**. Both compounds displayed a characteristic pattern of apoptosis as exemplified by the activation of caspases 3/7 and mitochondrial damage (Uche, Drijfhout, McCullagh, Richardson, & Li, 2016). A similar mechanistic investigation was also reported for isochondodendrine (**142**) and 2′-norcocsuline (**143**) which exhibited *in vitro* cytotoxicity against the tested cell lines with IC_{50} ranging from 3.5 to 17 μM and 0.8 to 6.2 μM, respectively. Furthermore, both alkaloids displayed IC_{50} values of 10.5 and 8.0 μM, respectively on normal human ovarian epithelial (HOE) (Uche et al., 2017).

Steroidal alkaloids, which are nitrogen derivatives of natural steroids, represent another important class of natural products as potential anticancer leads. The leaves of *Holarrhena floribunda* afforded, solamargine (**144**) which is the most active cytotoxic component of Solanaceae plant family against cancer displaying IC_{50} of 15.62 μg/mL, and 9.1-fold P-glycoprotein inhibition at 100 μg/mL against the SH-SY5Y neuroblastoma cell line (Burger et al., 2018). Steroidal alkaloids like holamine (**145**) and funtumine (**146**) showed selective dose-dependent cytotoxicity against the cancer cell lines compared to normal cell lines (KMST-6). These compounds further induced cell cycle arrest at the G0/G1 and G2/M phases in the toxic cell lines with a significant reduction in DNA synthesis (Badmus, Ekpo, Hussein, Meyer, & Hiss, 2019).

The cytotoxicity of various alkaloids will be appreciated according to the cutoff point established as follows: outstanding activity ($IC_{50} \leq 0.5$ μM), excellent activity ($0.5 < IC_{50} \leq 2$ μM), very good activity ($2 < IC_{50} \leq 5$ μM), good activity ($5 < IC_{50} \leq 10$ μM), average activity ($10 < IC_{50} \leq 20$ μM), weak activity ($20 < IC_{50} \leq 60$ μM), very weak activity ($60 < IC_{50} \leq 150$ μM), and not active ($IC_{50} > 150$ μM) (Kuete, 2025). It has also been established that the degree of resistance (D.R.) < 0.9 defines collateral sensitivity, whilst D.R. between 0.9 and 1.2 defines normal sensitivity. The cross-resistance is noted if the cytotoxic agent is more active in the sensitive cell line than its resistant subline, with D.R above 1.2 (Efferth et al., 2020; Efferth et al., 2021; Mbaveng, Kuete, & Efferth, 2017). Collateral sensitivity or normal sensitivity of resistant *vs* sensitive cancer cell lines should be achieved for compounds with ability to combat cancer drug resistance (Efferth et al., 2020; Efferth et al., 2021; Kuete and Efferth, 2015; Mbaveng et al. 2017). This basis of classification will be used to discuss the cytotoxicity of alkaloids isolated from African medicinal plants. From the data summarized in Table 1, it appears that outstanding cytotoxic activities were obtained with compounds **61** against INA-6 cells, compounds **96** and **97** against CCRF-CEM cells, and compound **66** against HeLa and

PANC-1 cells; excellent cytotoxic potency was found for compounds **49, 50, 59, 99**, **100,** and **171** against CCRF-CEM cells, **61** against CEM/ADR5000 cells, compounds **64** and **65** against HeLa cells, **65** against PANC-1 cells, **143** against Igrov-1 cells, **158** against A549 cells, and **171** against U87MG.Δ*EGFR* cells, **143** against Ovcar-8 cells and Igrov-1 cells; very good cytotoxic effects were obtained with compounds **21, 41, 44, 51, 132,** and **172** against CCRF-CEM cells, compounds **59, 60,** and **62** against INA-6 cells, **64** against PANC-1 cells, compounds **114, 124,** and **171** against CEM/ADR5000 cells, compounds **99, 113, 124,** and **172** against MDA-MB231-*BCRP* cells, compounds **99, 124,** and **171** against U87MG cells, **132** against U87MG.Δ*EGFR* cells, **124** against MDA-MB231-*pcDNA* cells, compounds **138** and **143** against A2780 cells, **158** against MCF-7 cells, MDA-MB-231 cells, and Hs578T cells, **165** against SK-MEL cells and KB cells, **165** against BT-549 cells, SK-OV-3 cells, and LLC-PK11 cells, compounds **124** and **171** against HCT116 $p53^{+/+}$ cells and HCT116 $p53^{-/-}$ cells, **167** against SK-MEL cells, KB cells, and LLC-PK11 cells; good cytotoxic effects were obtained with compounds **41, 51, 59, 99, 100** against CEM/ADR5000 cells, **48** against CCRF-CEM cells, compounds **69, 74,** and **76** against PANC-1 cells, compounds **99, 114, 171** and **172** against MDA-MB231-*pcDNA* cells, compounds **100, 114,** and **172** against U87MG cells, compounds **111, 113, 114,** and **172** against HCT116 $p53^{+/+}$ cells, compounds **112, 113,** and **114** against U87MG.Δ*EGFR* cells, compounds **113, 114,** and **172** against HCT116 $p53^{-/-}$ cells, compounds **114** and **171** against MDA-MB231-*BCRP* cells, compounds **125, 127** and **129** against HCT116 cells, compounds **140** and **148** against Ovcar-8 cells, compounds **140** and **141** against A2780 cells, compounds **140, 141** and **142** against Igrov-1 cells, compounds **142** and **143** against Ovcar-4 cells, **147** against DU145 cells, **157** against A549 cells, MCF7 cells, and Hs578T cells, **158** against HCT-15 cells, **159** against HeLa cells, **166** against LLC-PK11 cells, **167** against SK-OV-3 cells, compounds **169** and **170** against A2780 cells. These compounds could be useful for the management of malignant diseases. Interestingly, collateral sensitivities CEM/ADR5000 cells *vs* sensitive CCRF-CEM cells to compound **114** (D.R.: 0.11), MDA-MB231/*BCRP* cells *vs* MDA-MB231-*pcDNA* cells to compounds **99** (D.R.: 0.62), **113** (D.R.: 0.33), **114** (D.R.: 0.87), **171** (D.R.: 0.92), and **172** (D.R.: 0.67), HCT116 $p53^{-/-}$ cells *vs* HCT116 $p53^{+/+}$ cells to compounds **111** (D.R: 0.39), **113** (D.R: 0.97), and **132** (D.R: 0.87), and U87MG.Δ*EGFR* cells *vs* U87MG cells to compounds **112** (D.R: 0.06), **113** (D.R: 0.51), **132** (D.R: 0.97), and **171**

(D.R: 0.38) were achieved (Table 1). This is an indication that these alkaloids are convincing cytotoxic compounds that could be used to fight cancer drug resistance.

6. Conclusions

In this Chapter, information on alkaloids from African medicinal plants that have been reported by various research groups to have *in vitro* and some *in vivo* anticancer potential against cancer cells that are both drug-sensitive and drug-resistant. A review of the literature uncovered 152 distinct alkaloids, spread throughout 27 species and 8 families, including Ancistrocladaceae, Rutaceae, Amaryllidoideae, Apocynaceae, Menispermaceae, Fabaceae, Phyllanthaceae, and Araceae. Notably, the bulk of the reported alkaloids were discovered in plants that were assumed to have a long history of usage in folk medicine, indicating that knowledge of traditional medicine may be highly beneficial in efforts to find new anticancer medications in African medicinal plants. The majority of African flora's anticancer alkaloids to date have come from naphthyisoquinolines (42.76%) found in the different tissues of *Ancistracladus* species that grow in the Democratic Republic of the Congo and Ivory Coast. The diversity in habitats found in Africa supports a large range of plants with various species which makes it an optimal place to search for novel anticancer agents. In addition to the extensive ongoing research on the isolation and assessment of the anticancer potential of alkaloids from medicinal plants of Africa, additional studies are required to clarify their anticancer activity mechanisms. Furthermore, the possibility of new clinical investigations necessitates in-depth animal model-based studies of identified alkaloids and their analogs. It is significant to note that, despite the reviewed alkaloids' promising *in vitro* anticancer potential, there is inadequate information on *in vivo* research, toxicological data, and cancer targets, which are essential for pre-clinical research and eventual therapeutic development. Future research should therefore focus on toxicological investigations to determine the negative effects and upper limits of exposure to these alkaloids and their derivatives. The establishment of a targeted library of alkaloids from the African flora with promising anti-cancer characteristics is worthwhile to facilitate the process of lead/hit discovery of possible anticancer drugs *via* virtual screening. This chapter has compiled data that could be crucial in directing researchers interested in investigating native African plant alkaloids for use in cancer treatment.

References

Abbas, Z., & Rehman, S. (2018). An overview of cancer treatment modalities. *Neoplasm, 1*, 139–157.

Abebe, B., Tadesse, S., Hymete, A., & Bisrat, D. (2020). Antiproliferative effects of alkaloids from the bulbs of *Crinum abyscinicum* Hochst. ExA. Rich. *Evidence-Based Complementary and Alternative Medicine, 2020*, 2529730.

Alipour, M., Khoobi, M., Moradi, A., Nadri, H., Moghadam, F., Emami, S., & Shafiee, A. (2014). Synthesis and anti-cholinesterase activity of new 7-hydroxycoumarin derivatives. *European Journal of Medicinal Chemistry, 82*, 536–544.

Amoa, O. P., Ntie-Kang, F., Lifongo, L. L., Ndom, J. C., Sippl, W., & Mbaze, L. M. A. (2013). The potential of anti-malarial compounds derived from African medicinal plants, part I: A pharmacological evaluation of alkaloids and terpenoids. *Malaria Journal, 12*(1), 1–26.

Anand, P., Kunnumakara, A. B., Sundaram, C., Harikumar, K. B., Tharakan, S. T., Lai, O. S., ... Aggarwal, B. B. (2008). Cancer is a preventable disease that requires major lifestyle changes. *Pharmaceutical Research, 25*(9), 2097–2116.

Ancolio, C., Azas, N., Mahiou, V., Ollivier, E., Giorgio, C., Di, ... Balansard, G. (2002). Antimalarial activity of extracts and alkaloids isolated from six plants used in traditional medicine in Mali and Sao Tome. *Phytotherapy Research, 16*(7), 646–649.

Andima, M., Coghi, P., Yang, L. J., Wai Wong, V. K., Mutuku Ngule, C., Heydenreich, M., ... Derese, S. (2019). Antiproliferative activity of secondary metabolites from *Zanthoxylum zanthoxyloides* Lam: *In vitro* and *in silico* Studies. *Pharmacognosy Communications, 10*(1), 44–51.

Awale, S., Dibwe, D. F., Balachandran, C., Fayez, S., Feineis, D., Lombe, B. K., & Bringmann, G. (2018). Ancistrolikokine E_3, a 5,8′-coupled naphthylisoquinoline alkaloid, eliminates the tolerance of cancer cells to nutrition starvation by inhibition of the Akt/mTOR/autophagy signaling pathway. *Journal of Natural Products, 81*(10), 2282–2291.

Badmus, J. A., Ekpo, O. E., Hussein, A. A., Meyer, M., & Hiss, D. C. (2019). Cytotoxic and cell cycle arrest properties of two steroidal alkaloids isolated from *Holarrhena floribunda* (G. Don) T. Durand & Schinz leaves. *BMC Complementary and Alternative Medicine, 19*(1), 1–9.

Benamar, M., Melhaoui, A., Zyad, A., Bouabdallah, I., & Aziz, M. (2009). Anti-cancer effect of two alkaloids: 2*R* and 2*S*-bgugaine on mastocytoma P815 and carcinoma Hep. *Natural Product Research, 23*(7), 659–664.

Bisai, V., Saina Shaheeda, M., Gupta, A., & Bisai, A. (2019). Biosynthetic relationships and total syntheses of naturally occurring benzophenanthridine alkaloids. *Asian Journal of Organic Chemistry, 8*(7), 946–969.

Bodede, O., Shaik, S., Singh, M., & Moodley, R. (2017). Phytochemical analysis with antioxidant and cytotoxicity studies of the bioactive principles from *Zanthoxylum capense* (Small Knobwood). *Anti-cancer Agents in Medicinal Chemistry, 17*, 627–634.

Bringmann, G., Mutanyatta-Comar, J., Greb, M., Rüdenauer, S., Noll, T., & Irmer, A. (2007). Biosynthesis of naphthylisoquinoline alkaloids: synthesis and incorporation of an advanced 13C2-labeled isoquinoline precursor. *Tetrahedron, 63*(8), 1755–1761.

Bringmann, G., Steinert, C., Feineis, D., Mudogo, V., Betzin, J., & Scheller, C. (2016). HIV-inhibitory michellamine-type dimeric naphthylisoquinoline alkaloids from the Central African liana *Ancistrocladus congolensis*. *Phytochemistry, 128*, 71–81.

Bringmann, G., Teltschik, F., Schäffer, M., Haller, R., Bär, S., Robertson, S., & Isahakia, M. (1998). Ancistrobertsonine A and related naphthylisoquinoline alkaloids from *Ancistrocladus robertsoniorum*. *Phytochemistry, 47*(1), 31–35.

Bunsupa, S., Yamazaki, M., & Saito, K. (2017). Lysine-derived alkaloids: overview and update on biosynthesis and medicinal applications with emphasis on quinolizidine alkaloids. *Mini Reviews in Medicinal Chemistry, 17*(12), 1002–1012.

Burger, T., Mokoka, T., Fouché, G., Steenkamp, P., Steenkamp, V., & Cordier, W. (2018). Solamargine, a bioactive steroidal alkaloid isolated from *Solanum aculeastrum* induces non-selective cytotoxicity and P-glycoprotein inhibition. *BMC Complementary and Alternative Medicine, 18*(1), 1–11.

Choi, G. S., Choo, H. J., Kim, B. G., & Ahn, J. H. (2020). Synthesis of acridone derivatives via heterologous expression of a plant type III polyketide synthase in *Escherichia coli. Microbial Cell Factories, 19*(1), 1–11.

Choi, K., Morishige, T., Shitan, N., Yazaki, K., & Sato, F. (2002). Molecular cloning and characterization of coclaurinen-methyltransferase from cultured cells of *Coptis japonica. Journal of Biological Chemistry, 277*(1), 830–835.

Chou, C. J., Lin, L. C., Chen, K. T., & Chen, C. F. (1994). Northalifoline, a new isoquinolone alkaloid from the pedicels of *Lindera megaphylla. Journal of Natural Products, 57*(6), 689–694.

Cragg, G. M., & Newman, D. J. (2005). Plants as a source of anti-cancer agents. *Journal of Ethnopharmacology, 100*(1–2), 72–79.

Dey, P., Kundu, A., Kumar, A., Gupta, M., Lee, B. M., Bhakta, T., ... Kim, H. (2020). *Analysis of alkaloids (indole alkaloids, isoquinoline alkaloids, tropane alkaloids). In. Recent advances in natural products analysis.* Elsevier, 505–567.

Efferth, T., Kadioglu, O., Saeed, M. E. M., Seo, E. J., Mbaveng, A. T., & Kuete, V. (2021). Medicinal plants and phytochemicals against multidrug-resistant tumor cells expressing ABCB1. *ABCG2, or ABCB5: a synopsis of 2 decades. Phytochemistry Reviews, 20*(1), 7–53.

Efferth, T., Saeed, M. E. M., Kadioglu, O., Seo, E. J., Shirooie, S., Mbaveng, A. T., et al. (2020). Collateral sensitivity of natural products in drug-resistant cancer cells. *Biotechnology Advances, 38*, 107342.

Elgorashi, E., Zschocke, S., & Van Staden Eloff, J. (2003). The anti-inflammatory and antibacterial activities of Amaryllidaceae alkaloids. *South African Journal of Botany, 63*(3), 448–449.

Evans, W. (2009). *Trease and Evans' pharmacognosy.* Elsevier Health Sciences.

Eze, F. I., Siwe-Noundou, X., Isaacs, M., Patnala, S., Osadebe, P. O., & Krause, R. W. M. (2020). Anti-cancer and anti-trypanosomal properties of alkaloids from the root bark of *Zanthoxylum leprieurii* Guill and Perr. *Tropical Journal of Pharmaceutical Research, 19*(11), 2377–2383.

Fayez, S., Cacciatore, A., Sun, S., Kim, M., Aké Assi, L., Feineis, D., ... Bringmann, G. (2021). Ancistrobrevidines A-C and related naphthylisoquinoline alkaloids with cytotoxic activities against HeLa and pancreatic cancer cells, from the liana *Ancistrocladus abbreviatus. Bioorganic and Medicinal Chemistry, 30*, 115950.

Fayez, S., Feineis, D., Aké Assi, L., Kaiser, M., Brun, R., Awale, S., & Bringmann, G. (2018b). Ancistrobrevines E-J and related naphthylisoquinoline alkaloids from the West African liana *Ancistrocladus abbreviatus* with inhibitory activities against *Plasmodium falciparum* and PANC-1 human pancreatic cancer cells. *Fitoterapia, 131*, 245–259.

Fayez, S., Feineis, D., Assi, L. A., Seo, E. J., Efferth, T., & Bringmann, G. (2019). Ancistrobreveines A-D and related dehydrogenated naphthylisoquinoline alkaloids with antiproliferative activities against leukemia cells, from the West African liana: *Ancistrocladus abbreviatus. RSC Advances, 9*(28), 15738–15748.

Fayez, S., Feineis, D., Mudogo, V., Seo, E. J., Efferth, T., & Bringmann, G. (2018a). Ancistrolikokine I and further 5,8′-coupled naphthylisoquinoline alkaloids from the Congolese liana *Ancistrocladus likoko* and their cytotoxic activities against drug-sensitive and multidrug resistant human leukemia cells. *Fitoterapia, 129*, 114–125.

Ferlay, J., Colombet, M., Soerjomataram, I., Parkin, D. M., Piñeros, M., Znaor, A., & Bray, F. (2021). Cancer statistics for the year 2020: An overview. *International Journal of Cancer, 149*(4), 778–789.

Frédérich, M., Bentires-Alj, M., Tits, M., Angenot, L., Greimers, R., Gielen, J., ... Merville, M. P. (2003). Isostrychnopentamine, an indolomonoterpenic alkaloid from *Strychnos usambarensis*, induces cell cycle arrest and apoptosis in human colon cancer cells. *Journal of Pharmacology and Experimental Therapeutics, 304*(3), 1103–1110.

Funayama, S., & Cordell, G. (2014). *Alkaloids derived from phenylalanine and tyrosine. Alkaloids*. Elsevier, 21–61.

Gezici, S., & Şekeroğlu, N. (2019). Current perspectives in the application of medicinal plants against cancer: novel therapeutic agents. *Anti-cancer Agents in Medicinal Chemistry, 19*(1), 101–111.

Graham, J., Quinn, M., Fabricant, D., & Farnsworth, N. (2000). Plants used against cancer–an extension of the work of Jonathan Hartwell. *Journal of Ethnopharmacology, 73*(3), 347–377.

Hagel, J., & Facchini, P. (2013). Benzylisoquinoline alkaloid metabolism: a century of discovery and a brave new world. *Plant and Cell Physiology, 54*(5), 647–672.

Han, N., Yang, Z., Liu, Z., Liu, H., & Yin, J. (2016). Research progress on natural benzophenanthridine alkaloids and their pharmacological functions: A review. *Natural Product Communications, 11*(8), 1181–1188.

Hashmi, M., Khan, A., Farooq, U., & Khan, S. (2018). Alkaloids as cyclooxygenase inhibitors in anticancer drug discovery. *Current Protein and Peptide Science, 19*(3), 292–301.

Hesse, M. (2002). *Alkaloids: Nature's curse or blessing?* Wiley & Sons.

Huang, L., Zhe-Ling, F., Yi-Tao, W., & Li-Gen, L. (2017). Anticancer carbazole alkaloids and coumarins from *Clausena* plants: A review. *Chinese Journal of Natural Medicines, 15*(12), 881–888.

Huang, W., Wang, Y., Tian, W., Cui, X., Tu, P., Li, J., ... Liu, X. (2022). Biosynthesis investigations of terpenoid, alkaloid, and flavonoid antimicrobial agents derived from medicinal plants. *Antibiotics, 11*(10), 1380.

Ibrahim, S., & Mohamed, G. (2015). Naphthylisoquinoline alkaloids potential drug leads. *Fitoterapia, 106*, 194–225.

Ka, S., Masi, M., Merindol, N., Di Lecce, R., Plourde, M. B., Seck, M., ... Evidente, A. (2020). Gigantelline, gigantellinine and gigancrinine, cherylline- and crinine-type alkaloids isolated from *Crinum jagus* with anti-acetylcholinesterase activity. *Phytochemistry, 175*, 112390.

Kaigongi, M. M., Lukhoba, C. W., Yaouba, S., Makunga, N. P., Githiomi, J., & Yenesew, A. (2020). In vitro antimicrobial and antiproliferative activities of the root bark extract and isolated chemical constituents of *Zanthoxylum paracanthum* kokwaro (Rutaceae). *Plants, 9*(7), 1–15.

Karou, D., Savadogo, A., Canini, A., Yameogo, S., Montesano, C., Simpore, J., ... Traore, A. S. (2005). Antibacterial activity of alkaloids from *Sida acuta*. *African Journal of Biotechnology, 4*(12), 1452–1457.

Kavatsurwa, S. M., Lombe, B. K., Feineis, D., Dibwe, D. F., Maharaj, V., Awale, S., & Bringmann, G. (2018). Ancistroyafungines A-D, 5,8′- and 5,1′-coupled naphthylisoquinoline alkaloids from a Congolese *Ancistrocladus* species, with antiausterity activities against human PANC-1 pancreatic cancer cells. *Fitoterapia, 130*, 6–16.

Kiplimo, J. J., Islam, M. S., & Koorbanally, N. A. (2011). A novel flavonoid and furoquinoline alkaloids from *Vepris glomerata* and their antioxidant activity. *Natural Product Communications, 6*(12), 1847–1850.

Kittakoop, P., Mahidol, C., & Ruchirawat, S. (2014). Alkaloids as important scaffolds in therapeutic drugs for the treatments of cancer, tuberculosis, and smoking cessation. *Current Topics in Medicinal Chemistry, 14*(2), 239–252.

Krane, B. D., & Shamma, M. (1982). The isoquinolone alkaloids. *Journal of Natural Products, 45*(4), 377–384.

Kuete, V. (2014). *Health effects of alkaloids from African medicinal plants. Toxicological survey of African medicinal plants*. Elsevier, 611–633.

Kuete, V. (2025). Chapter Four-African medicinal plants and their derivative as the source of potent anti-leukemic products: rationale classification of naturally occurring anticancer agents. *Advances in Botanical. Research; a Journal of Science and its Applications,* 113. https://doi.org/10.1016/bs.abr.2023.12.010.

Kuete, V., & Efferth, T. (2015). African flora has the potential to fight multidrug resistance of cancer. *BioMed Research International, 2015,* 914813.

Kuete, V., Fouotsa, H., Mbaveng, A. T., Wiench, B., Nkengfack, A. E., & Efferth, T. (2015a). Cytotoxicity of a naturally occurring furoquinoline alkaloid and four acridone alkaloids towards multi-factorial drug-resistant cancer cells. *Phytomedicine, 22*(10), 946–951.

Kuete, V., Sandjo, L. P., Mbaveng, A. T., Zeino, M., & Efferth, T. (2015b). Cytotoxicity of compounds from *Xylopia aethiopica* towards multi-factorial drug-resistant cancer cells. *Phytomedicine, 22*(14), 1247–1254.

Kuete, V., Sandjo, L. P., Seukep, J. A., Zeino, M., Mbaveng, A. T., Ngadjui, B., & Efferth, T. (2015c). Cytotoxic compounds from the fruits of *Uapaca togoensis* towards multifactorial drug-resistant cancer cells. *Planta Medica, 81*(1), 32–38.

Laines-Hidalgo, J., Muñoz-Sánchez, J., Loza-Müller, L., & Vázquez-Flota, F. (2022). An update of the sanguinarine and benzophenanthridine alkaloids' biosynthesis and their applications. *Molecules (Basel, Switzerland), 27*(4), 1378.

Li, J., Seupel, R., Bruhn, T., Feineis, D., Kaiser, M., Brun, R., ... Bringmann, G. (2017b). Jozilebomines A and B, Naphthylisoquinoline dimers from the Congolese Liana *Ancistrocladus ileboensis,* with antiausterity activities against the PANC-1 human pancreatic cancer cell line. *Journal of Natural Products, 80*(10), 2807–2817.

Li, J., Seupel, R., Feineis, D., Mudogo, V., Kaiser, M., Brun, R., ... Bringmann, G. (2017a). Dioncophyllines C2, D2, and F and related naphthylisoquinoline alkaloids *from* the Congolese Liana *Ancistrocladus ileboensis* with potent activities against *Plasmodium falciparum* and against multiple myeloma and leukemia cell lines. *Journal of Natural Products, 80*(2), 443–458.

Lichman, B. (2021). The scaffold-forming steps of plant alkaloid biosynthesis. *Natural Product Reports, 38*(1), 103–129.

Liu, Y., Harinantenaina, L., Brodie, P. J., Slebodnick, C., Callmander, M. W., Rakotondrajaona, R., ... Kingston, D. G. I. (2013). Structure elucidation of antiproliferative bisbenzylisoquinoline alkaloids from *Anisocycla grandidieri* from the Madagascar dry forest. *Magnetic Resonance in Chemistry, 51*(9), 574–579.

Lombe, B. K., Feineis, D., & Bringmann, G. (2019). Dimeric naphthylisoquinoline alkaloids: polyketide-derived axially chiral bioactive quateraryls. *Natural Product Reports, 36*(11), 1513–1545.

Mansoor, T. A., Borralho, P. M., Dewanjee, S., Mulhovo, S., Rodrigues, C. M. P., & Ferreira, M. J. U. (2013b). Monoterpene bisindole alkaloids, from the African medicinal plant *Tabernaemontana elegans,* induce apoptosis in HCT116 human colon carcinoma cells. *Journal of Ethnopharmacology, 149*(2), 463–470.

Mansoor, T. A., Borralho, P. M., Luo, X., Mulhovo, S., Rodrigues, C. M. P., & Ferreira, M. J. U. (2013a). Apoptosis inducing activity of benzophenanthridine-type alkaloids and 2-arylbenzofuran neolignans in HCT116 colon carcinoma cells. *Phytotherapy, 20*(10), 923–929.

Mansoor, T. A., Ramalhete, C., Molnár, J., Mulhovo, S., & Ferreira, M. J. U. (2009a). Tabernines A-C, β-carbolines from the leaves of *Tabernaemontana elegans. Journal of Natural Products, 72*(6), 1147–1150.

Mansoor, T. A., Ramalho, R. M., Mulhovo, S., Rodrigues, C. M. P., & Ferreira, M. J. U. (2009b). Induction of apoptosis in HuH-7 cancer cells by monoterpene and β-carboline indole alkaloids isolated from the leaves of *Tabernaemontana elegans. Bioorganic and Medicinal Chemistry Letters, 19*(15), 4255–4258.

Masi, M., Koirala, M., Delicato, A., Di Lecce, R., Merindol, N., Ka, S., ... Evidente, A. (2021). Isolation and biological characterization of homoisoflavanoids and the alkylamide *N-p*-coumaroyltyramine from *crinum biflorum* rottb., an Amaryllidaceae species collected in Senegal. *Biomolecules, 11*(9), 1–21.

Masi, M., Van Slambrouck, S., Gunawardana, S., Van Rensburg, M. J., James, P. C., Mochel, J. G., ... Evidente, A. (2019). Alkaloids isolated from *Haemanthus humilis* Jacq., an indigenous South African Amaryllidaceae: Anticancer activity of coccinine and montanine. *South African Journal of Botany, 126*, 277–281.

Mbaveng, A. T., Bitchagno, G. T. M., Kuete, V., Tane, P., & Efferth, T. (2019b). Cytotoxicity of ungeremine towards multi-factorial drug resistant cancer cells and induction of apoptosis, ferroptosis, necroptosis and autophagy. *Phytotherapy, 60*, 152832.

Mbaveng, A. T., Kuete, V., & Efferth, T. (2017). Potential of Central, Eastern and Western Africa medicinal plants for cancer therapy: spotlight on resistant cells and molecular targets. *Frontiers in Pharmacology, 8*, 343.

Mbaveng, A. T., Noulala, C. G. T., Samba, A. R. M., Tankeo, S. B., Abdelfatah, S., Fotso, G. W., ... Efferth, T. (2021b). The alkaloid, soyauxinium chloride, displays remarkable cytotoxic effects towards a panel of cancer cells, inducing apoptosis, ferroptosis and necroptosis. *Chemico-Biological Interactions, 333*, 109334.

Mbaveng, A. T., Noulala, C. G. T., Samba, A. R. M., Tankeo, S. B., Fotso, G. W., Happi, E. N., ... Efferth, T. (2021a). Cytotoxicity of botanicals and isolated phytochemicals from *Araliopsis soyauxii* Engl. (Rutaceae) towards a panel of human cancer cells. *Journal of Ethnopharmacology, 267*, 113535.

Mbaveng, T. A., Damen, F., Çelik, İ., Tane, P., Kuete, V., & Efferth, T. (2019a). Cytotoxicity of the crude extract and constituents of the bark of *Fagara tessmannii* towards multi-factorial drug resistant cancer cells. *Journal of Ethnopharmacology, 235*, 28–37.

McCalley, D. V. (2002). Analysis of the Cinchona alkaloids by high-performance liquid chromatography and other separation techniques. *Journal of Chromatography. A, 967*(1), 1–19.

Michael, J. (2017). *Acridone alkaloids. The Alkaloids: Chemistry and Biology, Vol. 78*, Elsevier, 1–108.

Mohammed, M. M. D., Ibrahim, N. A., Awad, N. E., Matloub, A. A., Mohamed-Ali, A. G., Barakat, E. E., ... Colla, P. L. (2012). Anti-HIV-1 and cytotoxicity of the alkaloids of Erythrina abyssinica Lam. growing in Sudan. *Natural Product Research, 26*(17), 1565–1575.

Morishige, T., Tsujita, T., Yamada, Y., & Sato, F. (2000). Molecular characterization of the S-adenosyl-L-methionine: 3′-hydroxy-N-methylcoclaurine 4′-O-methyltransferase involved in isoquinoline alkaloid. *Journal of Biological Chemistry, 275*(30), 23398–23405.

Moyo, P., Shamburger, W., Van Der Watt, M., Reader, J., De Sousa, A., Egan, T., ... Birkholtz, L. (2020). Naphthylisoquinoline alkaloids, validated as hit multistage anti-plasmodial natural products. *International Journal for Parasitology: Drugs and Drug Resistance, 13*, 51–58.

Nair, J. J., & Van Staden, J. (2014). Cytotoxicity studies of lycorine alkaloids of the amaryllidaceae. *Natural Product Communications, 9*(8), 1193–1210.

Nganou, B. K., Mbaveng, A. T., Fobofou, S. A. T., Fankam, A. G., Bitchagno, G. T. M., Simo Mpetga, J. D., ... Tane, P. (2019). Furoquinolines and dihydrooxazole alkaloids with cytotoxic activity from the stem bark of *Araliopsis soyauxii*. *Fitoterapia, 133*, 193–199.

Nouga, A. B., Ndom, J. C., Mpondo, E. M., Nyobe, J. C. N., Njoya, A., Meva'a, L. M., ... Wansi, J. D. (2016). New furoquinoline alkaloid and flavanone glycoside derivatives from the leaves of *Oricia suaveolens* and *Oricia renieri* (Rutaceae). *Natural Product Research, 30*(3), 305–310.

Nwodo, J. N., Ibezim, A., V. Simoben, C., & Ntie-Kang, F. (2015). Exploring cancer therapeutics with natural products from african medicinal plants, part ii: alkaloids, terpenoids and flavonoids. *Anti-cancer Agents in Medicinal Chemistry, 16*(1), 108–127.

Omosa, L. K., Mbogo, G. M., Korir, E., Omole, R., Seo, E. J., Yenesew, A., ... Efferth, T. (2021). Cytotoxicity of fagaramide derivative and canthin-6-one from *Zanthoxylum* (Rutaceae) species against multidrug resistant leukemia cells. *Natural Product Research, 35*(4), 579–586.

Omosa, L. K., Nchiozem-Ngnitedem, V. A., Mukavi, J., Atieno Okoko, B., Ombui Nyaboke, H., Hashim, I., ... Spiteller, M. (2022). Cytotoxic alkaloids from the root of *Zanthoxylum paracanthum* (mildbr) Kokwaro. *Natural Product Research, 36*(10), 2518–2525.

Othman, L., Sleiman, A., & Abdel-Massih, R. M. (2019). Antimicrobial activity of polyphenols and alkaloids in middle eastern plants. *Frontiers in Microbiology, 10*, 911.

Pacifici, G. (2016). Metabolism and pharmacokinetics of morphine in neonates: A review. *Clinics (Sao Paulo, Brazil), 71*, 474–480.

Paterna, A., Gomes, S. E., Borralho, P. M., Mulhovo, S., Rodrigues, C. M. P., & Ferreira, M. J. U. (2016a). Vobasinyl-iboga alkaloids from *Tabernaemontana elegans*: Cell cycle arrest and apoptosis-inducing activity in HCT116 colon cancer cells. *Journal of Natural Products, 79*(10), 2624–2634.

Paterna, A., Gomes, S. E., Borralho, P. M., Mulhovo, S., Rodrigues, C. M. P., & Ferreira, M. J. U. (2016b). 3′R)-hydroxytabernaelegantine C: A bisindole alkaloid with potent apoptosis inducing activity in colon (HCT116, SW620) and liver (HepG2) cancer cells. *Journal of Ethnopharmacology, 194*, 236–244.

Pauli, H., & Kutchan, T. (1998). Molecular cloning and functional heterologous expression of two alleles encoding (S)-N-methylcoclaurine 3′-hydroxylase (CYP80B1), a new methyl jasmonate. *The Plant Journal, 13*(6), 793–801.

Rainsford, K. D., & Alamgir, A. N. M. (2017). *Therapeutic use of medicinal plants and their extracts: volume 1, Vol. 1*. Springer International Publishing AG.

Roy, A., Ahuja, S., & Bharadvaja, N. (2017). A review on medicinal plants against cancer. *Journal of Plant Sciences and Agricultural Research, 2*(1), 1–5.

Samoylenko, V., Jacob, M. R., Khan, S. I., Zhao, J., Tekwani, B. L., Midiwo, J. O., ... Muhammad, I. (2009). Antimicrobial, antiparasitic and cytotoxic spermine alkaloids from *Albizia schimperiana*. *Natural Product Communications, 4*(6), 791–796.

Sandjo, L. P., Kuete, V., Tchangna, R. S., Efferth, T., & Ngadjui, B. T. (2014). Cytotoxic benzophenanthridine and furoquinoline alkaloids from *Zanthoxylum buesgenii* (Rutaceae). *Chemistry Central. The Journal, 8*(1), 1–5.

Sawadogo, W., Schumacher, M., Teiten, M., Dicato, M., & Diederich, M. (2012). Traditional West African pharmacopeia, plants and derived compounds for cancer therapy. *Biochemical Pharmacology, 84*(10), 1225–1240.

Schmidt, A., & Liu, M. (2015). *Recent advances in the chemistry of acridines. Advances in heterocyclic chemistry*. Elsevier, 287–353.

Segun, P. A., Ismail, F. M. D., Ogbole, O. O., Nahar, L., Evans, A. R., Ajaiyeoba, E. O., & Sarker, S. D. (2018). Acridone alkaloids from the stem bark of *Citrus aurantium* display selective cytotoxicity against breast, liver, lung and prostate human carcinoma cells. *Journal of Ethnopharmacology, 227*, 131–138.

Singh, S., Pathak, N., Fatima, E., & Negi, A. (2021). Plant isoquinoline alkaloids: Advances in the chemistry and biology of berberine. *European Journal of Medicinal Chemistry, 226*, 113839.

Talaty, N., Takáts, Z., & Cooks, R. (2005). Rapid in situ detection of alkaloids in plant tissue under ambient conditions using desorption electrospray ionization. *Analyst, 130*(12), 1624–1633

Tao, H., Zuo, L., Xu, H., Li, C., Qiao, G., Guo, M., & Lin, X. (2020). Alkaloids as anticancer agents: A review of Chinese patents in recent 5 years. *Recent Patents on Anti-cancer Drug Discovery, 15*(1), 2–13.

Tshitenge, D. T., Bruhn, T., Feineis, D., Schmidt, D., Mudogo, V., Kaiser, M., ... Bringmann, G. (2019). Ealamines A-H, a series of naphthylisoquinolines with the rare 7,8′-coupling site, from the Congolese liana *Ancistrocladus ealaensis*, targeting pancreatic cancer cells. *Journal of Natural Products, 82*(11), 3150–3164.

Tshitenge, D. T., Feineis, D., Mudogo, V., Kaiser, M., Brun, R., Seo, E. J., ... Bringmann, G. (2018). Mbandakamine-type naphthylisoquinoline dimers and related alkaloids from the Central African liana *Ancistrocladus ealaensis* with antiparasitic and antileukemic activities. *Journal of Natural Products, 81*(4), 918–933.

Uche, F. I., Abed, M. N., Abdullah, M. I., Drijfhout, F. P., McCullagh, J., Claridge, T. W. D., ... Li, W. W. (2017). Isochondodendrine and 2′-norcocsuline: Additional alkaloids from *Triclisia subcordata* induce cytotoxicity and apoptosis in ovarian cancer cell lines. *RSC Advances,* 7(70), 44154–44161.

Uche, F. I., Drijfhout, F. P., McCullagh, J., Richardson, A., & Li, W. W. (2016). Cytotoxicity effects and apoptosis induction by bisbenzylisoquinoline alkaloids from *Triclisia subcordata. Phytotherapy Research,* 1533–1539.

Unger, M., & Stöckigt, J. (1997). Improved detection of alkaloids in crude extracts applying capillary electrophoresis with field amplified sample injection. *Journal of Chromatography. A, 791*(1–2), 323–331.

Wang, M., Carrell, E. J., Ali, Z., Avula, B., Avonto, C., Parcher, J. F., & Khan, I. A. (2014). Comparison of three chromatographic techniques for the detection of mitragynine and other indole and oxindole alkaloids in *Mitragyna speciosa* (kratom) plants. *Journal of Separation Science, 37*(12), 1411–1418.

Wansi, J., Devkota, K., Tshikalange, E., & Kuete, V. (2013). *Alkaloids from the medicinal plants of Africa. Medicinal Plant Research in Africa.* Elsevier, 557–605.

World Health Organization. (2012). *Cancer Fact Sheets*. Globocan 2012 (Iarc). ⟨https://www.who.int/news-room/fact-sheets/detail/cancer⟩.

Xu, M., Bruhn, T., Hertlein, B., Brun, R., Stich, A., Wu, J., & Bringmann, G. (2010). Shuangancistrotectorines A–E, dimeric naphthylisoquinoline alkaloids with three chiral biaryl axes from the Chinese Plant *Ancistrocladus tectorius. Chemistry–A European Journal, 16*(14), 4206–4216.

Printed in the United States
by Baker & Taylor Publisher Services